Maize Cobs and Cultures: History of *Zea mays* L.

John E. Staller

Maize Cobs and Cultures: History of *Zea mays* L.

 Springer

Professor John E. Staller
Department of Anthropology
The Field Museum
Chicago IL, USA
E-mail: jstaller@earthlink.net

ISBN: 978-3-642-04505-9   e-ISBN: 978-3-642-04506-6
DOI 10.1007/978-3-642-04506-6
Springer Heidelberg Dordrecht London New York

Library of Congress Control Number: 2009938015

Cover illustration: Maize varieties around Pisac, Peru are still grown on terraces constructed by the Inca.
Pisac is in the Urubamba Valley, renowned for its unusual maize landraces

Cover design: WMXDesign GmbH, Heidelberg, Germany

Printed on acid-free paper

Springer is part of Springer Science+Business Media (www.springer.com)

Acknowledgments

This book is a product of long-term research interest in maize and domestication in general, and is in part a result of interactions and discussions on the origins of maize, plant domestication, and cultivation in general with numerous colleagues and collaborators. I am particularly indebted to Michael Blake (University of British Columbia) and John E. Terrell (The Field Museum). Over the years, these scholars and other colleagues have shared their valuable insights, thoughts and kindly passed along their research and other published data in the biological and social scientific literature on these topics. I also express my sincere gratitude to Bruce Smith (Smithsonian Institution) for providing me critical information on the maize cobs from Tularosa Cave at Field Museum, and for the inspiration his research on maize and domestication have provided. Many thanks to John Tuxill (Western Washington University) and Michael D. Carrasco (Florida State University) for extending the courtesy and use of some images. I also extend my thanks to Stephen E. Nash (Denver Museum of Nature and Science) and Scott Demel (The Field Museum) for granting access to the ceramic and botanical collections at Field Museum. Special thanks as well to Christine Giannoni (Field Museum Library) and her staff, for allowing me to photograph images from the pre-Linnaean and colonial herbals in the rare books collection. I want to express my sincere thanks to Tom Zuidema (University of Illinois-Urbana) and Sabine P. Hyland (St. Norbert College) for the many insights they have provided over the years into New World ethno-history and using sixteenth century accounts. All interpretations and statements of fact are my own, and any errors of fact or misinterpretation of the data should not in any way reflect on these scholars.

Chicago, September 2009   John E. Staller

Contents

Chapter 1
An Introduction to Maize Cobs and Cultures

The importance of maize (*Zea mays* L.) has long been critical to our understanding of the development of pre-Hispanic cultures in the New World. Our perceptions and conceptions regarding its roles and importance to ancient economies are largely the product of scientific research on the plant itself, this developed, for the most part, out of botanical research and scholarship in plant biology and its recent role as one of the most important economic staples in the world. The mutability of the plant and its ability to adapt and reproduce in a wide variety of environmental circumstances led to the previously untested assumption that its central economic role to sociocultural development was at the very basis of its transformation from its wild progenitor *Zea mays* ssp. *parviglumis* to domesticated corn (Matsuoka et al. 2002). The morphological and botanical research surrounding maize has also had a profound influence upon archaeological interpretation, since those biological data set the limits of what was possible regarding the origins of maize and provided a basis for understanding its biogeography in different regions of the Americas. Anthropological research in the early part of the last century based largely on the historical particularistic approach of the Boasian tradition provided the first evidence that challenged the assumptions about the economic importance of maize to sociocultural developments for scholars of prehistory. These and subsequent ethnographic studies showed that the role of maize among Native American cultures was much more complex than just as a food staple.

Multiple roles and uses of maize were also implied from early linguistic studies in which the data suggested that its meanings and referents went beyond the preparations as food and also crossed over to the religious association with primary deities and traditional folk healing. However, the later emphasis on historical linguistics and its association with certain language groups favored an emphasis on the movement of maize economies with certain language groups (Sauer 1952). The shift in emphasis was, in part, influenced by Old World scholarship, which largely perceived by the spread of agriculture as moving with farming populations from the Near East to different regions of western Europe as a kind of wave of advance (see e.g., Childe 1935, 1939, and 1954). The introduction of agricultural grains, like wheat and barley, spread with an associated tool kit to different areas of

J.E. Staller, *Maize Cobs and Cultures: History of Zea mays L.*,
DOI 10.1007/978-3-642-04506-6_1, © Springer-Verlag Berlin Heidelberg 2010

the European continent, and the presence of such artifacts provided the earliest direct evidence of agriculture outside the southwestern Asia. Such models and theories of agricultural origins became increasingly popular in American archaeology during and after the 1930s. Previous cultural historical evidence generated out of the Boasian school still continued to influence methodological approaches in the field of archaeology. New World Archaeologists attempted to explain the biogeography and adaptive shift to a greater dependence upon agriculture, particularly maize agriculture, in terms of its earliest presence in the archaeological record and, due in part to a lack of organic preservation, the appearance of associated processing tools, e.g., ceramics, grinding stones as the evidence of a formative economy (see e.g., Ford 1969). Even as recently as a half-a-century ago, the appearance of such materials in archaeological deposits was largely perceived of as synonymous with an agricultural adaptation and, in the Neotropics, with early maize agriculture. After the advent of radiocarbon in 1950, research at a number of cave sites in various regions of Mexico attempted to document the earliest appearance of maize in the archaeological record and suggested that domesticated varieties appeared long before the associated processing tools (see e.g., MacNeish 1958, 1962, 1967a–c; 1985, and1992). However, these data did not hinder the general predisposition by New World prehistorians to perceive the spread of ceramic innovation as synonymous with a maize agricultural economy (see e.g., Ford 1969; Lathrap 1970, 1974; Lathrap et al. 1975).

The archaeological reconstructions of the post-radiocarbon era were greatly influenced by botanical and genetic research carried out by luminaries such as George Beadle, Paul Mangelsdorf and, more recently, Jack Harlan and Hugh Iltis. During the late 1930s, a series of articles were published on the domestication of Old World cereal grains, wheat and barley, which provided internally consistent lines of evidence for the history of domestication of these food crops. Biological scientists in the New World unwittingly attempted to apply similar approaches to understanding and explaining the origins of maize (Iltis 1971; Harlan 1975). One of the most influential scholars on the genetics, biology, and morphology of corn was Paul Mangelsdorf. In a coauthored study, in1939, they presented genetic and biological evidence to suggest that maize evolved from an as yet unknown wild ancestor, and the hybridization of a second wild species of grass (Mangelsdorf and Reeves 1939). Their results, subsequently referred to as the 'Tripartite Hypothesis', were immediately challenged by Beadle (1939) who suggested that teosinte was the wild ancestor of corn and that domestication involved the genetic mutation of four or five genes. The Tripartite Hypothesis proposed by Mangelsdorf and Reeves had far reaching implications for archaeological interpretations. This research also played a major role surrounding archaeological interpretations of the spread of maize races or varieties throughout the Americas. These data provided a basis from which to assert multiple domestication events.

Teosinte was so different, morphologically, from domesticated maize that the vast majority of prehistoric scholars doubted an evolutionary association, and pursued research, which would document earliest presences. Moreover, the Tripartite Hypothesis was more closely attuned to what was known about the domestication of Old World cereal grains. Teosinte is quite different from the projenitors of Old World cereals. Teosinte grains are inedible, encased in a hard, nonremovable cupulate fruitcases and have generally been absent from early Archaic Period caves and rockshelter sites (cf. Iltis 2006; see also Flannery 1968a–d). It was not until later genetic evidence demonstrated a biological association between varieties of teosinte and domesticated corn that archaeologists began to consider these evolutionary relationships more seriously. Hugh Iltis (2000) proposed the provocative hypothesis that teosinte was not initially exploited as food, but rather for the sugar in the pith of its stalk were consumed as a condiment and used as an auxiliary catalyst to fermentation (see also Smalley and Blake 2003).

Most of the scientific research that has historically come from the biological sciences surrounding the origins and evolution of corn was funded and geared to making the plant more resistant to disease, insects and variations in climate – to increased productivity. Genetically modified varieties of corn have now become essential to the survival of most second- and third-world economies the world over. Moreover, the overwhelming literature on the genetic modification and scientific manipulation of the mutational properties of *Zea mays* L., have also centered around making this resilient plant more suitable to the feeding of livestock and to monocrop cultivation. These biological and generic studies and the data they generated, fostered the perception in archaeological circles that the current uses of maize could be projected into the past. Thus, the archaeological record has been, to a large degree, biased to the extent that methodologies and approaches to understanding the cultural aspects of the plant in the past have been addressed with the same detail, as its economic role to sustaining population densities, and to the development of sociocultural complexity. Moreover, these perceptions were largely supported by ethnohistoric accounts. Chroniclers generally perceived maize as the New World economic staple and from their observations in Mesoamerica,[1] that it was consumed as a grain, very similar in its role in the pre-Hispanic subsistence economy to wheat and barley in the Old World.

The second chapter begins with a consideration of the ethnohistoric evidence as these represent the only direct first-hand accounts of the role of maize in the Americas. However, the field of ethnohistory has, relatively, recent beginnings in the western hemisphere relative to the Old World. An awareness of primary sixteenth to seventeenth century documents spurred the movement toward the modern field of ethnohistory. As researchers began reading the output of these historians, the potential documentary evidence (for the most part literal readings) as a source for anthropological studies became the focus of interest for a cadre of dedicated scholars from both anthropology and history in the middle of the last century. This represents the beginning of the modern field of ethnohistory (Barber and Berdon 1998; Carmack 1973; Carmack et al. 1996). However, in the Americas, an agreed-upon definition of

---

[1] The culture area of Mesoamerica, is located in what are now the countries of southern and central Mexico, Guatemala, Belize, El Salvador, and the western parts of Honduras and Nicaragua. This culture area was the focus of complex, hierarchical cultures at the time of Spanish Contact, and was long theorized to be a hearth of early agriculture in the New World.

what was under the purview ethnohistory was not realized until the late 1970s. Nevertheless, the overall emphasis with regard to the historical documents was to use those accounts that emphasized the role of maize in the ancient subsistence economy and in the spread of polities into regions outside of their heartland centers (see e.g., Murra 1973, 1980, and 1982). Since most of the ethnohistoric literature on the topic of maize is derived from either Mesoamerica or Peru, particular emphasis is given by scholars of ethnohistory to how the plant functioned culturally and economically within, what is known about, indigenous economies and religions in these regions of the New World. This is also the case with regard to the ethnohistoric literature of Northern Mexico and the American Southwest. Historically, the emphasis in those regions has been to perceive cultural development and decline as directly related to the importance of maize to the prehistoric economy. Climatic degradation and the reduced carrying capacity that presumably resulted in the American Southwest as a by-product of maize agriculture are often seen as primary factors in the disappearance of Anazazi and related cultures (Willey 1964). The ethnohistoric evidence, however, suggests that the importance of maize went far beyond its role as an economic plant in this part of the hemisphere as well. The general reluctance of scholars in the social sciences, particularly, anthropology and archaeology, to address the ethnohistoric literature on the symbolic, ritual and religious significance of maize, have to varying degrees influenced the current debates coming out of the biological sciences and biochemistry, more specifically, stable carbon isotope research on paleodiet (see e.g., Staller 2006b, 2008b, and 2009).

The third chapter takes an historical approach to maize research initially focusing upon early studies in the archaeological and biological sciences and how those studies were used by later scholars and indirectly influenced the current debates. Research on plant domestication in the Old World also had a profound effect upon the early biological research upon maize. A review of the botanical and archaeological literature indicates that these early studies on domestication in general and primary economic staples in particular, had an important and often central influence upon later interpretations. These historical influences have led, in some cases, to erroneous assumptions surrounding the economic importance of plants such as maize in the ancient past.

Discussion of the economic importance and role of maize to prehistory would be incomplete without taking into account the early innovative research from the biological sciences, particularly by specialists in paleobotany and plant morphology. The research in paleobotany and plant morphology sustained the Tripartite Hypothesis as researchers continued to look for species of wild grass that had morphological characteristics, which were similar to those of maize (Mangelsdorf et al. 1967; MacNeish and Eubanks 2000; Eubanks 2001a,b). Some scholars even began to study at ancient pre-Columbian art and iconography seeking to identify botanical species, which would provide some clues as to maize's seemingly mysterious origins (see e.g., Eubanks 1999). Within the field of ethnobotany techniques involving the identification of silica bodies or phytoliths had a profound influence upon archaeological interpretations of the role of maize to ancient economies and sociocultural development. Most of these studies adhered to the Tripartite

Hypothesis, which assumes multiple origins or domestication events and pursued the identification of microfossils, both opal phytoliths and pollen, in archaeological soils. These studies focused particularly upon the earliest ceramic cultures, and in some cases preceramic cultures in Central America, outside of the natural spread of teosinte (Pearsall 1978, 1989, 1999, 2002; Pearsall and Piperno 1990; Piperno 1984, 1991, 2006; Piperno et al. 1985, 2007, 2009).

Similarly, early data on maize morphology was also predicated upon multiple origins or domestication events. The numerous designations given to the multitude of varieties or races throughout the Americas have taken on a life of their own and have been given scientific names and cultural designations, which have little or anything to do with their actual genealogy or phylogeny (see e.g., Anderson 1943; Zevallos et al. 1975; Ampuero 1982). These morphological changes are a product of conscious and unconscious selection by humans for larger cob and kernel size, increasing row number, or modifications due simply to adaptation to distinct environmental setting and climatic variations outside of their previous natural range. Maize divergence is also a result of anagenesis, the persistence of one or a suite of biological traits, which leads to varietal divergence over time, or cladogenesis, the development of evolutionary novelty through the eventual extinction of preexisting forms (Benz and Staller 2006, 2009). The emphasis upon anagenesis or cladogenesis will depend upon how the analyst perceives evolutionary diverge. The morphological data are complicated even more by the anthropological evidence on maize varieties. These data have clearly shown that distinct varieties have strong ethnic affiliations and therefore are sometimes taken as indirect evidence of long-distance interaction, as certain varieties are associated with particular regions and ethnic groups or societies (McK Bird 1966; McK Bird et al. 1988; see also Rivera 2006; Shady 2006). Many of these varietal designations are a direct outgrowth of the Tripartite Hypothesis as certain varieties or races were further split into subraces (Mangelsdorf et al. 1967; see also Huckell 2006, pp. 105–106). The prevalence of this botanical and morphological research and its widespread influence upon archaeological interpretations and modeling early maize behavior has created a formidable obstacle to understanding the biogeography of maize and the taxonomic relationship(s) of the various lineages (Huckell 2006, p. 105).

The fourth chapter explores the methodological and technological break-throughs in the study of maize over the past 30 years. Many of these more recent approaches have provided a different series of data sets from which to understand the prehistoric domestication and the role of maize to such developmental processes. The most recent debates surrounding the antiquity and location(s) of the original domestication event(s) have been greatly influenced by technological innovations. These techniques include direct dating of macrobotanical remains and plant microfossils. Recent archaeobotanical advances in the study of pollen, opal phytoliths from archaeological soils and more recently from carbon residues in ancient pottery has dramatically revised our understanding of the biogeography of and antiquity of maize in the Americas (Thompson 2006, 2007; Thompson and Mulholland 1994; Thompson and Staller 2001; Staller and Thompson 2000, 2002; Chavéz and Thompson 2006; Sluyter and Dominguez 2006; see also Lusteck 2006).

These approaches have, when dated by association, as in the case of pollen and phytoliths from archaeological soils, or directly AMS dated as in the case of microfossils from carbon residues, generated chronologies and culture histories that are much more ancient than what had been documented on the basis of macrobotanical evidence (Long et al. 1989; Benz and Long 1999; Blake 2006).

The data from the biological sciences has historically had a profound influence upon archaeologists working on domestication and the role of primary economic staples like maize, to such sociocultural processes. Perhaps, the most significant breakthrough has been through analytical techniques at the molecular level of plant DNA. The maize genome project and the breakdown of microsatellites at the level of DNA have not only produced compelling evidence for the origins of maize (see Matsuoka et al. 2002), but also the spread of maize lineages to different regions of the Americas (Freitas et al. 2003). Studies at the molecular level have also identified the existence of various alleles responsible for those characteristics such as starch production and sugar content, which are necessary for the manufacture of maize into flour for human consumption (see e.g., Whitt et al. 2002; Jaenicke-Després et al. 2003; Jaenicke-Després and Smith 2006). Early genetic research along these lines considered characteristics such as glume architecture, which is intrinsic to modifications associated with human selection and an increased interdependence between the maize and humans (Doebley and Wang 1997; Doreweiler 1996; Dorwieler and Doebley 1997; Dorweiler et al. 1993; Eyre-Walker et al. 1998; Benz 2001; Staller 2003; Thompson 2006, 2007; Hart and Matson 2009; Hart et al. 2007a,b).

In the past decade, direct dates on ancient cobs have indicated a more recent spread of maize through much of the Neotropics. AMS dates taken directly from macrobotanical maize cob samples recovered from the earliest levels of the Tehuacán caves in highland central Mexico, and the Guilá Naquitz rockshelter in highland Oaxaca have, in some specimens, produced younger dates by over two millennia than the initial radiocarbon assays from associated archaeological strata (Long et al. 1989; Piperno and Flannery 2001: Fig. 1). The morphological results from the most ancient cobs at Guilá Naquitz have also indicated that maize was not yet fully domesticated at 5450 B.P (Benz 2002; Staller 2003; see also Bellwood 2005; Blake 2006). These results indicate that in highland Oaxaca in the regions just outside of where maize was domesticated, all of the genetic mutations and modifications associated with fully domesticated maize were still undeveloped. The implications of these data for all future archaeological interpretations regarding the origins and early economic significance of maize to New World prehistory are profound. Some botanists and archaeologists have already begun to explore alternative explanations for the seemingly rapid spread of this important New World domesticate (Iltis 2000, 2006; Staller and Thompson 2002; Smalley and Blake 2003). The book concludes with a summation of the current state of research regarding the application of new groundbreaking approaches to understanding the maize at the molecular as well as the morphological and phenotypic levels and its role in the ancient diet. Case studies are provided which touch on some of the major implications brought about by recent methodological approaches and the importance of internally consistent lines of evidence to our understanding of maize evolution and biogeography.

Chapter 2
Ethnohistory: Impressions and Perceptions of Maize

2.1 Ethnohistoric and Ethnographic Perceptions of Maize

The sixteenth century documents, pictorial codices, and iconographic and hiero-glyphic texts are all evaluated to consider how earlier Indo-European perceptions of the New World influenced our current understanding of the roles and importance of maize to sociocultural development. Primary focus is given to the earliest primary and secondary ethnohistoric accounts regarding the role of maize to New World cultures. Since all the sixteenth century accounts were written to be part of history, they are generally narrative and descriptive (Carmack 1973). Their analytical and historical importance is not only that they provide a picture of relatively pristine native culture (see, e.g., Cortés 1963 (1485–1547?); 1991 (1519–1526); Díaz 1953 (1567–1575); and Landa 1975 (1566)), but also that they are a reflection of the sixteenth century New Word culture and their perceptions of the world around them. The only regions where native documents compare in ethnohistoric value to the Spanish sources are those written in Mexico and Guatemala during the sixteenth century[1] (Carmack 1973; Carmack et al. 1996; and Barber and Berdan 1998). Most of the preHispanic codices were destroyed in various campaigns to eradicate pagan idolatry (Acosta 1961 (1590); Durán 1971 (1581); Landa 1975 (1566); Las Casas 1992 (1552); and Sepulveda and Las Casas 1975 (1540)). Those codices produced after the conquest are largely commissioned by the Spanish nobility and illustrated by indigenous and mestizo scribes who had converted to Catholicism. Conse-quently, the content of most such colonial indigenous texts were conditioned to

---

[1]Spanish influence stimulated a large corpus of Contact Period native Quichean documents. Most were written during the first half of the sixteenth century. The fact that some Spaniards as well as Mesoamerican scribes were literate is important for the study of native culture and sixteenth century European culture —therefore of potential value to archaeological and anthropological reconstruction (Carmack 1973; Carmack et al. 1996; and Schwartz 2000).

J.E. Staller, *Maize Cobs and Cultures: History of Zea mays L.*,
DOI 10.1007/978-3-642-04506-6_2, © Springer-Verlag Berlin Heidelberg 2010

varying degrees by sixteenth century European perceptions and cultural biases (Staller 2009).

Mesoamerica is the only region in the New World in which a highly specialized native literary tradition already existed before the Contact Period (Anderson et al. 1976; Barber and Berdan 1998). It was primarily from these early accounts that western culture began to comprehend that much of what had been written in the scriptures did not take into account the existence of this New World, a world that was dramatically different both environmentally and culturally from their own. These sixteenth century accounts, were transcribed while these initial contacts were occurring, and have immediacy and freshness in their descriptions (Carmack 1973). They are dominated by personal impressions and details of special value to the archaeologist as well as the historian, for the purposes of studying Pre-Columbian foodways in general and the roles and uses of maize in particular.

Maize was unknown in Europe prior to the arrival of Columbus in 1492. By that time, Native Amerindians cultivated maize over much of the tropical and temperate portions of the western hemispere from southern Canada to south central Chile (Staller et al. 2006). The great adaptability and plasticity of maize are evidenced by the fact that today it represents the second most important food plant on earth and its current distribution is worldwide.

Much of what was initially known about maize comes from European explorers, primarily clergy, Conquistadores and mercenaries who came to this hemisphere seeking wealth, fortune, or to escape political and religious oppression and start a new life. Most of the earliest eyewitness accounts of the conquest and of the Native American cultures are from European clergy, such as colonial monks, priests, and scribes (Staller 2009). Clergy and members of the European aristocracy constituted the vast majority of the literate peoples of western Europe at the beginning of the fifteenth century. As apparent by most of the extant literature surrounding the conquest of the New World, European arrivals in the New World initiated processes of cultural change that were sometimes rapid and catastrophic, sometimes protracted and complex (see, e.g., Carrasco 1999; Schwartz 2000). The King of Spain provided Ferdinand Magellan and his navigator Juan Sebastian Elcano five ships to circumnavigate the globe in 1519. This, after the king of Portugal Manuel I, repeatedly turned down Magellan and his navigator. The commercial success of the voyage laid the foundation for the Pacific oceanic empire of Spain and the colonization of the Philippines (Elliot 1963). Combined with the Magellan expedition's circumnavigation of the globe in 1522, the rapid conquest of the civilizations of the New World convinced the Spanish Crown King Charles V of his divine mission to become the leader of the Christian world, a world that still perceived a significant threat from the forces of Islam (Elliot 1963; Carmack 1973; and Ife 1990). According to the *Capitulaciones*, the formal agreement between the Spanish Crown (Ferdinand and Isabella) and Christopher Columbus, the explorer would become the viceroy and governor-general of any and all lands and islands he discovered (Columbus 1970 (1492), p. 23; Ife 1990, p. xvi). The Spanish Crown would take 90% of all income generated from the territories under his jurisdiction[2] (Ife 1990, pp. xvi–xvii). In the prologue to the journal of the expedition itself, it is written;

> "Your Highnesses, as Catholic Christians and princes devoted to the holy Christian faith and the furtherance of its cause, and enemies of the sect of Mohammed and of all idolatry and heresy, resolved to send me, Christopher Columbus, to the said regions of India to see the said princes and the peoples and lands and determine the nature of them and of all other things, and the measures to be taken to convert them to our holy faith; and you ordered that I should not go by land to the East, which is the customary route, but by way of the West, a route which to this day we cannot be certain has been taken by anyone else." (Columbus 1970 (1492), p. 23).

The original documents of the agreement or collaboration between the Spanish Crown and the Conquistadores, therefore, sought the acquisition of territory and their inherent riches, while at the same time promote the religious principles under which most of the legal authority of the ruling nation states was sanctioned (Madariaga 1947, pp. 10, 12–14). The accounts provide considerable detail regarding religious rituals, because many ecclesiastics and political authorities were focused upon identifying "pagan idolatry" in whatever form they may find such activity (Staller 2009, pp. 26–27). Thus, accounts by religious clerics are generally rich sources of information for pre-Columbian scholars concerned with native ritual and religious belief (Carmack 1973; Barber and Berdan 1998). The accounts clearly indicate that one of the primary goals of the conquest of the New World was to convert its peoples. If the quest for wealth and power was what fueled the discovery and conquest of the New World, it was the conversion of indigenous populations that provided the religious and spiritual rationale for how these societies and cultures were to be integrated into the empire of New Spain (Madariaga 1947, p. 10). As the conquistador Bernal Díaz del Castillo stated, "We came to serve God and his Majesty, to give light to those in darkness, and also to acquire that wealth which most men covet" (Elliot 1963, p. 64; Díaz 1953 (1567–1575), p. 2).

2.1.1 Consequences of Conquest and Empire

The pursuit of wealth and power under the auspices of the Spain Crown led to the widespread destruction and oppression of indigenous populations, and some clerics and colonial officials protested against the Spanish Crown on behalf of the native populations (see, e.g., Las Casas 1992 (1552); Sepulveda and Las Casas 1975 (1540)). In 1550, Charles V convened a now famous conference at Valladolid, Spain in order to consider the morality of force used against the indigenous populations of the New World (Madariaga 1947; Martby 2002). Charles V was

---

[2]The *Capitulaciones* reserved certain rights to the Crown of Spain in newly conquered territories, while at the same time guaranteeing the expedition leader due *mercedes* or rewards for services rendered to the Crown (Elliot 1963, p. 58). Columbus also expected to enjoy the spoils of conquest, through the attainment of property and captive slaves, as well as to receive the grants of the land and title of noble (Elliot 1963, pp. 58–59; see also Madariaga 1947).

both king of Spain (King Charles I) and Holy Roman Emperor in 1519, and it was during the early period of his reign that the Aztec and Inca civilizations were conquered and their wealth brought to the Spanish Crown and various royal houses of Europe (Madariaga 1947; Elliot 1963; and Maltby 2002). Hernán Cortés destroyed the Aztec Empire of Motecuhzoma II with 600 soldiers and 16 horses; Francisco Pizarro, the Inca Empire with 37 horses and 180 men (Elliot 1963, p. 62). The conquests of these New World civilizations helped to solidify the rule of Charles V, providing the state treasury with enormous amounts of precious metals, jewels, and incomprehensible wealth. Although it was his predecessors that initiated the quest for a different route to the Indies, it was Charles V who sponsored most of the early expeditions and organized and appointed the early colonial officials who oversaw the governments of the kingdoms of New Spain (Madariaga 1947, pp. 10–11). When Charles V passed away in 1558, he had amassed unimaginable wealth and power for Spain through the expeditions he had sponsored into the West Indies, essentially creating the first empire in which the sun never set, an empire that spanned some 4 million square kilometers (Fig. 2.1). Charles V of Spain ruled over extensive domains in central, western, and southern Europe, as well as the Spanish colonies in the Americas (Elliot 1963). He is credited with the first idea

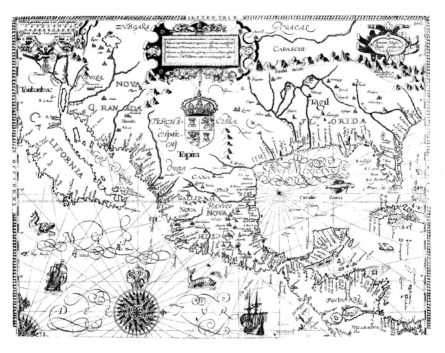

Fig. 2.1 An sixteenth century map by the English chart maker Gabriel Tatton engraved in 1600 showing the Empire of New Spain (*Nova Hispania*). European interests by this time appear to have shifted from Hispanola to Mexico, Nova Granada, California and the northern part of Florida, which was the focus of an expedition by Hernando de Soto in 1540–1543 (Courtesy of Library of Congress, Geography and Map Division)

Fig. 2.2 Charles V, King of Spain and Holy Roman Emperor on horseback. Painting by the Flemish artist Anthony Van Dyck c. 1620, oil on canvas

of constructing an American Isthmus canal in Panama as early as 1520 (Haskin 1913). From an historical perspective, his reign represents the pinnacle of Habsburg[3] power, when all the far-flung holdings of the royal house were united in the person of Charles V of Spain (Fig. 2.2).

The discovery of the New World also had a profound effect upon western Europe. On the one hand, it opened the way from the secular world we live in today by ultimately revealing to the world that *not* all of human history was contained

---

[3]The House of Habsburg (sometimes spelled *Hapsburg*) was an important royal house in Europe best known for supplying all of the Holy Roman Emperors between 1452 and 1740, as well as numerous rulers and ruling families of the Spanish and Austrian Empire (Evans 1979). Charles V was heir to the Habsburg's of Austria, The Valois of Burgundy, Trastamara of Castile and the House of Aragon (Elliot 1963, pp. 61–64; Evans 1979; Maltby 2002).

in Holy Scripture. Moreover, the Protestant Reformation began to threaten the power and authority of the Catholic Church just before and after the discovery of the New World. Many European Catholics were troubled by false doctrines within the church, particularly the teaching and sale of plenary indulgences (Tentler 1977). Soon after Charles V took power as Holy Roman Emperor, the city of Rome was sacked in 1527. Mutinous troops loyal to Charles V are said to have played a critical role in the victory over League of Cognac, which allied the Vatican with France, Florence, Milan, and Venice (Tentler 1977; Coe 1994b). The sacking of Rome marked a *crucial* early victory for the Holy Roman Empire over the Vatican and its allies. This victory was a harbinger of later historical events, which to varying degrees mark the overall decline of the political authority and power of the Roman Catholic Church throughout Europe in centuries that followed (Tentler 1977; Coe 1994b, p. 32). Another factor that fostered the Protestant Reformation was the spread of literacy, particularly the invention of the *printing press* by Johann Gutenberg c. 1439. Such advances had a direct effect upon how and to what detail discoveries in the New World were recorded and disseminated (Fig. 2.3). The Holy Roman Empire also determined where and how such information regarding the flora and fauna of New Spain was to be recorded and disseminated.

2.1.2 Western Perceptions of New World Cultures

The earliest contact of Europeans with such cultures occurred in the Caribbean (Hispanola), Mesoamerica (New Spain), and Andean South America (Perú). As these newly discovered territories became incorporated into part of the empire of New Spain, they became independent kingdoms that were ultimately under the subject and political authority of the Spanish Crown (Elliot 1963). The ruler who was most responsible for the creation of the Spanish American empire was Charles V, who ruled Spain from 1516–1556. It was in these years that the Aztec and Inca civilizations were conquered by the conquistadores in little more than a decade (Madariaga 1947, pp. 8–9; Elliot 1963, p. 61). During the fifteenth and sixteenth centuries, the Church in Rome sanctioned much of the political authority enjoyed by the European aristocracy through a symbolic, and in some cases literal, association with those royal families (Staller 2009). The ruling aristocracy attained great wealth, and in some cases, absolute power over their subjects through a divine right to rule. Before the sixteenth century and for several centuries later, Holy Scripture provided the basis for European perceptions of all things great and small, as well as the creation of heaven and earth. Coincidentally, writings of the Classical Age appeared at the end of the fourteenth century, when there was a general revival of learning associated with the Italian Renaissance (De Vorsey 1991, p. 17; Maltby 2002). Within these writings of the Classical Age was the *Geographia*, a complete cartographers handbook, originally written in the second century in Greek by Claudius Ptolemy. He described a method for producing maps of a curved surface; in other words, in latitude and longitude (De Vorsey 1991, p. 71). Columbus (1930 (1507)) applied the Ptolemaic concept of a west-to-east extent and of a

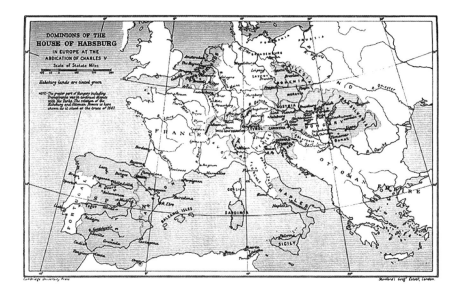

Fig. 2.3 (**a**) Map of the territorial possessions of the House of Habsburg in 1556 and at the time of the abdication of Charles V, King of Spain and Holy Roman Emperor. Habsburg lands are shaded (from Ward et al. 1912). (**b**) Flag and Coats of Arms of the Holy Roman Empire under Charles V, from 1519 to 1558. The Emperors used the double-headed eagle as a symbol of their authority. Individuals from the Habsburg dynasty were Holy Roman Emperors from 1452 until 1740 with Charles VI

habitable world or "*oecumene*" to be 180° or half the earth's circumference, thus extrapolating the known world from western Europe to eastern India and China (see also Staller 2009).

Since New World societies were not found in the Old or New Testaments, many European clerics and scholars searched for a biblical explanation of their existence until the eighteenth century, with the Age of Enlightenment (Staller 2009). These never-beforecivilizations and cultures, plants, and animals fascinated as well as puzzled and troubled the European explorers (Sauer 1969). The plants, animals, and cultures were, in fact, totally alien to anything written or spoken about by the Church in Rome. Europeans nevertheless tried to fit these alien entities into a cultural and religious framework consistent with what was known and familiar to them (Carrasco 1999, pp. 11–13). Many Spanish chroniclers in Mesoamerica, mainly Mexico, Guatemala, and parts of Honduras, Nicaragua, and El Salvador speculated that the indigenous cultures and civilizations of these regions were descendants of the Lost Tribes of Israel (see, e.g., Durán 1994 (1588?), pp. 3–7). Christopher Columbus and Amerigo Vespucci believed that the native Americans were the result of the great biblical flood (Columbus 1990 (1492); Traboulay 1994). A widespread belief that the Amerindians were part of the 10 tribes of Israel was based upon readings from the *Book of Genesis* and persisted in various forms into the nineteenth century and even later (see, e.g., Durán 1994 (1588?), pp. 8–10). Jose de Acosta (1962 (1590), pp. 45, 54) believed the native Americans to have come from Africa, although he was a bit more realistic and scientific in his assessment stating, "*we would like to know how and why they came*" (cf. Durán 1994 (1588?), p. 3).

Spanish chroniclers in Andean South America, in fact, speculated that native Andeans were descendants of the ancient Chaldeans who once lived on the Plain of Sennaar in the Persian Gulf (Valera 1968 (1594), pp. 153–154; cf. Hyland 2003, p. 96). The Andeans worshiped natural features such as mountains, lakes and springs, as well as celestial bodies in the night sky much like the Caldeans mentioned in the Bible. The discovery of the New World revealed to Europeans what could never have been imagined–a world of untold-riches, strange customs and wondrous sights. The Spaniards, and later to a lesser degree, the French, English, Dutch and others recorded their impressions of the native peoples from the arrival of the first conquistadores or pilgrims to the retreat of the last European official during the periods and wars of independence. Indeed, the propensity of the European colonial officials to keep detailed records of their affairs in their colonies has been of considerable value to students and scholars of the colonial period in the New World (Carmack 1973; Spores 1980).

2.2   Using Sixteenth Century Accounts

The earliest colonial accounts consist of primary sources, such as the Conquistadores writings, and *relaciones*, which were first-hand accounts of what natives told colonial officials in the context of legal claims, tribute, and religious practices (Gadacz 1982). Primary ethnohistoric accounts are valuable sources of information for archaeological interpretation because they are direct observations of nearly untouched or pristine native culture. Primary sources have in common the fact that as a whole, they are generally unsympathetic toward native culture, and most of the earliest *relaciones* were predicated by an underlying desire to gain privilege from their patrons, primarily Charles V, and through him the European aristocracy or so-called royal houses, and of course, the Holy See (Elliot 1963; Innes 1969; Carmack 1973; Newsom 1996; Ife 1990; and Barber and Berdan 1998). The accounts left to us by the conquistadores, explorers, and soldiers of fortune who conquered the New World were relegated to the dusty archives of history, or to monastic repositories and libraries in the former colonies of France, England, Portugal and particularly Spain, until archaeologists and historians rediscovered them at the turn of the twentieth century. Ethnohistoric sources have been appearing with increasing frequency in the archaeological literature since the early eighties (Spores 1980). Our current perceptions of the economic role that maize played in the development of civilization in this hemisphere, was largely influenced by our interpretations and study of such early primary ethnohistoric accounts, particularly colonial botanicals (see, e.g., Staller 2009). The general tendency of the conquistadors to evaluate what they observed in their initial encounters with Amerindians in terms of relative size and similarity to their own civilization are also useful for understanding how maize became to have a role in the colonial cultures of the New World and Europe (Oviedo 1959 (1526)).

Although New World archaeologists have been using ethnohistoric accounts for their research on ancient sites for over 30 years, such data and research have only recently begun to play a central role in archaeological and anthropological research (Spores 1980). The only region in the New World where Spanish documents can complement, compare in ethnohistoric value to the native sources are those written in Mexico and Guatemala during the sixteenth century. Scholars who have written on those regions have noted that Spanish accounts, as a body of information about native life, fairly well with what the natives had to say about themselves in some aspects of the culture (Sahagún 1963 (d. 1590)). The fact that the Mesoamericans' were literate as well as complex societies appears to play a role in how the chroniclers describe and discuss the native Mesoamerican societies (Carmack 1973). Many later European expeditions to the New World were focused upon recording the natural history, botany and biology of plant and animal species. The early colonial botanicals have been of value to botanists and plant taxonomists for many years (e.g., Oviedo 1969 (1526); Sahagún 1963 (d. 1590); Gerard 1975 (1633); see also Gerbi 1985). The application of such historical data to archaeological reconstruction is relatively recent compared to the biological sciences. When geneticists and botanists began to study the origins of maize, such ancient texts and botanicals were readily cited in the literature (e.g., Mangelsdorf and Reeves 1939).

Most of the information we have about maize from colonial accounts comes from what are referred to in the ethnohistoric literature as, *relaciones*.[4] Landa's (1975 (1566)) *relación* is the most detailed account of Maya writing and religious

---

[4]The colonial letters or *relaciones* that come most often to mind are Hernán Cortés' letters to Charles V of Spain such as "*la carta de Vera Cruz*" (Cortez 1991 (1519–1526), pp. 16–20, 33–34: Innes 1969), his Cartas y Documentos (Cortés 1963 (1485-1547). or Francisco Pizarro's accounts through his scribe Xerex, of the conquest of Peru (Xerex 1985 (1534); see also Pizarro 1921 (1517)), or Bernal Díaz' first-hand account of the conquest of Mexico (e.g., Díaz 1953 (1567–75)).

practices remaining from the early colonial period. *Relaciones* also include first-hand accounts [primary sources] of what natives themselves said to colonial officials (Sarmiento 1942 (1572); Arriaga 1968 (1621); Sahagún 1963 (d. 1590); Landa 1975 (1566)). However, most *relaciones* are records and notes taken at colonial administrative centers or churches about native customs, territorial disputes, land rights, grazing rights, cultivation and access to irrigation canals etc., (Valera 1968 (1594); Acosta 1961 (1590); Arriaga 1968 (1621); Sarmiento 1942 (1572)). Such colonial and native interactions were generally governed by self-interest, that may work both ways, such as when aborigines are making territorial claims, or when the information will be used by the Crown to determine the jurisdiction of colonial officials (Carmack 1973; Murra 1973, 1980). Territorial claims in both Mesoamerica and the Andean highlands often involved areas where maize cultivation was important to the local economies, or where certain highly valued varieties of maize were being cultivated (see Carmack 1973; Murra 1973, 1980; Morris 1993; and Morris and Thompson 1985).

The monastic orders that provided the most useful information on the conquest of the New World, are the Dominicans, Franciscans, and later the Jesuits. With respect to the Dominicans, Las Casas defended of the Indians of Guatemala at around 1540 and initiated his program of peaceful pacification in Veracruz (Las Casas 1971 (1527–1565); Sepulveda and Las Casas 1975 (c. 1540)). The sympathetic, ethnographic tradition of the Dominicans persisted for over two centuries. It is somewhat ironic therefore that it was the Dominican order who initiated the purges and inquisition earlier in Europe and then the later purges and extirpations of mestizo and indigenous populations in the New World colonies (e.g., Albornoz 1967 (1570–1584); Arriaga 1968 (1621); see also Traboulay 1994; Pérez 2005). In a symbolic demonstration of their domination over the Inca, they built a Dominican monastery on top of the Coricancha or Golden Enclosure, the palace complex and religious center of the Empire. The monastery was leveled along with most of the colonial architecture in the former Inca capital by a catastrophic earthquake in 1650, while the remaining Inca walls of the Coricancha withstood this quake (Cobo 1990 (1653)). Subsequently, a Dominican priory and the Church of Santo Domingo were constructed in the same location and likewise destroyed in another major quake in 1950. The systematic eradication of native religious architecture and expression has its historical basis in the Andes with the reign of the Viceroy Don Francisco de Toledo in the 1570s (Toledo 1940 (1571); see Pérez 2005; Homza 2006). The Franciscans, who first came to the New World in the 1540s, never developed an ethnographic tradition equal to that of the Dominicans. Nevertheless, their excellent dictionaries (Quiche and Cakchiquel) attest to an early interest in converting the native cultures (Ochoa and Jaime Riverón 2005).

Overwhelmingly, most ethnohistoric documents were written for administrative purposes: to report to higher officials on the general condition of the Indians, or to gather specific information about native culture for some immediately practical use. In either case, they were political instruments (Carmack 1973, p. 84). Diego de Landa's *Relación de las cosas de Yucatan* is the most detailed account of the ancient Maya surviving from the early colonial period, when some contact with the pre-Hispanic culture was still possible and the processes of acculturation were not yet very far advanced. Landa's *relación*, together with a handful of "native" pictographic and hieroglyphic writings in Yucatec Maya, and written down in the Latin many years after the Conquest is all that remains regarding the written evidence on the writing of a once-flourishing civilization. In early 1562, de Landa began to investigate incidents of idolatry and suspected countenance among the Maya populations of the province of Yucatán.[5] The *Relación* was composed while Diego de Landa lived in Spain, at around 1566, and may have been his written defense during an investigation by the Church of his inquisitorial activities into presumably idolatrous practices among the Yucatec Indians. Landa (1975 (1566), p. 18) was suspicious that the Maya acceptance of the Colonial religion was an error. He correctly observed that, in many cases, it was simply the addition of a Christian god to an already flourishing religious pantheon and tradition. These Maya populations were accused of idolatrous practices previously outlawed by the Franciscans. Many of these idolatries and ritual practices involved offerings of maize in various forms (Landa 1975 (1566)). Some general characteristics of the sixteenth century Spanish sources can be explained in terms of historical factors and the sociocultural conditions of Spain and the colonies (Carmack 1973; Pérez 2005; and Homza 2006). For example, the superiority of the accounts of religious clerics and friars over those of civil officials in providing useful early ethnographic observations on native life is a reflection of the strong humanistic tradition that the clergy brought to the New World. The humanistic component of the Renaissance came to Spain primarily through the Church, and was brought from there to the New World (Alt 2005). It is from such ethnohistoric accounts that most of the information on how maize was consumed and used in different aspects of native culture at the time of the conquest was first recorded (e.g., Landa 1975 (1566); Cobo 1990 (1653)).

Ethnohistoric accounts reveals a great deal about how ancient political economies were organized, and how food crops as well as other subsistence and utilitarian resources were managed, stored, and redistributed by the elite to their subject populations (see e.g., Berdan 1982; Berdan and Anawalt 1992 (1541–1542)). Present-day economic staples such as maize (*Zea mays* L.), peppers (*Capsicum* spp.), and squash (*Curcurbita* spp.) were unknown in Europe prior to the arrival of Columbus in 1492 (Schiebinger 2004). Trade and commerce of these domesticates and other food crops and economic plants into the Old World and Asia have had far-reaching effects upon agricultural economies, cuisines, and food habits worldwide (Coe 1994b; Schiebinger and Swan 2005).

---

[5]Most of the pre-colonial codices written by the Yucatec Maya were destroyed under the orders of Diego de Landa, in his lifelong quest to eradicate idolatry.

2.2.1   Early Pre-Linnaean Botanicals

Our general perceptions of maize (*Zea mays* L.) are conditioned to varying degrees by what was said about the plant when European settlers first arrived in the New World. Our more lasting preconceptions and misperceptions were all about of how maize and other New World cultigens were distributed and disseminated throughout Europe and northern Africa. With the invention of the printing press, information about the New and Old World flora and fauna could be systematically documented and recorded. As more European universities were created, such reference sources provided ever-wider access to both elite and later to literate commoner populations. The first natural history of the Indies was written by the Spanish aristocrat Gonzalo Fernández de Oviedo y Valdes (1959 (1526)), which was first issued in 1526 in Toledo in the form of a summary.[6] The so-called *Sumario* or *De La Natural Hystoria de Las Indias* represents the first systematic attempt at a natural history of the New World. European wood carvers and illustrators used Oviedo's descriptions of exotic plants and other objects from this book, and the later 1535 edition of this work and many became part of western folklore (see, e.g., Staller 2009). Oviedo's book described and illustrated New World plants such as the pineapple, cacao, chile peppers, tobacco and of course maize. Like many of these early accounts, they included fanciful depictions derived through secondary sources, or authors who copied and illustrated what was published in the first-hand accounts. A fanciful image of a hammock from the 1535 edition of *De La Natural Hystoria de Las Indias* (Fernández de Oviedo 1969 (1535)) forever changed long-distance travel on the Seven Seas, after European ship builders, sailors and explorers subsequently incorporated hammocks on their ships (Sauer 1969, p. 236). Oviedo did not restrict his observations of New World flora and fauna solely to what kinds of plants and animals were being consumed, but also included *how* native societies consumed them. Caribbean islanders were observed to only consume maize by roasting the ears or parching dry seeds (Sauer 1969, p. 236).

As more information came to western Europe from the Americas, many attempts were made at systematically documenting and describing the New World flora and fauna (Gerard 1975 (1633); Andrew et al. 2001 (1789–1794)). However, it not until the eighteenth century, with a florescence of exploration involved in recording the geography and natural history of the plants and animals that many of these initial descriptions were found to be more imagined than accurate. Scientific and botanical descriptions and illustrations of the natural biota of the Americas did not begin until early in the seventeenth century, at the same time when plant and animal specimens

---

[6]The first part of *De La Natural Hystoria de Las Indias* appeared in Seville during 1535. The complete work was not published until 1851–1855 for the Spanish Academy of History. Though written in a diffuse style, it embodies a mass of curious information collected first-hand. The incomplete Seville edition was widely read in the English and French versions published, respectively, in 1555 and 1556 (Fernández de Oviedo 1969 (1535)).

were being collected and studied in western museums (Andrew et al. 2001 (1789–1794)). Camerarius, a German botanist, began to systematically identify plants primarily based upon their reproductive parts (Fussell 1992, p. 61).

The discovery of the New World provided more than just information on a whole different array of exotic plants and animals, it introduced Western Enlightenment thought to a systematic way of describing and understanding the natural ecology. In the eighteenth century, a Swedish botanist, Carl Linnaeus began to hierarchically classify the natural world into families, genus and species, providing the basis for modern taxonomy (Schiebinger 2004, p. 194; Knapp et al. 2007, p. 261). Linnaeus believed that the foundations of scientific botany were classification and binomial nomenclature (Schiebinger 2004, p. 194). It is one of the greatest triumphs of western science that these early pioneers developed a way to name and refer to all described organisms on earth, and their fossil ancestors (Godfrey 2007, p. 259). Using what became the universal languages of science, Latin and Greek, Linnaeus classified maize as, *Zea*, a Greco-Latin term for "a wheat-like grain," and species *mays*, a Latinized term derived from the Taíno-Arawakan word *mahiz*, or "life-giver" (Las Casas 1971 (1527–1565); cf. Fussell 1992, p. 60). There were a number of spellings of the term "maize" when it first appeared in written documents in Europe, but in 1516, when the term first appeared in print in one of the Peter Martyr publications, it was spelled in the Latin, *maizium* (Weatherwax 1954, p. 5; Martire 1907).

Nomenclature can reveal a great deal about the culture history of plants; how knowledge of them spread through botanical networks, and European cultures understood their biogeography and use–as well as western botanists evaluated indigenous systems of knowledge (Schiebinger 2004, p. 195). Much of the ethnohistoric literature suggests that the earliest European explorers in the western hemispere saw maize as the cultural and economic equivalent of wheat and barley in the Old World (Oviedo 1959 (1526); Las Casas 1971 (1527–1565); Cieza de León 1998 (1553); Innes 1969; and Sauer 1969). Pre-Linnaean herbals published within a century after the conquest, give the general impression that among European explorers, clerics, and botanists maize represented a primary economic staple – a "staff of life" (Fuchs 1978 (1543), *fol.* 473; and Gerard 1975 (1633), pp. 81–82). This perception may also be related to later nomenclature – Corn, or more specifically, Indian corn, is the name applied to maize by the English explorers and consequently, the dominant name applied to *Zea mays* L. in English-speaking areas of North America (Weatherwax 1954). This term is taken from the sixteenth century German (*korn*), which they applied to wheat and the name was used in various other places in western Europe for oats, rye, barley, and even lentils (Cutler and Blake 2001 (1976), p. 2). The earliest descriptions of maize and all the flora and fauna witnessed upon the earliest voyages were by laymen, as no competent botanists and naturalists were on board in these initial explorations (Weatherwax 1954, p. 5). Subsequently, some botanists perceived the rise of Linnaean systematics as a form of "linguistic imperialism," a politics of naming, spread by, and at the same time, promoting imperial global expansion and colonization (Schiebinger 2004, p. 195; Andrew et al. 2001 (1789–1794)). Eighteenth century nomenclature served, both explicitly and implicitly, as an instrument of empire removing plants from their indigenous cultural moorings and placing them within a classificatory schema that was first and foremost comprehensive to Europeans (Schiebinger 2004, p. 224).

2.2.2   Earliest Sixteenth Century Accounts

On the October 12, 1492, sailing under the flag of Spain and sponsored by the Crown of Castile, Ferdinand and Isabella, the Italian explorer Christopher Columbus first set foot on Watling Island, a place he named *San Salvador*, in the central Islands of the Bahamas (Keegan 1989, pp. 31–34, 40). There, he found naked people, "...*very well made, of very handsome bodies and very good faces*" (Fig. 2.4). Columbus believed that he had found the outlying islands of East Asia, and this is why he called the inhabitants of the new lands "*Indios*," or Indians (Sauer 1969). The European Cultures that came to the New World after the arrival of Columbus in 1492 were searching for a passageway to India.[7] While western history has generally placed much significance on his first voyage of 1492, Columbus did not actually reach the mainland until his third voyage in 1498 when he explored the north coast of South America. Only 7 years after the death of Columbus on May 20, 1506, the conquistador Vasco Nuñez de Balboa trekked across Central America and gazed upon the Pacific–although he had already heard about the ocean by 1511 (Sauer 1969, p. 222). In a letter dated September 25, 1513, Balboa stated that while on a hill near the Gulf of San Miguel in present-day Panama, he observed the existence of a South Sea or Pacific Ocean (Blacker and Rosen 1960, pp.55–58). The so-called West Indies were not part of China after all, but "...*a New World more densely peopled and abounding in animals than Europe, or Asia, or Africa*" (Blacker and Rosen 1960, p. 59). Despite this observation and subsequent discovery of the western coastlines of Central and North America, fanciful depictions of New World cultures, plants and animals continued for many years, as Europeans attempted to connect them to Asia and the Far East (Milbrath 1989, pp. 184–185). For example, the pre-Linnaean herbal by John Gerard (1975 (1633), p. 81) has an image of the "Corne of Asia" suggesting that at this time some botanists believed maize may have come to Turkey from the Far East (Fig. 2.5a). In the same 1633, herbal different woodcuts of maize are called

---

[7]Columbus's voyages across the Atlantic Ocean began a European effort at exploration and colonization of the western hemispere. Columbus's four voyages (between 1492 and 1504) came at a time when there was growing national imperialism and economic competition among various European nation states. The national imperial expansion occurred under the rule of Charles V, who in 1516 took the throne of Spain upon his marriage to Princess Joanna, the second daughter of King Ferdinand and Isabella of Castile (Elliot 1963, pp. 2–3). Charles V, later, took power as Holy Roman Emperor in 1519, just before Cortés landed on the present-day Veracruz coast (Madariaga 1947; Maltby 2002). Many western ruling families in both the Christian and Islamic cultures began seeking wealth by establishing trade routes and colonies in other parts of the world (Madariaga 1947; Maltby 2002; and Schiebinger 2004).

Fig. 2.4 Wood carving of Christopher Columbus meeting the Taíno on the Island of Hispaniola. First-hand information on native dress was the basis for this depiction by the woodcarver. The Mediterranean rowing gallery indicates the wood cutter was familiar with what kinds of ships used in the expedition (from Giuliano Dati's version of a letter from Christopher Columbus dated to 1493) (from Milanich and Milbrath 1989)

"Turky Wheat" or foreign wheat. Maize was a source of wonder and fascination to Renaissance Europe (Staller 2009, Fig. 2.7). In sixteenth century Europe, most people believed Turkish traders spread maize to regional food markets throughout the Mediterranean, but the origins of the plant were shrouded in mystery (Weatherwax 1954). The term "Turkey" and/or "Turkish" came to refer to many animals and plants foreign to western eyes and cuisines[8] (Coe 1994b). Spanish traders initially spread maize, then called "Turkish Corn" or "Turkey Wheat" throughout the Mediterranean in the early sixteenth century and Venetian traders are

---

[8]The bird turkey (*Meleagris gallopavo* L.) derives its name from these early trading patterns. European explorers who first encountered turkeys in the New World believed the birds to be type of guineafowl, and because of their importation to Central Europe by Turkish traders, there was a general tendency at that time to attribute exotic animals and plants to far-off places (Coe 1994b). During the sixteenth century, anything "Turkey" or "Turkish" was synonymous with "foreign" (Madariaga 1947; Elliot 1963).

Fig. 2.5a (**a**) In sixteenth century Renaissance Europe, maize (*Zea mays* L.) fascinated elites and commoners alike. Many Europeans thought Turkish traders spread maize throughout the Mediterranean. In the Gerard (1975 (1633), p. 81) herbal, there is an image of the "Corne of Asia" suggesting some sixteenth century European botanists believed the plant may have been traded from the Far East (Courtesy of Field Museum Library, Chicago) (Photograph by John E. Staller). (**b**) Maize (*Zea mays* L.) was perceived by Europeans to represent the economic equivalent to wheat in the Old World. It was not only a food staple in the eyes of many Europeans, but also a "staff of life." The word maize is from the Taíno-Arawakan word *mahiz* or "life-giver," again reinforcing this association. In the 1633 *General History of Plants* by John Gerard, maize is called "Turky Wheat" or foreign wheat. Gerard appears to differentiate varieties or landraces on the basis of kernel color, although the description mentions kernel shape and ear morphology as well. During the sixteenth century, Europeans perceived maize as a grain and emphasized its economic and religious importance (from Gerard 1975 (1633), p. 82) (Courtesy of Field Museum Library, Chicago) (Photograph by John E. Staller)

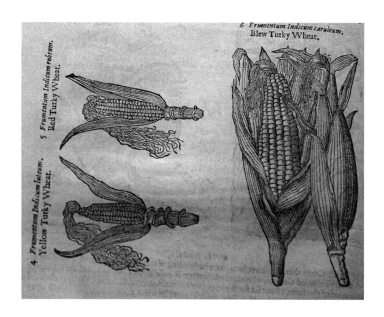

Fig. 2.5b (Continued)

believed to have subsequently traded it to the Far East (Fig. 2.5b) (Tannahill 1973, p. 246). This terminology later confused scholars who studied the ethnohistoric sources as many chroniclers referred to maize beer as "wine of wheat or maize" (Bruman 2000, p. 78; see also Weatherwax 1954). Such terminology may also have influenced how Europeans and Old World cultures incorporated the plant into their cuisines. Portuguese explorers and slave traders first introduced maize to Africa where it quickly integrated into the local economies. Maize was initially referred to by its common Taíno-Arawakan name, *mahiz* or "life-giver" (Weatherwax 1954; Sauer 1969). Such references and other references in early accounts and pre-Linnaean herbals appear to have later predisposed Europeans to emphasize its economic importance to the rise of New World civilizations.

Spanish colonial records from the early historic period mention some of the plants that were cultivated, venerated, and esteemed by the Taíno and other Caribbean indigenes. In many accounts, there is clear evidence that maize was venerated and central to indigenous cosmology and religion. Taíno-Arawak societies first encountered by Columbus were said to live in small- to moderate-sized villages of up to a 1,000 people. They were also said to have a strong economic dependence upon cassava [bitter and sweet manioc] and maize (Deagan 1989, p. 49). Columbus brought maize kernels back after his second voyage, which the Italian Scholar Pietro Martire d'Anghiera (Peter Martyr) understood to be called *maíz*, hence the modern term "maize" (Staller 2009). The earliest descriptions of maize in Europe were by Martyr who wrote the first accounts of explorations in Central and South America through a series of letters and reports, grouped in the original Latin publications of 1511–1530 into sets of 10 chapters called "*Decades*." The *Decades* were extensively copied and of great value to later scholars of the Age of Exploration. The second chapter of Martyr's *De Orbe Novo* [On the New World], was written to a high-ranking cleric and dated April 29, 1494, and includes an account of the first voyage (Martire 1907 (1516)). This important historical document is the first time that the term the New World appeared in print. Columbus just returned to Spain, in his letter he writes; "*The bearer* [of this letter] *will also give you in my name, certain white and black grains of wheat from which they make bread* [maize]" (Weatherwax 1954, p. 32). Martyr planted some kernels and provided western Europeans with the first description of the plant (Sauer 1969, p. 55). Martyr began his account in *De Orbe Novo Decades* saying; "*they make bread from a certain grain,*" noting this grain was snow-white inside (Sauer 1969, p. 55; Martire 1907 (1516)). A clear description of the maize plant by Como dating to the second voyage of Columbus and published in Latin by Scillacio on December 6, 1494 is noteworthy. He states; " ...*it is a grain of very high yield, of the size of the lupine* (Mediterranean white lupine), *of the roundness of the chick-pea* (*cicer*), *and yields a meal* (*farina*) *ground to a very fine powder*; ... *and yields a bread of very good taste*" (Coma, quoted in Sauer 1969, p. 55). Coma goes on to say that many "*Indios*" chewed the seeds when they were in need of nourishment.[9]

Oviedo (1959 (1526), Chap. 4) later wrote that Caribbean islanders mainly consumed maize by roasting the ears or parching dry seeds, and that it was not a major foodstuff. Many early explorers in North and South America mention that indigenous peoples as far a field as the Huron and the Andean peoples of the Sierra the Ancash prepared corn by soaking ears in water for as long as 2 or 3 months and cooking it (Cutler 2001, pp. 16–17). The earliest accounts appear to emphasize that maize was consumed as a grain, was an economic staple, and that Europeans, particularly the early explorers, seemed to prefer it, along with beans and squash, over other more exotic New World plant species such as cassava (*Manihot* spp.). Manioc (*Manihot esculenta* Crantz) was called *yuca*, in the 1633 British herbal by Gerard (1975 (1633), Chap. 155, p. 346) and as in the herbal, "*The root whereof the bread Casua or Cazava (cassava) is made.*" *Yuca* includes both bitter manioc, and sweet manioc and was according to chroniclers, intensively exploited. Manioc was also called cassava and *mogo* or *mandioca*. Manioc is a woody shrub of the Euphorbiaceae family, native to South America, and presently the third largest source of carbohydrates in the world. Tubers and root crops were very important to the pre-Columbian Caribbean and Middle American diet because they were not as susceptible to climatic perturbations (rains, hurricanes, etc.,) and able to withstand drought when maize crops would have been destroyed (Freidel and Reilly 2009).

The next earliest account of maize comes from Michele de Cuneo. He was a native of Savona, who traveled with Columbus on his second expedition. He wrote

---

[9]The last statement by Coma probably refers to parched maize.

Gerolamo Annari a letter dated October 15–28 1495, in reference to the abundance of green vegetables in the New World and that a certain "melic" (apparently maize) " ... *is not very good for us*" and that it "*tastes like acorns*" (Gerbi 1985, p. 32). Michele de Cuneo attempted to write a natural history of the New World by providing vague descriptions and tasting the fruits and vegetables to determine whether they were like those in Europe, "*to our taste*" or "*fit only for pigs*" (Gerbi 1985, p. 33). Most Spanish sailing expeditions from the West Indies were stocked with cassava bread and pigs (Innes 1969). His tasting and experimenting was the first step to methodologically empirical investigation, but in no way approaches the later *Sumario* of Gonzalo Fernández de Oviedo (1969 (1526)), which was the first serious scientific attempt at a natural history of the New World. Cuneo's opinion of maize was not generally shared by the Conquistadors and most of the other explorers or chroniclers (Weatherwax 1954, pp. 28–30). Chronicler accounts repeatedly state outright or in passing that they preferred maize over most other economic staples, and later demanded it in large quantities as tribute during the Colonial Period (Ochoa and Jaime Riverón 2005). Nevertheless, most of the early sources describe the most prevalent indigenous cultigens in the agricultural plots of the Caribbean as consisting of root crops and tubers (Fernández de Oviedo 1969 (1535), pp. 12–14). The preference for maize over other indigenous crops such as cassava becomes apparent in many later accounts and may have influenced archaeological perceptions of its economic role in the Neotropics.

It appears that these early accounts, to varying degrees, evoked a perception among Europeans to perceive maize as analogous to the Old World grains, wheat and barley. Maize was brought to be cultivated in Spain early on. In his *relaciones y cartas* of 1498, Columbus wrote that "*there was a lot of it* [maize] *in Castile,*" and earlier in 1493, Peter Martyr mentions that in the West Indies, the Indians "*produce without much trouble, a bread that is sort of a millet, similar to what is found in abundance around Milan and in Andalusia ...,*" indicating that maize was brought back to Europe after the first voyage or that Columbus confused maize with sorghum and millet (Weatherwax 1954, pp. 5, 29, 33; Gerbi 1985, p. 189; Martire 1907 (1516)). Las Casas (1971 (1527–1565), p. 110) mentions that in parts of what is today Haiti, they cultivated and harvested a plant throughout the year, or at least twice a year, "the grain *maize*, which the Admiral calls *panizo* ..." Columbus was in this case using a Spanish term for grain sorghum or millet (Weatherwax 1954, p. 29; Martire 1907 (1516)). The ethnohistoric literature provides detailed descriptions of how maize was consumed in different parts of the western hemispere at the time of contact. These descriptions suggest a much more complex role than has generally been emphasized and later reported by many scholars in the social sciences and humanities.

Oviedo (1959 (1526)) described pre-Hispanic agricultural plots as fertilized by cutting and burning vegetation to clear ground. He states that; "*The Indians first cut down the cane and trees where they wish to plant it* [maize] ... *After the trees and cane have been felled and the field grubbed, the land is burned over and the ashes are left as dressing for the soil, and this is much better than if the land were fertilized*" (Oviedo (1959 (1526)), pp. 13–14). Most of the early sources emphasize that *yuca* or cassava (bitter and sweet manioc; *Manihot esculenta* and *M. palmate*

Fig. 2.6 Most agricultural economies among New World cultures were dependent, to varying degrees, upon domesticated food plants. A European engraver created this carving based upon Oveido's description of the Arawaks small circular mounds or *conucos* upon which their cultigens were grown. Natives carry water directly to their fields, a form of pot irrigation (from Oviedo's *De la Natural Hystoria de Las Indias* c. 1535) (Courtesy of Field Museum Library, Chicago). (Photograph by John E. Staller)

aipí, respectively) and sweet potatoes (*Ipomea batatas* L.) as the most prevalent cultigens on the indigenous agricultural plots of the Caribbean (Fernández de Oviedo 1969 (1526), pp. 12–14; see also Las Casas 1971 (1527–1565), p. 110). Initial preparation of agricultural plots was in some areas followed by the construction of small earthen mounds or platforms, measuring one foot high and three to four feet in diameter. Upon such mounds they cultivated their various crops (Newsom 2006, p. 328). Ultimately, the fields, "*conucos*," consisted of a series of the small circular earthen mounds or platforms, measuring one foot high and three to four feet in diameter in some areas (Fig. 2.6). A variety of plants, what is generally referred to as, multiple cropping was common to various regions of the Caribbean and Mesoamerica and has been found to enhance environmental quality as well as cultivation, particularly in regions prone to flooding (Mt. Pleasant 2006, pp. 530–531; Newsom 2006, p. 328; see also Sauer 1969, pp. 51–54; Las Casas 1971 (1527–1565), pp. 110–111). Similarly, in Hispanola, what is today the Dominican Republic, there was not a shift to maize cultivation after the arrival of the Spaniards in Oviedo's time (1502–1509) as manioc was much more productive in the mountainous terrain and they supplied themselves with pork and beef as their primary sources of protein (cf. Sauer 1969, p. 157). The observation that there was a *shift* to the cultivation of maize after the arrival of the Spaniards is significant, in that our current perceptions of its importance to pre-Hispanic New World economies may in fact be a byproduct of early European perceptions regarding indigenous economies.

The early account of maize and its preparation in the *Sumario de la Natural Hystoria de Las Indias* is considered by many ethnohistoric scholars as the most accurate and precise. Oviedo based his observations in Cueva country–what is now Costa Rica (Fernández de Oviedo 1959 (1526), Chap. 4, 1969 (1535) pp.10–12).

"Its ears, more or less a geme-long [the span between thumb and forefinger], bear grains of the size of garbanzos [chick-pea]. The planting ground is prepared by clearing the canebrakes and montes, which are then burned, the ashes leaving the ground in better condition than if it had been matured. The Indians then make a hole in the ground by using a pointed stick, about as long as they are tall, and drop seven to eight grains into the opening, repeating the process as a step ahead is taken, and thus they proceed in a file across the clearing. The grain matures in 3–4 months... The Indian women grind the grain on a somewhat concave stone by means of another round one that they hold in their hands, using the strength of their arms, as painters are accustomed to grind their pigments. They add water little by little which is mixed with the grain as it is ground and results in a sort of paste resembling dough [masa]. They take some of this and wrap it in a leaf of some plant or in one of the maize. This is put into the hot embers and baked until it is dry and turns into something like white bread, with a crust on the outside, the inside of the roll being crumbly and somewhat softer ... such rolls may also be prepared by boiling but are inferior. This bread, baked or boiled, keeps only a few days. In four to five days it moulds and is no longer fit to eat."

This description refers to a floury white variety that is soft enough to be ground into meal, and does not need to be soaked in lime. The first printed botanical illustration of maize that made a lasting impression on Western culture was published in 1542 in the "*De historía stirpium commentarii insignes...*" first published in Latin. The German physician Fuchs' attempted to identify plants described by the classical authors of ancient Greek mythology, Plato Socrates, Homer etc., (Weatherwax 1954, p. 35; see also Fuchs 1542). The herbal[10] by Leonhard Fuchs was translated and published in German the following year and the illustration of what he named "Turkish Corn" is taken from a German wood-cutting that appears to depict a popcorn variety[11] (Fig. 2.7). Leonhard Fuchs was one of the founding fathers of botany. He initially set out to identify and illustrate plants described by classical authors. The 1543 German edition has descriptions of about 100 domesticated and 400 wild plant species as well as their medical uses (*Krafft und Würckung*) in alphabetical order (Fig. 2.8). Fuchs calls the maize plant *Frumentum Turcicum* believing it brought into Germany from Asia by Turks, who were said to have planted it when other grains were scarce (Fuchs 1978 (1543), *fol.* 473; Finan 1950, p. 159). Another woodcut, illustrated in a later Italian edition[12] of Oviedo's *Sumario de la Natural Hystoria*, in 1552 also appears to be depicting a popcorn variety (Fig. 2.9). These images of maize as a grain, a primary food crop,

---

[10]The first botanical illustration of maize appeared in 1535 in a little known book by Jerome Bock. Most early depictions were in *herbals*. *Herbals* were species lists popular in western Europe during the sixteenth and seventeenth centuries as they satisfied an interest in the flora and fauna that grew around many European societies. *Herbals* were essentially precursors to regional flora lists documented by modern ecologists and botanists (Weatherwax 1954, p. 35).

[11]Fuchs made no attempt at a natural system of classification, but the woodcuts were however based upon first-hand descriptions, as well as botanical samples and are anatomically and morphologically accurate. Many later sixteenth – eighteenth century herbals have copied images from this herbal. Fuchs' herbal includes 512 images of plants, largely locally grown, and printed from woodcuts. These include some of the earliest depictions of maize and chilli peppers in the Old World.

[12]Oviedo wrote two books about the natural history of the Americas, a shorter and longer version of the same theme, and an image of maize first appeared in an Italian translation, edited by Ramusio and first published around 1552 in a series of editions (Weatherwax 1954, p. 37, Fig. 14).

TVRCICVM
FRVMENTVM
Turcḟid̕ torn.

Fig. 2.7 One of the earliest and influential botanical depictions of maize "Turkish corn" in *De historía stirpium* by Leonhard Fuchs, 1542 (from Weatherwax 1954, p. 35)

are further enhanced by the descriptions of its preparation. The accounts state the dough or *masa* was shaped into rolls, wrapped in leaves then baked or boiled. This preparation is in contrast to what occurs in Mexico, where maize kernels are first soaked, then ground back and forth with grinding implements, i.e., *manos* and *metates*, and baked into thin cakes or *tortillas*[13] (Sauer 1969, p. 242; see also Díaz

[13]The Mexican today make tortillas (*tlaxcalli*) or maize bread by soaking the kernels in water with ashes and chunks of lime. When the kernels have softened, the mixture is heated to the boiling point, and then placed upon a metate (*metlatl*) or griddle where it is ground into flour from which small, thin cakes are padded into shape. The cakes are then baked on a dry griddle (Durán 1971 (1588), p. 136 *f*. 6).

Fig. 2.8 Leonhard Fuchs (1501–1566) was a German physician and generally considered to be one of the founding fathers of botany and the biological sciences. His interest in botany grew from his research on medicinal plants mentioned in the classic Roman and Greek literatures mainly in the *materia medica*. His botanical in Latin and the 1543 herbal printed in German had a profound influence upon botanists and plant biologists in later centuries

del Castillo 1953 (1467–1575), p. 90; Fernández de Oviedo 1969 (1524), p. 11). The chroniclers emphasize that two types of maize were grown in the Caribbean, popcorn and a floury (white) variety, and the ethnohistoric evidence suggests they had slightly different maturation periods (see Newsom 2006, p. 328, Fig. 23-2; see also Anderson 1947a). The following excerpt from Layfield (1995 (1598), p. 154) is an example:

> "Beyond cassava they depend on corn, from which a very fine bread is made, which they make much use of. There are two classes of corn; the smallest does not differ much from

Fig. 2.9 The earliest printed
depiction of a floury variety
maize, presumably a popcorn
in Oviedo's *De la Natural
Hystoria de Las Indias*,
Seville. Although this image
was believed to have been
published in the 1535 edition
of his work, many scholars
now maintain it first appeared
in the second edition of *De la
Natural Hystoria de Las
Indias* published in 1552
(from Finan 1950, Fig. 1)

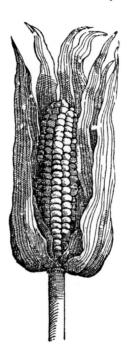

rice in proportion, size, and flavor. I never saw this [type] in crop or raw, but I have seen it in the bowl, and at first I took it for rice, except that I believed it was a little inflated. Those that ate it [said] it tasted like rice. I have seen the other class in fields, and it is the same, or appears very similar, to the grain that we call wheat. It grows with a knotted stalk like a cane with large scattered leaves. It grows up to a height of at least a fathom and a half, and the ear grows from the very tip[14]"

However, unlike cassava, corn bread does not store well, but cobs do, and they can be easily transported (Sauer 1969, p. 242; see also Raymond and DeBoer 2006). Diego de Landa states that the Yucatan Maya would make balls of maize dough that would, "*last several months and only become sour* [but do not go bad]. *From the rest they take a lump and mix it in a bowl made from the shell of a fruit which grows on a tree and by which God provided them vessels*" (1975 (1566), pp. 66–67). The vessels or containers may be referring in this case to coconuts (*Cocos nucifera* L.), but they could also have been calabashes or bottle gourds (*Lagenaria siceraria*).[15] Early Colonial Period observations emphasize gourds were used as containers and

---

[14]Translation by Lee Newsom (2006, p. 328).

[15]When gourds were introduced into Europe, in the sixteenth century, they were generally associated with Old World species of cucumbers and melons. Many New World introductions of the family Cucurbitaceae, which is predominantly distributed in the tropics, were confused with similar edible fruits from the tropical regions of the Old World. Gourds and melons were amongst the earliest cultivated plants in both the Old and New Worlds.

in fact continued to be widely used until recently when plastics appeared. New World cultures often planted squash with maize and beans, as the cornstalk provided support for the climbing bean-stalks, and also shade for the squash (Mt. Pleasant 2006). Because of the high yield of the plant per unit surface planted, maize became the primary staple for the Spaniards who could store cobs and transport them wherever they traveled (Sauer 1969, p. 242). Chronicler accounts throughout Central and South America repeatedly mention raiding indigenous maize fields and after taking control of regions forcing indigenous populations to grow races, which were of a higher yield, that is cobs with more rows and larger kernels (see, e.g., Huckell 2006; Rainey and Spielmann 2006; and Wright et al. 2005). Ethnohistoric accounts indicate that the sixteenth century Spanish diet in the New World was influenced by native subsistence practices, and aboriginal food ways were, in turn, modified by European contact (Chaney and Deagan 1989, pp. 180–181). Although maize is by far the most important New World food crop in the rest of the world, the discovery of the New World introduced a number of food crops and other plants, which are now staples worldwide (see Coe 1994b; McNeil 2006; Staller and Carrasco 2009; and Staller et al. 2006, 2009). Old World cultigens and domesticated livestock were soon integrated into native subsistence practices, as were metal hoes and other farm implements such as horse draw plows.[16] These factors had an important effect on the cultivation of maize as a primary subsistence crop, as did the imposition of religious clerics in forcing native populations into sedentary communities (Chaney and Deagan 1989, p. 181). The discovery of the New World touched off an unprecedented movement of flora and fauna to both sides of the Atlantic, that essentially restructured the world agricultural map (Schienbinger and Swan 2005, p.9). It is the height of irony that after the publications of Carl Linnaeus European efforts to scientifically classify the natural world coincided paradoxically with the large-scale alteration of flora and fauna worldwide through the military, economic, and political activities of western colonial powers (Schienbinger and Swan 2005, p. 8).

In 1509, Columbus's son Diego de Velásquez became governor of Española. The boom in gold mining there was as short-lived as the indigenous population, so the Spaniards colonized Puerto Rico, Cuba and Jamaica. Christopher Columbus had visited Jamaica on his second and fourth voyages, spending almost a year on that island. There was little by way of gold there, but all commentators marveled at its food supplies and its peaceful inhabitants. The indigenous populations of Jamaica grew cassava, maize, and cotton, which they traded out to towns in mainland Panama and Cuba. Colonization of Jamaica was similar to Española; due to disease, native populations catastrophically declined, and were soon replaced by cattle, pigs,

---

[16]New World crops such as maize, beans and squash replaced Old World legumes and cereal grains in the Spanish diet, but indigenous cultures also began to incorporate such European crops into their diets (Landa 1975(1566), p. 67; see also Staller and Carrasco 2009; Staller 2009). Particularly important European contributions to the neotropic diet were poultry, pork, beef, and to a lesser extent mutton. Native techniques for cultivating indigenous crops and hunting wild game and fish were also adopted by the Spanish (Chaney and Deagan 1989, p. 180).

sheep and goats from Europe (Traboulay 1994, p. 34). Archaeological evidence from the site of En Bas Saline suggests the now extinct Taíno-Arawak societies cultivated cassava and maize and were heavily dependant upon maritime resources. Chroniclers observed that these societies carried out rituals and religious activities that involved the veneration of idols or *Zemis* made from stone, ceramics, bone wood and cotton (Deagan 1989, p. 49). Newsom (2006, pp. 331, 333) states that the maize varieties from this region and at En Bas Saline were generally associated with high status individuals. The Caribbean evidence suggests that maize was integral to ritual feasting and communal-ritual activities, and may have been consumed as beer. It is more probable that the popcorn variety identified by Newsom was used for fermentation, while the floury variety was consumed in the manner previously described (Newsom 2006, pp. 331, 333).

2.3   Central America and Mexico

On Christopher Columbus' last voyage, they were traveling along the coast of what is today known as Costa Rica (Veragua), searching for gold. A youth of 14, named Ferdinand wrote, "*They have for their nourishment also much maize... from which they make wine and red wine, as beer is made in England, and they add spices to their taste by which it gets a good taste like sour wine ....*" Despite the fact that he knew maize was an economic staple in this region, the youth was more interested in the beer that was made from its maize kernels (Sauer 1969, p. 133). The societies of Costa Rica apparently did not make bread from cassava or manioc like those of the West Indies, but rather they made cakes from maize (Fig. 2.10). The Italian explorer Girolamo Benzoni recorded the three steps involved in making corn cakes, boiling the maize in lime, grinding husked corn to make dough, which was then patted into small cakes and then cooked on a *comal* (*comalli*) or griddle (Staller 2006b, Fig. 32-1B). Such griddles have been identified in archaeological sites throughout northern South America, the Amazon Basin, and eastern Central America. They continued to be used among indigenous populations in these regions until recent times (Fig. 2.11). The labor surrounding the preparation of maize for is different regarding gender. According to Landa (1975 (1566), p. 66), maize was the principal subsistence crop of the Yucatan. During the sixteenth century as now, men worked the *milpa,* performed appropriate rituals, and planted and harvested the maize crop, while Maya women performed much the same tasks for the preparation of maize, primarily grinding and cooking, as mentioned in the *Popol Vuh* and *Chilam Balam* (Tedlock 1985; Christenson 2003). Present-day Maya still perform their tasks regarding maize as they did in the sixteenth century; men labor in the *milpa*, while women would grind corn and make tortillas. Tortilla production is labor-intensive, a result of hours of cultivating and tending in the field and grinding with *mano* and *metate* or standing with a griddle in front of the hearth. Indigenous populations in Mexico still tend to take as much pride in their tortillas, as they do in maintaining ethnically distinct morphological maize varieties (Raven 2005; Benz

Fig. 2.10 Fanciful depiction of native Amerindians drinking as imagined by Theodore de Bry c. 1594. Women in the lower left corner initiate the fermentation process by chewing the kernels (maize?) with their saliva into the bowl. The mashed fermenting maize kernels were strained then boiled with lime, as depicted in the foreground at the right and center, respectively. Natives drink and smoke cigars in the central image, and males dance, drink, and shake rattles in the background. The woodcarvings of the Italian explorer Girolamo Benzoni who recorded the fermentation process while traveling in Central and South America between 1541and 1555 may have influenced this depiction (from Bradford 1973, p. 136)

et al. 2007). Landa goes on to state that, Maya women had to grind their husbands' corn by hand on the *metate* and that maize was the principal food crop among the Maya populations in the Yucatan Peninsula. Diego Landa (1975 (1566)) wrote that the Maya would

"... make various kinds of food and drink; and even when it is drunk [instead of being eaten] it serves them for both food and drink. The Indian women leave the maize to soak overnight in lime and water so that by the morning it is soft and therefore partly prepared; in this fashion, the husk and the stalk are separated from the grain. They grind it between stones ..." (pp. 66–67).

The Maya appear to have drunken more maize than other Mesoamerican cultures. In Santiago Atitlán highland, Maya would make a gruel called "*maatz*" and drink this during ceremonies associated with the planting of crops (Christenson 2001, p. 123; see also Staller 2009). They prepared *maatz* by making flour from toasted maize and then mixing this with ash and water (Carlsen and Prechtel 1991,

Fig. 2.11 Maya woman from San Antonio Agua Calientes, Guatemala in a *comal* (*comalli*) prepares tortillas on a griddle. Note the various griddles standing along the back wall on the left side of the image. (Courtesy of Michael D. Carrasco). (Photograph by Michael D. Carrasco)

p. 31–32). Landa (1975 (1566), p. 67) observed they used finely ground maize to "extract a milk," which they thickened over a fire to make a kind of porridge or gruel and also made a drink from "raw" ground maize kernels and bread in a number of different ways. The finely ground maize from which the milky substance was extracted symbolized mother's milk or semen among some Maya societies, and its consumption is connected to concepts of life-renewal, rebirth and regeneration from death (Freidel and Reilly 2009; Freidel et al. 1993, p. 180; Christenson 2001, p. 123; and Sachse 2008, p. 140). *Maatz*, central to certain feasts, was often served in elaborately painted ceramic vessels, which were themselves of significant ritual and political importance (Landa 1975 (1566), pp. 68–69; Coe and Coe 1996, pp. 43–54). The Yucatecan Maya would also toast and grind the maize and dilute it with chilli pepper (*Capsicum annuum* L.) and cacao (*Theobroma cacao* L.) to make a drink and that they made a foaming drink generally associated with the caciques and elite that was central to their ceremonial feasts (Landa 1975 (1566), p. 67; Díaz del Castillo 1953 (1467–1575)). Landa (1975 (1566), p. 67) states that, "*From ground maize and cacao they make a foaming drink with which they celebrate their feasts.*" Beverages made from maize and cacao, have clear and unambiguous associations to high-status individuals in Mesoamerica, as well as to mythological beings (Taube 1985, 1989; Coe and Coe 1996). Like maize, cacao is linked to fertility, the rebirth of ancestors, the feminine, in the iconography and death as funerary offerings (McNeil 2006, pp. 360–362, 2009).

Trade involving cacao and maize along the coast of the Yucatan, Tabasco, and Veracruz was largely focused upon providing the necessary ingredients to consume maize in the form of a beverage. Honey was also sometimes used in these drinks, particularly in the Yucatan Peninsula (Landa 1975 (1566); Díaz del Castillo 1953

(1467–1575)). Some chroniclers noted that maize flour could not be kneaded like wheat flour, and when indigenous peoples would sometimes make a loaf of bread from corn flour that "it is worthless" (Landa 1975 (1566), p. 67). Landa states that, *"They make bread in a number of ways; and it is a good and healthy bread; but it is bad to eat when cold so the Indian women go to pains to make it twice a day* (1975 (1566), p. 67). The Maya and other Mesoamerican societies would use maize when they made stews, which would sometimes include a variety of other vegetables, and aquatic resources, deer meat and/or wild and tame fowl (Landa 1975 (1566), p. 67). It is perhaps for this reason when Cortés and his army landed in Veracruz the various high-ranking lords they encountered would initially leave food as provisions (see also Staller 2009).

Many sixteenth century accounts provide detailed information regarding what was demanded and provided as tribute by rulers of the various city-states. The Aztecs were said to have had a basic plant diet consisting of maize, amaranth, beans, curcurbits, and chillies at the time of the conquest (Durán 1994 (1588?), 40n). The most important forms of tribute were maize (ears and flour) beans, sage or *chia* seeds, amaranth, pumpkin seeds along with various kinds of chilli peppers (see also Staller 2009, Fig. 2.16). Durán (1994 (1588?), p. 412) states that Tlatelolco warriors laid sacks of ground cacao before Motecuhzoma II as tribute, as well as sacks, *"of toasted maize, maize flour, bean meal, loads of maize bread [tortillas], loads of chillies and of pumpkin seeds."* Maize was also stored in large quantities in the royal storehouses of the various polities[17] (Berdan 1982; Berdan and Anawalt 1992 (1541–1542); and Staller 2009). These accounts imply and in some cases document that the redistribution of food was under the auspices of high ranking lords. When the Aztec Emperor Motecuhzoma II heard of the arrival of Cortés' ships in what is today Veracruz, he ordered them provisions which consisted of honey, turkeys, fish, eggs, maize bread [*tortillas* and *tamales*], and stressed that everything was to be provided in abundance (Díaz del Castillo 1953 (1467–1575), p. 90, 166–167; Durán 1994 (1588?), p. 509). Each of Cortés' soldiers was given a basket of *tortillas* for themselves and another for their horse (Durán 1994 (1588?), p. 510). The absolute power of rulers such as Motecuhzoma II is reflected by the fact that he could redistribute various foods as provisions and that he was aware of precisely where the Conquistadores were within his realm.

The various islands of the West Indies introduced Europeans to an alien world of exotic cultures and never-before-seen flora and fauna. However, it was the conquest of Mesoamerica and the Aztec civilization and later the Incas that brought them the riches and gold they truly sought (Fig. 2.12). In conquering the Aztec, capital of Tenochtitlan (now Mexico City), Hernán Cortés also conquered a large, indigenous empire. The empire extended from the Gulf Coast to the Pacific Ocean across Central Mexico, its northernmost boundary, the present-day Mexican state of

---

[17]A huge driving force of the Aztec state economy involved the destruction of large amounts of prestige goods and even ritually sacrificed humans. Such destructive activities and human sacrifice took place in the course of their state/court rituals and festivals. These cultural patterns suggest that both redistribution and ritual destruction of wealth was central to Contact Period political economies (Carrasco 1999, pp. 81–85).

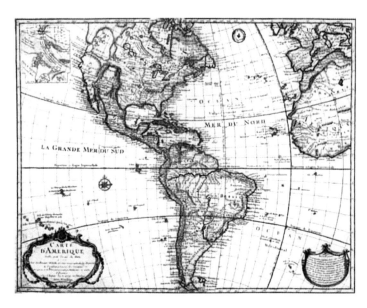

Fig. 2.12 A seventeenth century French map called the Carte D'Amerique was one of the first to show the entire western hemispere. Most of the gold and wealth accumulated by European aristocracy and Spanish explorers and conquistadores was derived from the conquest of the Aztec and Inca civilizations (from Azara 1809, p. 146) (Courtesy of the Library of Congress, Geography and Map Division)

Hidalgo, and its southward control extended to approximately the Mexico-Guatemala border (Baird 1993, p. 4; Schwartz 2000, pp. 4–6). The Aztec formed what is termed a triple alliance with the states of Texcoco, and Tlacopan (present-day Tacuba) thereby taking control of the Valley of Mexico (Fig. 2.13). The Aztec derived their name from their place of origin, Aztlan, which is located at a region called Chicomoztoc or seven caves (Schwartz 2000). Through warfare and alliances, they emerged as the preeminent polity of Mesoamerica for approximately 100 years. At the time Cortés and his army entered the valley of Mexico, the reigning monarch, Motecuhzoma II, in effect, ruled the entire area under the control of the Triple Alliance. Military force had placed and maintained Tenochtitlan in its supreme position (Baird 1993; Berdan 1982). The other great civilization of Mesoamerica was that of the Maya, which at the time of the conquest was made up of several confederated city-states.

2.3.1   Sixteenth Century Agriculture and Plant Cultivation in Mesoamerica

Mesoamerica refers to the areas that in the present encompass the countries of Mexico, Belize, and Guatemala, extending as far north as the Panuco River delta

Fig. 2.13 The Aztec Empire c. 1519 showing the various territories and allied and tributary city-states

and Lerma-Santiago River (Stross 2006, p. 577). This region of the Americas is characterized by great cultural and linguistic diversity, yet there are underlying cosmologies and worldviews, particularly with regard to maize, that unite this culture area (Fig. 2.14). In the Gulf coast and highland regions of central Mesoamerica, large populations were sustained by economies of scale that involved elaborate and complex forms of irrigation technology, floating gardens, and elaborate draining technologies. According to the *Codex Mendoza* (1540s) and the *Relaciones Geográficas* (1580s) and various other chronicler accounts, much of this technology was involved in the cultivation of maize and amaranth (*Amaranthus* spp.) (Berdan and Anawalt 1992 (1541–1542); Ochoa and Riverón, 2005). Amaranth seeds were consumed in the form of a gruel called *pinole*, and sometimes mixed with ground maize in tamales. Intensive agriculture was practiced in various regions of Mesoamerica, particularly along the Gulf Coast, Maya lowlands, and the Valley of Mexico. Although such cultivation involved various plant species, maize was a primary food crop in these regions during the Contact Period. The Aztec developed complex drainage and irrigation systems (*chinampas*) in the valley lakes of Zunpango, Xaltocan, Chalco, Xochimilco and Texcoco (Parsons 2006, pp. 11–15, Fig. 2.2; Staller 2009, Fig. 2.6).

The critical factors to plant cultivation in the Valley of Mexico were adapting to the complex hydrology of the valley, which was highly prone to erosion and maintained by restoration techniques that consisted primary of using organic plant and fish fertilizer and allowing fields to fallow. The other primary challenges

Fig. 2.14 Mesoamerica refers to western, southern, and central Mexico, Guatemala, Belize, El Salvador, and the western parts of Honduras and Nicaragua. This culture area was the focus of major New World civilizations at the time of Spanish Contact, and was long theorized to be a hearth of early agriculture in the New World

were climatic and involved periodic early frost in the higher elevations and drought. The Aztec built elaborate terraces on the surrounding mountain slopes, which covered most of the northern and central slopes in the fifteenth century (Sahagún 1963 (d. 1590)). Farmers maintained these terraces by living among their fields.[18] Familial and/or community identity was closely tied to their agricultural fields, particularly their cornfields or *milpas* (Stross, 2006, p. 581). Fray Bernardino de Sahagún (1963 (d. 1590)) states that in the Valley of Mexico maize was planted in April or early May and highly dependent upon rainfall and temperature. The *chinampas*, which means "fence of reeds" in Nahuatl, was the most complex irrigation technology in the Valley of Mexico (Parsons 2006). The *chinampas* or as the Spaniards referred to them, "floating gardens" were cultivated year-round and provided much of the food resources for the Aztec capital at Tenochtitlan and the great open-air markets there and also at Tlatelolco and Tetzococo (Fig. 2.15). Tlatelolco was at first a separate city on an adjacent island, but was eventually

---

[18]*Chinampas* were not owned by the farmers but the *calpulli*, which is a Nahuatl term meaning "Big house," and referred to the core territorial unit of Aztec social organization (Schwartz 2000, p. 254). *Calpullis* were associated to land units or *barrios* and could refer to a clan system, somewhat analogous to *ayllu* in the Andes (see below). Farmers and their families who worked the *chinampas* and could increase their landholdings if the family increased in size or the *calpulli*-owned vacant land.

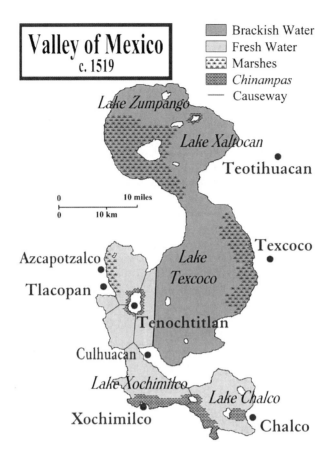

Fig. 2.15 The Valley of Mexico c. 1519 showing the various locations of major centers, artificial causeways, and agricultural features

incorporated as part of greater Tenochtitlan. It also had pyramids, palaces, and markets, but was most famous for its great marketplace (Schwartz 2000, pp. 7–8). Because of its location in the midst of the lake and its maze of canals, many of the first Spanish explorers who described Tenochtitlan compared the city to Venice (e.g., Díaz del Castillo 1953 (1567–1575), pp. 177–178). According to the chronicles, food crops, cotton, maguey and other primary resource commodities were differentiated and classified by indigenous populations on the basis of where they came from geographically (Sahagún 1963 (d. 1590)). Unlike the Andes, Mesoamerican indigenous economies were market-based, and resources were brought to important regional centers from vast distances.

As just outlined, various chronicler accounts emphasize that the Aztec Empire was sustained by intensive agriculture through the creation of rectangular plots or *chinampas*. *Chinampas* were stationary, artificial islands that measured roughly 30 by 2.5 m (Townsend 2000). *Chinampas* fields were used to cultivate maize, beans, squash, amaranth, tomatoes, chillies, and a diverse array of flowers, which were particularly important to festivals and feasts (see Fig. 2.15). The *chinampas* fields around the imperial capital are estimated to have provided enough food to feed one-half to two-thirds of the populace of the city of Tenochtitlan (Townsend 2000, pp. 80–84; Parsons 2006). The island location of the Aztec capital gave the center its peculiar character with the constant traffic of canoes carrying goods to and from the city. Fresh water was supplied by aqueducts, and away from the central precinct many Aztec commoners farmed "floating gardens" or *chinampas*, rectangular plots of silt on which multiple harvests could be made in a single year in a kind of hydroponic agriculture (Schwartz 2000, p. 8; Parsons 2006). The fifteenth century descriptions of the *chinampas* around Tenochtitlan also emphasize the importance of fowl and aquatic resources to the Mexican diet, but state that the difference of the *chinampas* maize fields was that some had "... *corn ripe for the picking, others nearly ripe, still others with corn just spouting, and some with grains just planted. In this way there could never be hunger in that land*" (Durán 1994 (1588?) p. 222). There is archaeological evidence from the Gulf Coast at sites pertaining to the Chontal Maya of irrigated fields in the Canderaria River (see, e.g., Siemens and Puleston, 1972). The chronicler Francisco López de Gómara (1943 (1554), p. 91) informed Cortés in 1519 of such raised fields stating, that their agricultural fields, "*both worked and in fallow ...*" are "*difficult to cross...*" that those on foot could, "*walk on a straight line, crossing ditches at each step*"(see also Siemens and Puleston 1972).

In the Gulf Coast region, the Cempoalan channeled water through a series of aqueducts that flowed from the river into storage tanks or cisterns (Ochoa Salas et al. 2005, p. 39). From these storage facilities, water was channeled to other cisterns through aqueducts until finally emptying into canals. They planted maize, beans, and cotton provided a large surplus that was stored in silos. The Totonac of Hueytlalpan province cultivated three crops of maize each year and one crop of cotton every other year without irrigation. Analysis of the *Relaciones Geográficas* by Lorenzo Ochoa and Olaf Riverón (2005, p. 42), however, indicates that in many regions indigenous populations only had one harvest of maize per year. This would explain why it was an important commercial crop involved in interaction networks among various regions of Mesoamerica. The most well-known form of intensive cultivation involved the floating gardens or *chinampas* (Parsons 2006). Such fields are important in the Contact Period because they were controlled by the state and sustained populations surrounding the Aztec capital of Tenochtitlan and the Chontal Maya in the Gulf Coast lowlands of southeastern Mesoamerica–both regions that were explored by Cortés and his armies.

Various chroniclers report that maize was an important staple crop in the Gulf Coast, but the most important commercial crop was cacao during both pre and post-Contact times (Brown 2005, p. 117). After cacao, the most important commercial crops are maize, beans, manioc, yam, squash, and plantain. The flooded landscape of the Chontal Maya along the Gulf coast of Tabasco represents a unique geography and ecology. Dominated by the Grijalva-Mezcalapa drainages, the undulating streams and lagoons in this region comprise one-third of all the wetlands of Mexico

(Olmsted 1993, p. 657). This is a region where societies had historically been dependent upon aquatic resources and have traded such resources along with plant crops throughout this region of the Gulf Coast (Brown 2005).

2.3.2 Maize and the Chontal Maya

The Chontal Maya had a profound political, economic, and cultural influence on the rest of Mesoamerica (Thompson 1970). During the Contact Period, this region was referred to as Acalan or Acalan-Tixchel and the inhabitants were referred to as the Putun in some accounts (Scholes and Roys 1968, p. 52). J. Eric Thompson (1970, p. 7) called the Chontol Maya the "Phoenicians of the New World" because of their watercraft and reputation as traders (Brown 2005, p. 117). In fact, the name Acalan is from the Nahua word *acalli* or "place of the canoes" (Scholes and Roys 1968, p. 50; Thompson 1970, p. 118). Acalan is situated between Tabasco, the Petén, and SW Yucatan. Cuzumel and Bacalar in the Caribbean coast, where important Putun trading centers were, and interaction was carried out by watercraft and over land with these surrounding regions (Scholes and Roys 1968). Cortés (1963 (1485–1547), p. 421–422) related in his letters that an entire sector of the town consisted of merchants under the supervision of the ruler's brother, and that some of the principal commodities included pine resin, cloth, cacao, and red shell beads (probably, Thorny Oyster or *Spondylus* spp.). He goes on to mention that maize and beans were also traded and that the region was "...very rich in food supplies and there is much honey." The Chontal Maya carried out three annual harvests, the first March-May, is referred to as *marceño* and is primarily focused upon the cultivation of maize (Brown 2005, p. 128).

There is some reason to suspect that intensive cultivation of maize in this region as well as the Yucatan and other regions of present-day Mexico and Central America is related to tribute demanded by the colonial authorities of New Spain (Scholes and Roys 1968, pp. 149–153, 240–241; Mitchem 1989, p. 105). On the other hand, Cortés (1963 (1485–1547?)) received maps from merchant traders from Tabasco, Xicalango and Acalan which showed trade routes in which large quantities of maize, chillies, and other plants, fish and fowl were transported to the Zoque region in Chiapas in early colonial times in exchange for salt and other status-related commodities.

These accounts suggest that even before the creation of the *encomienda* system, maize was an important trade item, particularly to regions such as Zoque where the climate was not ideal for the cultivation of cacao and where maize was largely obtained through interaction networks (Scholes and Roys 1968, p. 39). For example, in the jurisdiction of Villa de los Valles, the Franciscan order established a network of missions for the purposes of "pacifying" the Huastec and Pame Indians of Southern Veracruz, and ultimately used these populations to plant and harvest sugarcane, raise livestock, and engage in artisan crafting (Herrera Casasús 1989). The indigenous populations under the jurisdiction of the Franciscan missions were largely involved in the cultivation of maize and beans for their own sustenance (Zevallos 2005, p. 89). In addition, they grew sugarcane, made and sold *piloncillo*, and engaged in other activities. The populations of Tampico exchanged their fish and shrimp for maize and beans in the local markets. The overall impression given by the *relaciones* is that these populations were specifically adapted to coastal and inland resources, respectively and used markets to obtain resources not available locally.

While marching through Iztapa to Acalan, Cortés and his soldiers were cut off from their supply ships on the coast of Eastern Yucatan (Scholes and Roys 1968, p. 102). After traveling through the dense jungle for 2 days, Cortés and his soldiers were near death from starvation, they came upon an abandoned and burned ceremonial center and discovered a plentiful supply of maize and other foods (Díaz del Castillo 1953 (1567–1575); Cortés 1991 (1519–1526)). Díaz and Mexía formed an advance party for Cortés on this journey and they were instructed to bring him as much food as possible. They collected as much maize, fowl, and other supplies as possible and with 80 Maya carriers brought these things to Cortés and his army (Díaz del Castillo 1953 (1576–1575)). His soldiers, hungry and exhausted, consumed all the food that they brought leaving none for Cortés and his captains. Díaz anticipated that there would be a wild scramble for the food so he hid some of it in the forest, and agreed to share it with Cortés and a captain, Gonzalo de Sandoval (Díaz del Castillo 1953 (1567–1575); Scholes and Roys 1968, pp. 105–106, 111). During the Contact Period, many of the Chontal Maya were bilingual, able to speak both Chontal and Nahua, and Cortés made several early incursions into Acalan and through Cehuache territory (Díaz del Castillo 1953 (1567–1575), p. 90; Scholes and Roys 1968, p. 391). The current consensus is that Cortés's famous translator Doña Malínche (Marina), was Chontal, as this would explain her ability to translate Maya and Nahua, and also her detailed knowledge of the region and talent in cross-cultural negotiations (Schwartz 2000, pp. 64–65, 251, Fig. 5; Brown 2005, p.118) (Fig. 2.16). Doña Marina was a slave of Maya cacique, who was probably given to

Fig. 2.16 Indigenous depiction of Cortés disembarking on the Mexican coast of Veracruz and his interpreter Doña Marina interpreting to him (from Durán 1994 (1581), Chap. 71)

Cortés by the Chontal Maya after their defeat at Potonchan (Schwartz 2000, p. 251). Doña Marina could speak both Nahuatl and Yucatec Maya and she could communicate with Cortés's Maya captive Gerónimo de Aguilar, who had before this time been his principal translator (Díaz del Castillo 1953 (1567–1575), p. 90; Schwartz 2000, pp.66, 251). Doña Marina was crucial to Cortés when he and his army entered the Aztec capital of Tenochtitlan and furthermore she gave him a son Don Martin Cortés (Schwartz 2000, p. 66, 126–129; see also Brown 2005). When Cortés and his men later passed through Acalan in 1525, they were well received and provided with a daily supply of, "*honey, turkeys, maize, copal and a great deal of fruit*" (Scholes and Roys 1968, p. 391).

2.3.3 Storage and Redistribution: Mesoamerican Accounts

The importance of maize to the Spaniards is clearly evidenced in these accounts. One of the reasons that maize was central to the indigenous societies throughout Mesoamerica as well as the Spanish was that it could be easily carried and stored. Colonial administrators throughout New Spain demanded that Indian populations increase their maize production and required that maize be given to the state as tax and tribute. The accounts from tax assessors and colonial bureaucrats indicate that maize was not paid as tax or tribute in many regions of New Spain in the early years because local populations were only cultivating as much as they would consume in an agricultural cycle. Surplus production of maize appears to have been at least in part related to the tax and tribute demands of the indigenous emperors and *caciques* in pre-contact times and later by Spanish colonial administrators, particularly with the onset of the *encomienda* system (Scholes and Roys 1968, pp. 184, 240–243). There is also evidence from these accounts that increased population densities in many regions of Mesoamerica and later New Spain may be related to cumulative effects of increased maize production (Scholes and Roys 1968, p. 301).

The basic plant diet of the Aztecs at the time of the conquest was maize, amaranth, beans, curcurbits, and chillies (Durán 1994 (1588?), 40n; Sahagún 1963 (d. 1590)). The traditional Aztec meal at the time of the conquest consisted of tortillas, a dish of beans and a sauce made from tomatoes or peppers (Sahagún 1963 (1590)). Maize (cobs and flour) was also important as a form of tribute, along with beans, sage or *chian* seeds, amaranth, pumpkin seeds and various kinds of chile peppers (Durán 1994 (1588?), pp. 205, 317; see also Long-Solís 1986). For example, Durán (1994 (1588?), p. 412) mentions that Tlatelolco warriors laid before Motecuhzoma II as tribute, sacks of ground cacao, "*of toasted maize, maize flour, bean meal, loads of maize bread (tortillas), loads of chillies and of pumpkin seeds.*" Maize was also important as provisions for warriors during times of conflict and was stored in large quantities in the royal storehouses of the various city-states. The importance of maize to the Aztec diet is reflected by the complex vocabulary associated with maize cultivation (see, e.g., Berlin 1992). Sahagún (1963 (d. 1590)) states that the Aztec differentiated "tender maize stalk" by the term *xiutoctuli, compala* referred to "rotten maize ear, and so on. Such complex vocabularies were common among all Mesoamerican cultures at the time of the conquest and continue to the present (see, e.g., Berlin 1992; Berlin et al., 1974; Stross 2006, Alcorn 2006; Hill 2006; and Hopkins 2006).

When the Tepeaca rebelled against the Aztecs, Motecuhzoma II requested that neighboring allied city states prepare great quantities of maize cakes (*tamales*), toasted maize kernels, and maize flour (*tortillas*) as well as other important foods such as chillies, and salt (Durán 1994 (1588?), p. 153). Aztec warriors carried provisions when they set out to war. These provisions included, *"toasted maize kernels as well as maize flour, bean flour, toasted tortillas, sun-baked tamales ... great loads of chillies,"* and ground cacao that was formed into small balls (Durán 1994 (1588?), p. 350). Durán later goes on to say that such provisions came from great storehouses and immense bins that were under the control of the lords of the various city-states. Large-scale management and exploitation of wild plant and animal resources such as agave or maguey, as well as an array of aquatic insects and algae sustained and distinguished highland Mesoamerican civilization (Parsons 2006). Such resources were also provided in tribute to Mesoamerican lords (Staller 2009, Fig. 2.16).

In a later chapter, Durán recalls the 3-year drought, which began in the year 1454 called "One Rabbit" (*Ce Tochtli*) by the Aztecs (Durán 1994 (1588?), p. 238). The drought appears to have had a devastating effect upon Central Mexico, as people began to depopulate this region in search of food, Motecuhzoma I agreed with his allies in neighboring Tlacaelel to use the royal storehouses to avoid depopulation (Durán 1994 (1588?), pp. 238–239). He ordered that *"canoes filled with [maize] gruel . . .[and] maize dough was to be cooked in the form of large tamales each one the size of a man's head"* and that these foods were to be dispersed among the Aztec population (Durán 1994 (1588?), p. 239). He ordered, under the pain of death that no grains or maize cobs were to be carried off to other parts. Twenty canoes of *tamales* and another ten of gruel, made from maize kernels and mixed with *chian* (*Salvia hispanica*) seeds entered the city for a period of 1 year, after which the supplies in the storehouses ran out (Durán 1994 (1588?), p. 239).

These accounts suggest that large quantities of maize were kept in such store-houses and that corn was depended upon by these societies as a buffer against climatic fluctuations, brought on as in this case by widespread famine and drought. It is also clear from the descriptions that unlike the Caribbean cultures, the civilizations of Central Mexico depended upon maize for their sustenance and that it was used economically and symbolically by the ruling elite as a means by which to justify their rule and status among their support populations. Moreover, these accounts clearly emphasize the importance of maize consumption in the form of a grain (flour), rather than a vegetable as appears to be the case in other regions of the Americas at the time of contact (Sauer 1950, 1952; see also Staller 2006b). These accounts reinforce more recent research that emphasizes the role of eco-nomic staples in defining and redefining status and class, and the roles of such consumables in reinforcing status differences through its redistribution (Freidel and

Reilly 2009). Control over land, tribute, and economic resources are a pretext of the right to rule and to the production and reproduction of socioeconomic differences and hierarchical relationships. Many of the pictographic and hieroglyphic documents left to us from Central Mexico and other subregions of Mesoamerica deal either directly or indirectly with storage, and keeping track of the movements and redistribution of various commodities, consumables as well as status related items (see, e.g., Berdan and Anawald 1992 (1540–1541)).

Despite the importance of maize to indigenous economies in these regions, it was also a critical food staple to the Spanish explorers, who commonly pillaged and destroyed agricultural fields to obtain maize for their survival. Durán (1994 (1588?), pp. 177–178) mentions that when the Aztec waged war they would pillage the maize fields of their enemies, and kill all the turkeys and domesticated dogs (edible) they came upon, "*just as our own Spaniards do today unless they are controlled*" (Durán 1994 (1588?), p. 178). When such wars were waged, the people in the towns and cities being attacked would hide their "*maize, chillies, turkeys ...and all their possessions*" (Durán 1994 (1588?), p. 178). Rations given to soldiers included "*a large handful of toasted tortillas and another of toasted maize kernels*" (Durán 1994 (1588?), p. 178). When the Aztecs sought to punish the people of Xochimilco the Aztec ruler Itzcoatl ordered five officers and five soldiers to "*...tear out ears of corn and destroy the plants ...*" of their biggest maize field ...(and to) "*devastate the field entirely ...*" (Durán 1994 (1588?), p. 107). When Motecuhzoma called his stone-cutters to carve his image on a rock in Chapultepec he repaid them for their services with "*...loads of maize, beans, chillies, also mantles and clothing for their wives and children*" (Durán 1994 (1588?), p. 481).

These accounts suggest that maize was central to tribute and as payment for services rendered to the state. Its critical importance to local and state economies is reflected in the fact that when conflict and destruction were inflicted among competing Mesoamerican polities, maize fields were often destroyed or pillaged as a form of punishment or retribution (Staller 2009). The preference of maize as a food source among Europeans is also implied by the apparent continuation of such practices by colonial officials upon rebellious or aggressive indigenous population resisting their right to rule. The Spaniards obtained their sustenance either by plunder or barter. By the time of the Spanish conquest of Perú, many of these seasoned explorers and soldiers understood the richness of these exotic New World food crops and how they were consumed (Ceiza de León 1998 (1553), p. 86, n.6).

2.4 First Impressions of Andean Civilization

The chronicles from the Andes, like those from, Mesoamerica, the Caribbean and the Gulf Coast of Mexico, repeatedly mention how the Spanish explorers were seeking maize for their sustenance. These accounts suggest a clear European preference for maize, beans, and squash over other indigenous crops. Mesoamerica, however, is the region in which maize originated and where it was first domesticated. When Charles V sent conquistadores to Andean South America, they encountered one of the most extreme environments in the world, a region where there were different food crops and cuisines than those they had observed and tasted further to the north (Fig. 2.17). Maize also maintained an economic importance among the indigenous Andean cultures, but not to the extent it did in Mesoamerica. Maize was primary consumed as a vegetable in South America,[19] and also as a

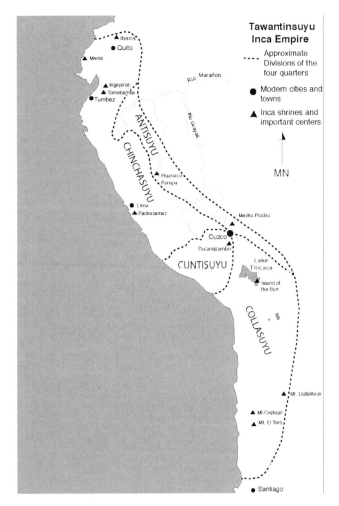

Fig. 2.17 Land of the Four Corners: Map of Tawantinsuyu, the Inca Empire showing the approximate boundaries of the four quarters and their territorial extent

[19]Rather than grinding maize into flour to make cakes or tortillas as in Mesoamerica, the ways it is prepared in the Andes are parching (*chojllo*), popping (*rosetas*), boiling (*chocohoka*), and roasting (*kamcha* or *hamka*) (Valcárcel 1946, pp. 477–482).

fermented intoxicant (Valcárcel 1946, pp. 477–478; Sauer 1950, p. 494; and Staller 2006b, pp. 449–450). Instead of using maize as an economic and symbolic way of justifying their rule over commoner populations and other city-states as in the case of Mesoamerica, maize was used in the context of traditional forms of reciprocity for sealing social and economic alliances as well as a recognition of hierarchy and status (Staller 2006b, pp. 454, 462–464). Despite these differences, the Spaniards had already become accustomed to the food crops in other regions of the Neotropics and sought out maize as a food staple, in part because it could be stored and carried while traveling (Raymond and DeBoer 2006).

While sailing along the eastern Pacific by what is today the Esmeraldas coast of Ecuador (Fig. 2.18), the Spaniards went ashore at a place called, Tacácmez, searching for food, and after much suffering, came upon Atacames[20] (Cieza de León 1998 (1553), p. 84). There they found, "*plenty of maize and other foods to eat...*" He goes on to say,

> "The Spaniards delighted with all the maize they had found, ate leisurely because where there is want, if men have maize, they do not feel it; indeed, a very good honey [maize beer or chicha] can be made from it, as those who have made it know, and as thick as they want, because I have made some when I was there" (Cieza de León 1998 (1553), p. 84).

These comments were made by a Conquistador, in this case, one of the first to write a global history of the Andes. Pedro Ceiza de León had already spent time in other regions of the Americas. This account infers that just a little over a quarter-of-a-century after European contact, there was a clear food preference by the Spaniards for maize over other indigenous crops. Moreover, the cultivation of maize along the coastal streams and on artificial terraces on the sides of mountains further reinforced the European impression that in the Andes, as in New Spain (Mesoamerica) it was a primary food staple, a grain, analogous to wheat and barley in the Old World.

Coastal societies told the Spaniards of a great civilization ruled by a highland culture called the Inca, which spanned most of the Andean cordillera and coast between present-day Santiago, Chile and Quito, Ecuador. In 1531, when the ships carrying Francisco Pizarro and his army landed on the shores of what is today Peru they were amazed at the stark desolation along this desert coast and the nearby mountainous terrain, and at the same time, the complex ecological and cultural diversity high in the cordillera. How could such coastal societies have developed urban centers and constructed such enormous pyramids? The accounts suggest that they believed it was no doubt related to maize cultivation. Maize, cotton and other indigenous root crops such as the potato (*Solanum tuberosum* L.) as well as legumes such as the peanut[21] (*Arachis hypogaea* L.) were grown along the coastal streams

---

[20]Tacácmez refers to the area between the mouth of what is today the Esmeraldas River and Cape San Francisco (Ceiza del León 1994 (1553), 86 *f*. 1).

[21]Peanuts may have originally been domesticated in Peru as early as 7,600 years ago (Dillehay et al. 2007, p. 1890). Their cultivation spread in pre-Hispanic times to Mesoamerica where they were sold in the markets in Tenochtitlan (Cieza de León 1998 (1553)).

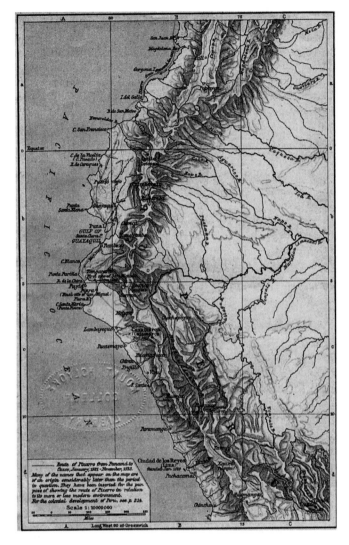

Fig. 2.18 Conquest of Peru. Map showing the early route of Francisco Pizarro and his army. Their first landing was on the northern coast of Ecuador just east of the Esmeraldas River

and seemingly every place where water could be channeled by the vast irrigation networks constructed by the indigenous populations.

When Pizarro and his men arrived in coastal Peru at a place called Peruquete, named after a chief or *cacique*, Pedro Cieza de Leon (1998 (1553) p. 49) states, *"they found nothing but some maize and those roots that they* (the Indians) *eat* (manioc or sweet potato)." Speaking of the coastal valleys of Peru, Ceiza de León (1998 (1553), p. 57) observed that the villages were dispersed and that the Andeans, *"cut the forest on the slopes and sow their maize and other food stuffs."* The Andean economy was on the one hand, based upon local food crops, potatoes, manioc, tubers of various kinds, raising camelids, and large-scale state-sponsored cultivation on artificial terraces build for and controlled by the Inca State (Murra 1973; Morris 1993). Most of the state-sponsored cultivation involved maize, which was consumed in various ways and generally redistributed as a form of reciprocity to subject communities in exchange for labor tasks carried out for the Inca state (Staller 2006b, p. 464). Traditional Andean forms of reciprocity provided the basis for tribute in the form of labor exchanges (*mit'a*).[22] (Valera 1968 (1594); Guaman Poma 1980 (1583–1615); and Murra 1973, 1980). *Mit'a* represented a cultural and symbol association that was ultimately beneficial to all parties, and organized and carried out on a rotating basis by communities (*ayllus*), organized by their community leaders (*curacas*) and is a reflection of the non-market basis of Andean economy (Vega 1966 (1609); Urton 1985; MacCormack 2004; and Cummins 2002).

During the Colonial Period, *mit'a* labor was used to carry out tasks for the interests of the Spanish Crown, who were only able to sustain an altered form of this traditional system of corvée labor by recognizing the authority of Andean *curacas* or community leaders (Cummins 2004, p. 4; Staller 2006b, pp. 252, 262). Subsistence cycles were based in part upon such labor exchanges and were organized by *ayllus*[23] on a rotating basis, while their leaders provided food and beer for the community several times a month as a form of reciprocity (Vega 1966 (1609); Guaman Poma 1980 (1583–1615); and Murra 1973, 1980).

The Spaniards were informed about this enormous empire called land of the four corners or Tawantinsuyu, despite being located in one of the most desolate and extreme landscapes on earth (Fig. 2.19). What was even more remarkable was that this vast empire was connected by a system of paved roads referred to as the Inca highway or *camino real* (Hyslop 1990) Along the *camino Real* were numerous Inca administrative centers that in some cases contained huge storehouses of maize, that were dispersed throughout the highlands and coast (Morris and Thompson 1985; Hyslop 1990). Maize was central to Inca political economy and closely associated with the cult of the Sun and veneration of their dynasty (Staller 2006b, p. 464). John Murra (1973, 1980) concluded on the basis of his reading of the sixteenth century accounts, that the emphasis upon maize as a prestige crop by the state created a split economy based upon local food crops, potatoes, tubers of various kinds, raising camelids, at the local level and large-scale state-sponsored cultivation. State-sponsored cultivation of maize consumed in various forms and generally redistributed as

---

[22]*Mit'a* is a Quechua term used to refer to a form of corvée labor provided to the state for services rendered or commodities provided. In pre-Hispanic, non-market Andean economies, it represented a form of reciprocity between the polity and the community of laborers (Staller 2006b, p. 449).

[23]Quechua term for a landholding collectivity, self-defined in kinship terms, including lineages, which derive their wellbeing and identity from the same locality or place, and through this identity set apart as a distinct social unit (Staller 2006b, p. 453). *Ayllu* is similar in certain respects to the Aztec *calpulli* although in the case of the *calpulli* with reference to barrios in different parts of the city (see e.g., Carrasco 1999; Schwartz 2000).

Fig. 2.19 The Incas had a complex system of roads that spanned the entire length and width of the empire. They also manufactured bridges made of fiber cord, which joined mountain passes and steep ravines. Here the Governor of the bridges (*vedor de puentes*) makes a gesture of counting with his left hand, perhaps to organize the passing on the bridge (from Guaman Poma 1980 (1583–1615), p. 328, *fol.* 356)

a form of reciprocity to subject communities primarily involved the manufacture of maize into beer or *chicha*. Chroniclers in South America noted early on that maize was predominantly consumed as *chicha*, the term they used to designate both alcoholic and non-alcoholic beverages made from a variety of plants and prepared in diverse fashions (La Barre 1938, p. 224; Staller 2006b, p. 449–450). The Quechua term for maize beer is *aqha*; however, the term *chicha* is now commonly used throughout the Andes with reference to maize beer.

Carl Sauer (1950) suggested that the word *chicha* had Arawakan roots, derived from "*chichal*" or "*chichiatl*," "*chi*" meaning "with," and "*chal*" denoting "saliva" or "with saliva," or "to spit," while *chichia* and *atl* mean "to fertilize" and "water," respectively (Nicholson 1960, p. 290; see also Staller 2006b, p. 449). The etymology of the term *chicha* was more recently identified as pertaining to Nahuatl, not Arawakan, derived from the Nahuatl "*chichiya*," which according to the chronicler Molina (1989 (1575)) means, "to become sour, bitter" referring in this case to the results of the fermentation process (Staller 2006b, pp. 449–450). Indigenous Maya populations in various regions of Guatemala still use "*chicha*" to refer to maize-based fermented intoxicants (Staller 2006b, pp. 449–450). Present-day Andeans use the term to refer to maize beer, and in Bolivia, to maize and more rarely to fermented intoxicants made from the Peruvian pepper tree (*Schinus molle* L.). The fruits of *Schinus molle* are bright pink in color and often sold as "pink peppercorns," although *S. molle* is not related to the true pepper (*Piper nigrum* L.), which is an Asian species, native to southern India.

In the Andes, the manufacturing process was generally under the exclusive domain of women (*chicherías*) in the cordillera, while men (*chicheros*), primarily *mit'a* workers, processed maize beer in the desert coast (Arriaga 1968 (1621), p. 12; see also Moore 1989; Rostworowski 1999). *Chicha* was made by allowing the kernels to spout in the shade until the spouts were about an inch in length. The spouted kernels or *joras* were then chewed, to initiate the fermentation process and formed into small balls that are then flattened and laid out to dry. Naturally occurring enzymes in the saliva catalyze the breakdown of starch in the kernels into maltose. The mashed fermented kernels were then strained and boiled and stirred for many hours with lime (Staller 2006b, Fig. 32-1A). Maize beer has a pale straw color, a slightly milky appearance, and a slightly sour aftertaste, a flavor reminiscent of hard apple cider. It is drunk either young and sweet or mature and strong and contains only a small amount of alcohol, about 1–3% (La Barre 1938). Oviedo described the preparation of *chicha* in Panama, pointing out that, as in the Andes, the kernels are allowed to spout (*joras*) and are then boiled in water which contained certain herbs for a considerable amount of time (Fernández de Oviedo 1969 (1535), p. 136). Once the *joras* and herbs are allowed to cool, it is left to sit for 3 or 4 days and then consumed (Bruman 2000, p. 97). However, there is *no* mention of the addition of saliva or chewed maize added to the fermenting mixture. Most chroniclers emphasize the importance of pineapple and plantain to the manufacture of fermented intoxicants in this region of Central America (Bruman 2000, p. 98). There was no indigenous culture in the hemisphere that had the vast system of storehouses, or received tribute in the scale of the Inca Empire.

2.4.1 Storage, Tribute, and Redistribution in the Andes

Inca storehouses were distributed throughout their empire and for the most part held maize, but in the higher *puna* and *altiplano* environmental zones, they stored freeze-dried ducks and tubers such as potatoes or oca (*Oxalis tuberosa* Savign) (Murra 1980). Oca is one of the important staple crops of the Andes, only the potato (*Solanum tuberosum* L.) is more important as a food staple, due primarily to its easy propagation and tolerance for cold and poor soils. The place where the Inca perhaps had their most extensive granaries, warehouses, and storehouses was in the Imperial Capital, Cuzco. These buildings were said to have been built under the orders of their first rulers, who also dictated how the stored commodities were redistributed (Coe 1994b, pp. 195–196). The conquistador Juan de Betanzos 1968 (1551), p. 35, stated that the Inca ruler,

> "... ordered that all the lords and chiefs who were there would meet in his house on a certain day, and when they had come together as he had ordered, and being in his house, he told them that it was necessary that there be in the city of Cuzco warehouses of all the foodstuffs: maize, chile, beans, tarwi, chichas, quinoa, and dried meat, and all other provisions and preserved foods that they have, and therefore, it was necessary that he order them to bring them from their lands. And then Inca Yupanqui showed them certain slopes and mountainsides around the city of Cuzco and visible from it and ordered them to build granaries there, so that when the food was brought there would be somewhere to put it. And the lords went to the sites that the Inca showed them, and got to work, and built the granaries..."[24]

The chronicler Felipe Guaman Poma de Ayala states that Inca law also placed strict regulations on how food crops should be grown, as well as what should be eaten,

> "We command that there be an abundance of food throughout the kingdom, and that they plant very much maize, potatoes, and ocas; and make caui, kaya, chuna, and tamas [various preserved roots], and chochoca [lightly boiled and sun-dried maize]; and quinoa, ulluco, and masua... That they dry all the foods including yuyos [greens] so that there will be food to eat all year round, and that they plant communally... maize, potatoes, chile... Let them make up the accounts each year; if this is not done, the tocricoc [royal official] will be cruelly punished in this kingdom. We command that nobody spill maize, or other foods, or potatoes, nor peel them, because if they had understanding they would weep while being peeled, therefore do not peel them, on pain of punishment" (Guaman Poma 1980 (1583–1615), I p. 164).

The Inca had sumptuary laws forbidding luxury in daily dress and status-related accoutrements such as adornments made of precious metals and gems and did away with banquets and meals (Coe 1994b, p. 200). The Inca put considerable technological effort into transforming the mountainous Andean landscape through the construction of vast irrigation terraces. Inca territorial expansion was based, in part, upon large-scale maize production associated with such widespread terrace construction on mountain slopes both in the highlands and the coast (Murra 1980). The transformation of steep mountainsides into garden terraces has its beginnings in pre-Inca times, but the Inca took this highland Andean practice to an unprecedented scale (Morris 1993). In order to harness the corporate labor necessary to maintain and create such terraces and their associated administrative facilities, the state employed traditional forms of reciprocity, what was called *mit'a*, in which

---

[24]Translation by Sophie Coe 1994b, pp. 195–196.

corporate labor was demanded in exchange for commodities provided by the state in the form of *chicha* and consumables (Murra 1980; Morris 1993). The widespread modification of the natural landscape in the form of mountainside terraces, such as those near Pisac in the Urubamaba Valley grew maize that was manufactured into beer or *chicha* at various Inca administrative centers along the Inca highway (Staller 2006b, pp. 463–464; see also Morris and Thompson 1985). The Inca and other highland cultures provided the Spanish clerics and colonial administrators information on territorial claims and how such consumables were redistributed by the state, as well as mythic histories dealing with Inca dynasty and the legends surrounding the various rulers (see, e.g., Toledo 1940 (1570); Sarmiento 1942 (1572); and Acosta 1961 (1590)).

Mythological history recorded in sixteenth century *relaciones* states that the ruler Inca Pachacuti Yupanqui instituted banquets and ritual feasting as a form of reciprocity for *allyus* involved in corporate labor projects (Sarmiento 1942, (1571)). Research by Thompson and Morris (1985) at the Inca administrative center and *ushnu* at Huanuco Pampa uncovered archaeological evidence of such feasting, banquets, and ceremonial activity (Staller 2006b, Fig. 32-9). Huanuco Pampa was a provincial capital directly beside the Inca highway, in which certain buildings and areas of the site were set aside for *chicha* brewing and ritual feasting and drinking (Staller 2006b, p. 462).These areas included large structures around the main plaza, which were full of organic remains associated with cooking, feasting, and broken drinking vessels or keros – remains from ritual drinking of maize beer. Sophie Coe (1994b, p. 192) states quite rightly that if Inca civilization were to be categorized by Europeans in the present, they would probably be seen as an empire of accountants. Thirty years after the conquest, chroniclers marveled that the Inca rulers and their accountants (*quipucamayus*) could still account for every grain of maize consumed by the armies of the emperor Huayna Capac during his campaigns into northern Ecuador (Fig. 2.20). Jose de Acosta states that the Inca state was,

> "...an admirable and provident government, because being neither religious, nor Christian, the Indians in their own manner reached this high perfection, having nothing of their own, yet providing everything necessary to everybody, and sustaining so generously the things of religion, and those of their lord and master." (Acosta 1961 (1590), p. 196)

The chronicler Martín de Murua pointed out that maize and a host of other food crops had close association to the ruling elite (see also Staller 2006b, 2008b). The Inca promoted laws in which the poor, destitute, blind, crippled and maimed were provided for with food and cloth from the public storehouses (Coe 1994b, p. 200). The emperor Huayna Capac was described by Pedro Pizarro (1978 (1571), p. 49) as,

> "...a great friend to the poor, and ordered that especial care be taken of them in all his dominions...They say that he drank more than three Indians put together, but nobody ever saw him drunk, and his captains and great lords asked him how he could drink so much and not get drunk, he said that he was drinking for the poor, because he was concerned with their sustenance."

These accounts appear to suggest that of all the consumables cultivated and stored by the state, maize is the plant with the closest symbolic association with the

Fig. 2.20 Felipe Guaman Poma's depiction of Collca the Inca storehouses. He writes "Topa Ynga Yupanqui" at the top of the far left storehouse, indicating the man represents the Inca emperor Topa Inca Yupanqui. The emperor gestures to the administrator of the storehouses Poma Chaua, his identity written on the far right storehouse. Poma Chaua holds the knotted cords or *quipus* which Inca *quipucamayus* used to keep track of the items stored in this facility (from Guaman Poma 1980 (1583–1615), p. 309 *fol.* 335)

elite. For example, the principal wife or *coya* of the first Inca emperor Manco Capac and her successors were said to have consumed maize in various forms. Indigenous Inca informants informed Murua what the first *coya* ate;

> "Her daily food was usually maize either as locros anca (stew), or mote (boiled maize grains), mixed in various manners with other foods, cooked or otherwise prepared. For us these are coarse and uncouth foods, but for them they were as excellent and savory as the softest and most delicate dishes put on the tables of the rulers and monarchs of our Europe. Her drink was a very delicate chicha, which among them was as highly esteemed as the fine vintage wines of Spain. There were a thousand ways of making this chicha... and the maidens of her household took great pains with it." (Murua 1962 (1590), p. 29)

The accounts emphasize that *chicha* was central to maize consumption in the Andes. The primary food was described in one relación as "*maize and chile and greens*" and little protein, which was derived mainly from maritime or aquatic resources, and "*for this reason they are so fond of drinking chicha, because it fills their bellies and nourishes them*" (cf. Coe 1994b, p. 205). The Andean cultures also derived considerable meat protein from camelids, particularly llama and alpaca, however even in this case such mammals were commonly offered as sacrifice (Guaman Poma 1980 (1583–1615)).

In summary, the primary consumption of maize in the Andes during the Contact Period appears – in contrast to Central and Mesoamerica – to have been in the form of beer[25] rather than as bread or *tortillas* (Staller 2006b, p. 449). When consumed as a fermented intoxicant, it had a profound significance to religious rituals and rites, as well as a symbolic significance to forms of traditional reciprocity, tribute and to *mit'a* obligations associated with corvée labor (Cutler and Cardenas 1947; Nicholson 1960; Murra 1973, 1980; and Morris 1993). Maize beer played a fundamental role in most all Andean fertility rites and was perceived as linking humans to the spiritual realm through the fecundity of the earth (Otero 1951; Coe 1994b). The ancient cultural and symbolic association between *chicha* and the earth (*pachamama*) is reflected in Andean customs, when noting that if it is spilled accidentally the rationalization or explanation is that, "the earth is thirsty" (Jimenez Borja 1953) Andean girls in some regions spill *chicha* on the earth during planting to make it more "fertile" (Nicholson 1960). Cutler and Cardenas (1947, p. 33) suggest that mildly alcoholic brews were so common in some regions of Bolivia and Peru, they may have contributed substantially to the ancient diet. An appreciation of the role of such customs requires some recognition of the inter-relatedness of the cultural and natural world to Andean worldview and cosmology (Rostworowski 1986; Staller 2008a). At the time of the conquest, it was common for Andean cultures to construct platforms and sacred oracles in the form of cut or sculpted stone on the summits of mountains and hills; these modifications of the natural landscapes are a reflection of a widespread concept of a sacred landscape (Staller 2008b; see also Townsend 1992)

2.4.2   Maize and Andean Political Economy

The Inca emphasized maize as a prestige crop, rather than as an economic staple, as appears to have been the case in Mesoamerica. Maize was also used to fostering social alliances in the context of long-distance interaction, and chroniclers state that it was redistributed by the Inca as a form of reciprocity in exchange for corveé labor (Morris 1979, 1993; Murra 1980; and Rostworowski 1986). The significance of maize to indigenous Andean culture is attested by numerous ethnohistoric accounts,

---

[25]The Quechua word for maize beer is *akka* or *aqha* while the Aymara term is *kufa* (Nicholson 1960). Indigenous Andean populations in the present generally refer to maize beer as *chicha*.

which state that when it was consumed in various forms it was an acknowledgement of rank and power (Betanzos 1987 (1551); Valera 1968 (1594); and Cobo 1990 (1653)). The economic significance of maize in the Andes was also closely tied to the Inca Solar deity (*Inti*) and to veneration of former rulers and nobles of high rank and status. Its significance to the ruling class and to dynastic imperatives is evident by its presence in the form of gold maize stalks in the Garden of the Sun located in a courtyard of the Coricancha, the most sacred temple in the imperial capital Cuzco (Cobo 1990 (1653), pp. 48–50; see also Vega 1966 (1609); Ceiza de León 1998 (1554); and Betanzos 1996 (1557)).

The consumption of maize beer in Andean culture played a fundamental role in exchange and particularly in sealing alliances between two or more social groups (Staller 2006b, pp. 454–456). Ethnohistoric accounts indicate that the sharing and drinking *chicha* was not only a gesture of hospitality, but also contained explicit references to hierarchy and power (Cummins 2002; Staller 2006b). *Chicha* offered with the left hand inferred the recipient was of inferior status, while beer offered with the right hand indicated that those individuals were of equal or superior status (Vega 1966 (1609); Guaman Poma 1980 (1583–1615). Maize beer was, therefore, well suited as a social lubricant to interactions and transactions between first-time partners, particularly when the individuals or groups are of unequal status. Ritual feasting and drinking furthermore couches interactions that may include the redistribution of goods as well as corvée labor, and labor exchanges (*mit'a*) into a cultural and symbol association that is ultimately beneficial to all parties (Morris 1979; Murra 1973, 1980; and Staller 2006b).

The symbolic exchange and consumption of maize beer was also common to Andean customs between humans and *huacas* or sacred places in the landscape, i.e., certain mountains, caves, rivers, natural springs, etc. Ceremonial centers, temples and oracles were customarily given maize beer as offerings (Betanzos 1996 (1557)). Sacred Places, material manifestations of the sacred, such as architecture or modifications to the landscape, and even in some cases departed ancestors, were customarily given *chicha* as offerings (Fig. 2.21). Maize beer was of critical importance to rituals and rites associated with the Cult of the Dead and to customs surrounding ancestor veneration, as were textiles and llamas, which were commonly burned in veneration and as ritual offerings (Zuidema 1973; see also Staller 2006b). Andean communities commonly had *huacas* that is, stone or wooden sculptures worshipped as its mythological ancestor (Classen 1993). Cobo (1990 (1653)) observed the Inca toasting their idols, and pouring out much *chicha* in their honor. These idols and oracles or *huacas* were fed and offered drink in return for good health, propitious weather, and bountiful crops (Vega 1966 (1609)). When members of a community or lineages (*ayllus*) exchanged drink or participated in feasting and imbibed fermented intoxicants, it was a reminder of the obligation they had to one another to aid in personal and communal work to sustaining a shared livelihood (Cummins 2002; Staller 2006b). Maize beer played a fundamental role in such activities, in maintaining alliances and carrying out communal labor tasks.

Maize beer was central to Andean political economy as evident from the chronicler accounts, which often emphasized that Inca nobility in Cuzco used it

Fig. 2.21 The emperor or *Sapa* Inca and the Queen (*Mamakilla*) drinking and making offerings of *chicha* to a previous ruler and queen. The deceased ruler sits on a low stool and the mummified queen kneels to the right of him. Mummies of previous rulers and their principal wives were commonly brought out during ritual festivals and rites of passage (from Guaman Poma 1980 (1583–1615), p. 262, *fol.* 287)

to extend and maintain their authority over their vast empire (Morris 1979; Murra 1973). In fact, *aqha* or *chicha* appears to have played a major role to Inca political expansion. Unlike Mesoamerica, all forms of interaction and exchange in indigenous Andean culture were non-market based, and involved in traditional customs surrounding reciprocity and redistribution (Staller 2006b, pp. 455–456; see also Murra 1975; Rostworowski 1977; and Morris 1979).

Redistribution and reciprocity in Andean culture are social and economic aspects of interaction and corvée labor similar to gift-giving in that the "value" of a commodity is culturally associated to the status of the individuals involved in the transaction (Morris 1979; Rostworowski 1977; and Murra 1980). The value of an object or commodity is not only dependent upon *what* is exchanged, but also *who* is exchanging it (Morris 1979) and in some cases even *where* it came from, e.g., *huacas* or oracles, important ceremonial centers (Staller 2006b, 2008b). Exchanges represent a form of reciprocity. Reciprocity and the customs and traditions surrounding this concept permeate all aspects of Andean social structure (Murra 1975; Vega 1966 (1609); Zuidema 1983; and Rostworowski 1999).

Evidence on traditional forms of Andean interaction comes from ethnohistoric documents and the Spaniards generally depended upon local leaders (*curacas*) to carry out orders and perform labor tasks. However, the Spaniards were not particularly predisposed to recognize the subtleties of native customs surrounding hierarchy and interaction. This becomes readily apparent when they attempt to curb native drinking customs and drunkenness (Rostworowski 1977; Staller 2006b). When the Spanish noble Gregorio González de Cuenca outlawed the *curacas* right to redistribute and dispense *aqha* or *chicha*, he almost immediately rescinded the order (Cummins 2002, pp. 42–43). *Curacas* customarily provided their workers drink in exchange for labor or services rendered and without reciprocity in the form of beer or other fermented intoxicants, the communities and their associated *ayllus* felt no obligation to bow to the will of their leaders (Cummins 2002, p. 43; Staller 2006b, p. 462). Subsistence cycles in the Andes were based, in part, upon labor exchanges (*mit´a*) (Murra 1982). Corvée labor and tribute to the state was organized by *ayllus* on a rotating basis (Cummins 2002). Community leaders (*curacas*) provided food and maize beer several times a month in reciprocity (Urton 1985; Cummins 2002; and Staller 2006b). The Jesuit chronicler Blas Valera, noted that the Inca reinstated a custom whereby *curacas* were commanded to hold feasts two or three times a month in their central plazas to make certain that the needs of the sick, lame, and widows were attended (Cummins 2002). Other chroniclers emphasize that such practices and customs were of great antiquity (Vega 1966 (1609); Guaman Poma 1980 (1583–1615)). The sociopolitical relationships between Inca rulers and local *curacas* were fermented by maize beer or *chicha* and multi-varied. Customs surrounding reciprocity and redistribution facilitated the movement of both consumables and sumptuary items throughout the empire in the highlands as well as the coast.

2.5   Maize and Ancient Religion

The sixteenth century accounts from various regions of the Neotropics clearly emphasize the importance of maize to indigenous ritual practices and religious beliefs. It is, however, in Mesoamerica, where maize appears to have played the largest role in ancient religion; this impression is based, in part, on the iconographic and epigraphic evidence from ancient texts and monuments. The Maya carved various plants, both real and imagined on stone monuments, modeled them into figurines, painted them on pottery and murals, and illustrated them in various codices (Taube 1989; Berdan and Anawalt 1992 (1541–1542)). Maize was of such religious and symbolic importance and a powerful icon to the Chontal Maya, that maize was and still is perceived of as the cultural ideal of beauty in its manifestation as the young Maize Lord[26] (Taube 1985, p. 181, 1996). The delicate features and Xoc Monster-*Spondylus* medallion of the young Maya Lord also reflect a feminine nurturing quality, often associated with tending corn among some contemporary Maya. Among societies of lowland Veracruz, maize was also of great religious, ritual and iconographic importance as evidenced by images of the maize deity (young Maize God) in a variety of Mesoamerican cultures from this region (Alcorn et al. 2006, p. 599). Sacred indigenous texts such as the Maya *Popol Vuh*[27] and various Aztec documents convey a clear impression that maize was central to the mythological origins, ethnic identification and very existence of the Mesoamerican people (Stross 2006, p. 578; see also Tedlock 1985; Durán 1994 (1588?); Alcorn et al. 2006).

The sacredness of maize is apparent when noting that Mexican shaman still use maize to interpret omens, make prophecies, divine future events, and its ancient divinity is clear from iconographic and hieroglyphic associations with primary deities and origin myths (Stross 2006, p. 588). The young Maize God of the Maya civilization, like most Mesoamerican deities, was anthropomorphic, in this case in the form of a young male ruler (Fig. 2.22a). In ancient codices like the Dresden Codex,[28] which dates to the eleventh or twelfth century, the young Maize God is shown conversing with Itzamnaaj, the Creator God, a deity often associated with the celestial realm, and with a bountiful harvest (Aveni 2000, p. 221). Ancient Maya codices were usually doubled in folds in an accordion-like fashion in the form of folding-screen texts (Fig. 2.22b). Like most prehistoric codices, pages of the Dresden Codex are made of flattened Amatl paper, that is, "*kopó*," fig-bark covered with a lime plaster or stucco stand some eight inches high, and are as a long as eleven feet in length (Aveni 2000, p. 221). Ancient and colonial period Mayan codices usually contained painted hieroglyphics and images of primary deities on both sides of each page. Most of the four remaining pre-Hispanic codices deal primarily with mythology, history, or astronomy. On page nine of the Dresden Codex, the hieroglyphic text above the central panel on page 9 of the Dresden Codex reads, u-nu-chu po-lo ITZAM-na UT?-?-li "Maize God," UK' WE'. The term "*u-nu-chu po-lo*" refers to a "discussion" or literally translated "putting heads

---

[26]Karl Taube (1985, p. 181) points out, "*his elongated tonsured head mimics the long-tasseled cob. Maize grain, at times infixed into his head, is an identifying feature of his personified nominal glyph.*" Jade ornaments generally associated with the necklace worn by the young Maize Lord evoke verdant, precious qualities of the living plant (Karl Taube (1985, p. 181)).

[27]The *Popol Vuh* is an oral narrative of Quiche Maya mythological origins thought to have been written in the middle of the sixteenth century by Maya speakers (Stross 2006, p. 584).

[28]The conquistador Hernán Cortés sent the Dresden Codex as tribute to King Charles V in 1519 (Madariaga, 1947). The book was then lost and rediscovered later in Vienna in 1739. Since that time, it has been housed in the Royal Library in Dresden, Germany. It was partially destroyed by the firebombing of that city during World War II. Despite the damage inflicted upon the manuscript, the Dresden Codex is considered the most complete of the four remaining American codices (Aveni 2000).

Fig. 2.22a (a) Maya
sculpture of the young Maize
God, commissioned by the
13th ruler "18-Rabbit"
(*Waxaklajuun Ub'aah
K'awiil*). This was one of
eight that were set on the
cornice of Structure 22 at the
ceremonial center of Copan.
Structure 22 was built in A.D.
715 to commemorate the 20th
anniversary of his accession
to the throne. The young
Maize God represents the
Mayan ideal of beauty, and
features prominently in Maya
art during the Classic period
(200 B.C.–A.D. 900). He
personifies the agricultural
cycle and is associated with
abundance and prosperity.
The head dress in this
sculpture is a stylized ear of
maize, his hair, the silk of the
cob (from Weatherwax 1954,
Fig. 75). (**b**) Pre-Hispanic
Maya codices with images of
the young Maize God include
the Dresden Codex. Dating to
the eleventh or twelfth
century, the Dresden Codex
comes from the Maya
ceremonial center of Chichén
Itzá. On page nine, the central
panel shows a seated young
Maize God (*left*) speaking
with the creator god,
Itzamnaaj (*right*), to bring a
bountiful harvest. Itzamnaaj
is teaching the ancestral hero
twins about their dead father's
(One Hunahpu) severed head,
presumably how to resurrect
their father's "seed" as a
spouting young maize god
sprouting young Maize God.
Mayan codices were
generally painted on both
sides and primarily dealt with
mythology or history

a

b

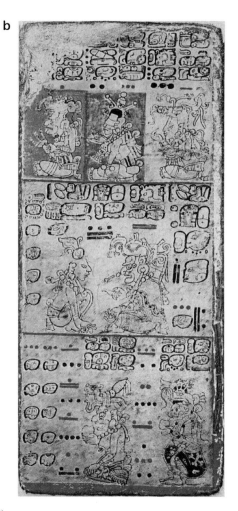

Fig. 2.22b (Continued)

together." The augury for this has been interpreted as "bountiful food," while "*uk we*" refers to feasting (Michael D. Carrasco, personal communication 2009). Iconographic interpretation on Classic Maya stelae and architecture indicated that the young Maize God was the "First Father," while the Quiche Maya term *Qanan* or corn literally means "Our Mother" (Taube 1985, p. 181; Freidel et al. 1993, p. 55; Stross 2006, p. 583; see also Freidel and Reilly 2009). Maize, therefore, represented a mythological lord, as well as "Our Mother," in this case through a metaphorical reference to the fertility and fecundity of the earth. The implication of these symbolic and metaphorical associations is that maize represents a transcendence of a binary duality or pairs of opposites, male/female. Among the Chontal Maya, maize was venerated, along with the sun, moon, rain, and the wind as a god (Schules and Roys 1968, p. 58; see also Carrasco and Hull 2002).

The close ethnic identity to maize fields or *milpas* is evident when noting that among the Yucatec Maya. A Yucatecan man's identity is defined by his *milpa*, the destruction of maize fields by indigenous nobility in times of conflict and by Spanish conquistadors strongly suggests that such cultural identities and associations harkens back to the colonial period and to prehistoric Mesoamerica (Stross 2006, p. 578; Carrasco 2009). The Yucatec Maya were noted for providing their deities with numerous offerings of maize and incense, while their lords were given a drink of toasted (parched) maize (Landa 1975 (1566), p. 101). Diego de Landa (1975 (1566), p. 101) also states that the Maya priest would incense an idol called Chacuuayayab with "fifty grains of ground maize and their incense which is called *zacah*." And that they would give Maya lords a drink "of 380 toasted maize cobs."

The spiritual power of maize in Mesoamerican culture is inferred anthropological evidence and sixteenth century accounts by its use in rituals. Maize kernels were usually cast, sometimes lots with pieces of thread during Aztec rituals (Durán 1971 (1588?), p. 118, 1994 (1588?), p. 493). Casting of maize kernels is related to the indigenous belief that the maize kernels will "jump" towards the culpable object or protagonist causing the illness or provide clues to the question being divined (Lipp 1991; Stross 2006). In the Bacalar region of the Yucatan indigenous tribes rebelling against seventeenth century Spanish authority replaced hosts and wine in their mass services with tortillas and maize gruel, emphasizing the importance that maize continued to have among Maya populations long after they had converted to Catholicism (Schules and Roys 1968, p. 346; Landa 1975 (1566), p. 62). The association of maize with mythological beings has its first expressions with the earliest civilizations of Mesoamerica (Taube 1996).

2.5.1 Maize and Religious Uses and Rites

Maize was crucial to ritual and sacrificial offerings to major deities during calendric rites and therefore, central to Mesoamerican religion (Carrasco 2009; Stross 2009). During the dry season, the Aztecs would make offerings of maize bread and fowl to the feathered serpent *Quetzalcoatl*[29] to bring on the coming rains and fertility of the crops. Offerings consisted of small plates on which were tamales, "*as large as fat*

---

[29]*Quetzalcoatl* literally means the "plumed or feathered serpent" and is considered by Mesoamerican scholars to have been the primary deity associated with priests and merchants, and revered for bringing language and civilization to Mexico. This deity has also been found to be associated with the Maya pantheon, where it was called *Kukulucan*. In both cases, offerings of maize in various forms were central to the rituals, rites and as offerings associated with this deity (Schwartz 2000).

melons" and upon these tamales were pieces of fowl and male turkeys (Durán 1971 (1588?), p. 136). Such offerings were made to ward off sickness (particularly, respiratory illnesses) or to predict the coming rainy season. Various forms of syncretism continue to the present, and maize continues to have a great importance to ethnic identity, ritual, and in the local economy as a food staple (Christenson 2008; see also Tuxill et al. 2009).

Aztec deities have, generally, been grouped into three broad categories: Those associated with the agricultural fertility and the earth, creator deities believed to be responsible for the beginnings and ends of cosmological and world cycles, and those, like their patron deity, Huitzilopochtli, associated with the cult of war and human sacrifice (Schwartz 2000, p. 9; Carrasco 1999, p. 23). Religious rituals surrounding the Festival of Toxcatl were dedicated to the Aztec deity the Lord of the Smoking Mirror, Tezcatlipoca.[30] Bunches of ears of corn as well as pine bark, turkey, quail, and fish were provided as offerings to the Smoking Mirror. The Temple of Tezcatlipoca had maidens who maintained the edifice and during the festival dedicated to the deity, would make offerings and carry out processions clad in blouses and skirts covered in strings of toasted maize. Their faces were also covered with strings of toasted maize, and on their heads they wore crowns of maize. Youths participating in this festivals had necklaces and garlands of toasted maize (Durán 1971 (1588?), pp. 104–105). The Lord of the Smoking Mirror Tezcatlipoca was also associated with the surface of the earth and in the east, his color was yellow referring to the rising sun and the fruitfulness of the maize plant. Hence, the strings of toasted maize kernels worn by celebrants with the festival of Toxcatl (Burland and Forman 1975, p. 56). Diego Durán (1971 (1588?), p. 107) observed that after much celebration dancing and playing of drums the Aztec returned to their homes at sunset and brought large platters filled with *tzoalli*, maize dough mixed with honey, covered with cloths decorated with skulls and crossbones, another reference to the underworld. The honeyed tamales were eagerly snatched up or carried away as relics.

In honor of the deity *Xipe Totec*, or the Lord of the Flayed Skin, the Aztec would eat a ritual food consisting of *tortillas* and *tamales* made of corn flour that was mixed with honey and beans. Durán (1971 (1588), p. 182) states that it was forbidden to eat any other bread on the day honoring this deity. Such descriptions imply that certain food prohibitions and fasting may have been practiced before the arrival of Europeans to these shores.

The deity that represented maize in the Aztec culture was the goddess *Chico-mecoatl* (serpent of the seven heads) and *Chalchiuhcihuatl* or Woman of Precious

---

[30]Tezcatlipoca literally means, "smoking mirror" in the Nahua language and he was considered the deity of rulers, sorcerers, shaman and warriors, as well as the lord of the night sky and divination (Schwartz 2000, p. 256). Taube and Miller (1993, p. 164) maintain that despite his many associations and symbolic referents, Tezcatlipoca was the "embodiment of change through conflict."

Fig. 2.23a (a) Aztec depiction of the Maize Goddess, Chicomecoatl (Seven Serpent), holding tasseled maize cobs, and wearing a necklace of young ears of corn. (from Durán 1971 (1588?), Chap. XIV, p. 222, Plate 23). (b) Other depictions of the Maize Goddess, Chicomecoatl (Seven Serpent), in her traditional crown and costume as well as her manifestation as Seven Serpent (right). In both images, the indigenous scribe shows her holding tasseled maize cobs, and wearing the head dress and holding the sun shield emblematic of her power to bring forth life (from Durán 1880 (1588?), Chap. XI, Plate 47)

Stone (Figs. 2.23a, b). She was the deity of the harvest and of all grains and plant species of the state (Durán 1971 (1588?), p. 222, 1880 (1588?), Chap. 1, Plate 47). Ritual celebrations made to the maize deity were held on September 14, preceded by 7 days of ritual fasting (Durán 1971 (1588?), p. 223). The Aztec made sculptures of maize goddess of finely carved wood, and she was often depicted holding carved maize cobs in her hands and had a necklace of golden ears of maize held by a blue ribbon (Durán 1971 (1581), Plate 23; Carrasco 1999, pp. 197–200). She was clothed in fine all-red garments and wore a tiara of red paper. Sometimes, the ears of maize in her hands were imitated in fine feather work or garnished in gold (Durán 1971 (1581), pp. 222, 1880 (1588?)). This idol was housed in a chamber in the temple of

chicomecoatl

Fig. 2.23b (Continued)

Huitzilopochtli "all to her greater honor and glory"[31](Durán 1971 (1588?), pp. 222–223). The chamber or hall that housed this idol was decorated and made green with numerous strings of corn-cobs, chile peppers, squash, flowers and offerings of amaranth seeds and seeds of various plants, which covered the floor (Durán 1971 (1588?), pp. 223).

The Aztec told chroniclers that she was part of them, their flesh and livelihood, and when they planted kernels in the new growing season they would cry, as though they were sacrificing part of themselves (Carrasco 1999; Schwartz 2000). Davíd Carrasco (1999, p. 200) has observed that in Aztec botanical thought, maize kernels were believed to be composed of the visible seed and the invisible "heart of maize." Once maize kernels were planted they were immersed into the underworld, a place called Tlalocan, in a colossal receptacle enclosed in the cosmic mountain. Aztecs believed that the only way for maize kernels to become active seed was for that seed to be united with the "heart" (Carrasco 1999, p. 200). The first name of this deity

---

[31]*Huitzilopochtli*, the blue hummingbird of the south, a form of sun god, was the patron deity of the Aztec and its temple located beside the main plaza or *zocalo* in the sacred center of Tenochtitlan was the site of much ritual human sacrifice (Schwartz 2000, p. 5; Carrasco 1999, pp. 199–200). Durán's reference emphasizes the great importance of the maize goddess through her association with the patron deity of the Aztec culture.

(*Chicomecoatl*) has metaphorical reference to the catastrophic destruction brought on by frost, drought and famine and the second name (*Chalchiuhcihuatl*) was in reference to her ability to bring forth life, fertility and fecundity (Carrasco 1999, p. 200). With reference to the destruction wrought by drought or famine, the Aztec also used the term "*tecuani*" to sting or bite, when referring to a maize field frozen or destroyed by drought and insects. It was common to say the fields were *tecuani* or eaten by frost, anything that stings or bites was *tecuani*, thus a metaphorical reference of the first name. The seven-headed serpent name may be indirect reference to the cardinal directions and the tripartite cosmos. In other words, to all plant life and to those things in the climate and environment that affect their survival, i.e., rain from the sky, dust from the wind, and the past and future events associated with such phenomena (Durán 1971 (1581), pp. 222–223).

During the fourth month of the Aztec calendar, the people prepared special maize bread, which was left as ritual offerings and only eaten during the ritual celebrations marking this month. This ritual bread was prepared in the following manner: a small quantity of ground maize kernels was mixed with toasted amaranth seeds, and then kneaded together. Honey, instead of water, was then mixed into the flour (Durán 1971 (1588?), pp. 422–423). They called this bread *tzocoyotl*, while the Spanish referred to it as *bollitos* or little loaves, and it was only eaten on this day. Maize appears to have been a central ingredient to foods consumed in a number of important rituals, and also had clear associations with certain parts of the annual cycle (see Staller 2009).

2.5.2 Maize: Religious Significance to Mesoamerican Civilization

It is evident from the ethnohistoric accounts that the color of kernels was also significant to the preparation of ritual offerings for certain festivals and to their association with certain deities. During the 17th month of the Aztec calendar called *Tititl*, they venerated *Camaxtli*, the god of hunting, with offerings of sour bread (*xocotamalli*), and a bitter porridge made of purple maize. Ritual offering of the bread and porridge were made to the civic shrine and temple in honor of this god as well as at domestic shrines in some houses. The bitter bread and porridge were also consumed by celebrants during the rituals surrounding this feast (Durán 1971 (1588?), p. 463). Blue maize in the form of kernels and flour were offered to the waters of the *chinampas* that were channeled into the raised fields (Durán 1994 (1588?), p. 368). The stepped or raised field system of maize cultivations as described here is also apparent in an engraving dated to 1585 from North America (Hale 1966).

The religious and ritual importance of maize to Mesoamerican civilization was also apparent by "broken pieces of maize bread" or tortillas, "*which were hung in the temple chambers, like those of the maize goddess, where they were strung on cords; this was like bread of oblation. They offered these to the captives ...*" and then the Aztec priests would address the prisoners with prescribed chants preparing them for death by sacrifice (Durán 1994 (1588?), p. 157). Maize was apparently consumed in these instances to create a symbolic link between those who were being sacrificed and those who were sacrificing them to the deities. The Aztec and other Mesoamerican cultures also believed that maize offerings were able to protect those who were in battle and to facilitate their safe return. While Aztec warriors were off to war, the native women would make offerings of *tortillas*, wine (*tezvin* or *pulque*), and all kinds of foods to the gods (little statues), and "*seated before these images they would weep and moan chanting prescribed prayers*" for such rituals, beseeching the deities to bring their husbands and sons home safely (Durán 1994 (1588?), pp. 350–351).

Durán (1994 (1588?), pp. 151–152) described funerary rites for fallen warriors in the conflict against the Tepeaca as initially involving 4 days of singing, chanting, and clapping to the playing of drums. On the 4th day, they set fire to bundles of pines and used the resultant ashes to wipe the faces of the war dead (Durán 1994 (1588?), p. 151). The widows of the deceased were then ordered to make offerings for 5 days. Funerary offerings consisted of "*prepared breads and bowls of toasted maize*," which were taken to the burned pine bundles. These offerings along with the mantles and breechcloths worn by the deceased were then all burned. After only ashes were left, they took "*wine of the earth* (*tezvin* or *pulque*) *and poured it all over the place the clothing had been burned*" (Durán 1994 (1588?), p. 152). Having completed these rituals for the war dead, widows were required to make the same offerings of food 80 days later, where they were again burned and "wine" was poured on the ground. It is important to emphasize here that commoners were not trained as warriors, only sons of the Aztec nobility, thus, these rituals and funerary rites were organized and made by the ruling class for their own departed children (Carrasco 1971, p. 372).

The ritual use of maize was also believed to have an impact on meteorological phenomenon. As the rainy season drew near, the Aztec performed rites and ritual sacrifices to summon the life-giving rains. During the ritual feasts called "Coming Down of the Waters" ceremonies included offerings of *tamales* as well as other vegetables and blood let from various parts of the body. These offerings were believed to help bring on the coming rainy season (Durán 1971 (1588?), p. 462). The rituals and rites associated with the Coming down of the Waters represented a metaphorical connections between the sacred center of Tenochtitlan and natural features in the surrounding landscapes, in this case the mountain of Tlalocan,[32] named after Tlaloc, the deity associated with rain, lightning, thunder, and in Maya and Toltec culture also with wind. The idol of Tlaloc, like the idol to the Maize Goddess was kept in the Temple of Huitzilopochtli, in the central precinct of

---

[32]Tlalocan means "place of Tlaloc," and had a temple dedicated to this deity on its summit (Durán1971 (1588?), p. 156). The temple on the summit of the mountain also had a centrally placed idol dedicated to the deity and smaller idols dedicated to the surrounding mountains and hills (Durán 1971 (1588?), p. 156).

Tenochtitlan[33] (Durán 1971 (1588?), p. 156). Tenochtitlan was to the Aztec, "the foundation of heaven," the political, symbolic, and ritual center of their universe (Schwartz 2000, p. 6). While the Aztec emperor and the rulers from the surrounding city states made offerings to the idol in the sacred center of Tenochtitlan, religious specialists, members of the elite and soldiers from all over Mesoamerica made ritual processions and pilgrimages to the mountain top and carried-out solemn rites and offerings during the Feast of Tlaloc, which occurred every year on the April 29 (Schwartz 2000, pp. 158–160). Durán (Schwartz 2000, p. 156) goes on to state that the purpose of the Feast of Tlaloc was for a good maize harvest "...*since all the maize which had been sown had now spouted.*" These rituals and ceremonial rites included human sacrifice and many objects of value, cloth, food, newly manufactured pottery and figures, all centered upon controlling the forces of nature, particularly meteorological phenomenon such as frost, hail, lightning etc., which had a direct impact upon agricultural fields, particularly the maize crop. The Festival of Tlaloc was one of regeneration and rebirth of the agricultural cycle in which maize was of critical importance to the survival and sustenance of Mesoamerican commoners and rulers alike.

The Festival of Tlaloc was particularly important to the Aztecs because they reckoned their cultural origins with water. According to the Aztec codices, their pictographic representations and oral history, which are a mixture of legend, myth, and history, they migrated into the Valley of Mexico around the year A.D. 1250 and by A.D. 1325 settled themselves in the midst of the lake (Schwartz 2000, p. 5). Their mythological origins are also associated with water, an island at a place called Aztlan, a Nahuatl term that means "place of whiteness or herons." The exact geographic location of this place is unknown, but believed to be northwest of the Valley of Mexico, in the Chichimeca region (Schwartz 2000, p. 5; Sachse 2008, p. 127). The Aztecs depicted their place of mythological origin as a cave called Chicomoztoc or "place of the seven caves" (Fig. 2.24). From these modest beginnings, the Aztecs came to rule the Valley of Mexico, exacting tribute from many distant regions of Mesoamerica, and their rise to power is according to their own pictographic representations symbolically linked to maize cultivation (Durán 1880 (1588?), Chap. 1, Plate 1; Sachse 2008, pp. 132–134).

Both colonial accounts and indigenous pictographic representations and codices emphasize the importance of maize to Mesoamerican religious rituals, calendric rites, tribute and economy, and particularly associations with the ruling class. The Spanish accounts strongly infer that it did not hold the economic importance to Contact Period cultures in Mesoamerica that it does today. Many accounts indicate that, the indigenous diet was highly diverse and complex. In the Valley of Mexico,

---

[33]The central temple precinct was dominated by a great 60-meter-high pyramid with twin temples (the Great Teocalli), one to Tlaloc, and another to Huitzilopoctli, the patron deity of the Aztecs. Around the central precinct were 70–80 other palaces and temples, including the ruler's residence and the school for the priesthood. Beyond the enclosing wall were other palaces, temples, markets, and the adobe residential buildings, some of them two stories high with gardens on their roofs (Schwartz 2000; see also Díaz del Castillo 1953 (1567–75), pp. 177–182).

Fig. 2.24 Pictographic representation of the Aztec origin site, Aztlan, the "Land of Heronsm," that the Aztec called Chicomóztoc, the Seven Caves. Other chroniclers say it was an island and the exact geographic location of this place is unknown, but thought to be northwest of the Valley of Mexico, in the Chichimeca region (from Durán 1880 (1581) Chap. 1, Plate 47)

Gulf Coast, and Maya lowlands, indigenous societies depended, upon aquatic and maritime resources, and a whole host of plants and fungi, both wild and altered by human selection, which were available locally or acquired through interaction (Sahagún 1963 (d. 1590); Durán 1880 (1588?), 1994 (1588?); see also Schules and Roys 1968; Thompson 1970; Parsons 2006).

The Spanish accounts also reflect a preference for maize, beans, and squash by
the early explorers, as well as a perception of maize as the equivalent of wheat and
barley in the Old World (Staller 2009). Even in regions such as the Caribbean,
where maize was clearly not consumed as a grain, the accounts appear to emphasize
those regions where it was consumed in this manner. Its apparent rapid introduction
into Western Europe immediately after the discovery of the New World further
created the impression of maize as a primary economic grain, and the basis for
complex sociocultural development in this hemisphere. With the later introduction
of the *encomienda* and plantation system, New World populations were encouraged
and at times forced by the colonial authorities to cultivate maize at ever-increasing
scale. These accounts show a clear ignorance of the importance of maize varieties
to various ethnic groups under their domain and a general trend to cultivate varieties
that were more productive, i.e., more rows and larger kernels (Huckell 2006;
Newsom 2006; and Rainey and Spielmann 2006). The predominant use of maize
in ceremonial rites and rituals associated with the Mesoamerican calendars, primary
deities, and ethnic identity is clear and abundant (Durán 1880 (1588?), 1994
(1588?); Sahagún 1963 (1590)). These accounts and descriptions strongly emphasize the sacredness of maize to Mesoamerican cultures and its symbolic importance
to the transference of divinity and spiritual power. The complexity in which it was
prepared and used in this manner among Mesoamerican cultures is extraordinary,
particularly in the light of how it was used and consumed in other regions of the
hemisphere.

2.5.3 Early Accounts on Maize Alcohol Consumption

In the area west of present-day Panama City, referred to as Coiba by the chroniclers,
toasted ground maize was stirred into water and drunk, identical to the drink *pinole*
of Mexico, and "wine" or maize beer was similar to what the Mexicans referred to
as *tezvin* (Sauer 1969, p. 271). In the interior of the region of Coiba, the indigenous
populations made "wine" or maize beer from a small-grained and floury lineage of
maize known as Early Caribbean (Newsom 2006, p. 330, Fig. 23-2; Newsom and
Deagan 1994, Table 13.1, pp. 215–216; see also Brown 1953, 1960; Bretting et al.
1987). This landrace or lineage of maize is believed to have been one of the earliest
types introduced into Europe and has been suggested to be only distantly related
to any known varieties in either Central or South America (Brown 1960, p. 161;
cf. Newsom 2006, p. 330).

The consumption of maize in the form of beer in Mesoamerica has generally
been absent in the archaeological record, but various *relaciones* and chronicler
accounts, particularly in the northern and western parts of Mesoamerica, suggest
"*vino de maíz*" or maize beer was widely consumed in the state of Morelos and in
regions to the north and west–more specifically among the Tarascans, Huichol and
indigenous cultures in the states of Michoacán, Duárngo, Chihuahua, Jalisco and
Nayarit (Bruman 2000, pp. 37–46). Maize beer is referred to as *tesgüino* in these

regions, which is a Uto-Aztecan term, probably Cazcan, and closely akin to Nahuatl (Bruman 2000, p. 39). Maize beer is also referred to by this term and consumed in eastern parts of Guerrero on the south coast, as well as Central Chiapas, and the eastern regions of coastal El Salvador (Bruman 2000, Map 5). José Tudela (1977 (1541)) states maize beer was central to various Tarascan ceremonies. In a later *relación*, Beaumont (1932 (1700s), vol. 3, p. 462) states, *"These Indians* (in Michoacán) *made many alcoholic beverages by the fermentation of the grains of this plant"* (cf. Bruman 2000, Chap. 4, *f*. 2). In the Memorias of Sinaloa of the *Carta Anua* of 1593 a Jesuit states, *"They make also wine from the same maize, and at times have a solemn drink festivals for which the whole town congregates, although they do not allow youths* (mozos y gente nueva) *to drink"* (Carta Anua 1593, p. 26; cf. Bruman 2000, p. 39). Henry Bruman (2000, pp. 40–41) states that the ethnohistoric literature on the varieties of maize used to make the strongest beer or *tesgüino* is lacking, however, his informants suggest a yellow variety (*maíz amarillo*) was preferred, and that it may have been a flint landrace.[34] Maize beer in Mesoamerica was made in various ways. Initially they malt the maize kernels by moistening them in the shade on a reed mat and covering them in a moist until they produce colorless spouts of about an inch or so long. The moistening of the kernels releases enzymes called "diastases" which convert them into fermentable sugars through the process of saccharification (Bruman 2000, pp. 40–41). Once the malt is ready it is ground and then diluted with water and boiled for over twenty four hours with continuous stirring into homogeneous syrup. The syrup is put into an olla and then left in the shade to ferment. It is usually ready to drink in twenty four hours time (Bruman 2000, p. 41).

2.5.4 Maize Beer and Pulque in Mesoamerica

The Nahua term for cornstalk and/or maize beer *"tesqüino*[35]*"* is generally used in the central highlands, western, and southern coastal regions of Mesoamerica. The Nahuatl term *"chicha"* was brought by the Spaniards from the Maya highland societies to different regions of the Americas to refer to fermented intoxicants in general (Staller 2006b, pp. 449–450). In the Andes, this term generally refers to maize beer and more rarely intoxicants made from the Peruvian pepper tree (Staller 2006b, p. 449). Mesoamerican cultures also made beer from cornstalk[36] (see, e.g., Beadle 1972; Smalley and Blake 2003). Cornstalks were crushed to collect the juice

---

[34]If the Mesoamerican accounts are accurate, this would be in contrast to the Andes where popcorn varieties were most commonly used to make maize beer (see below).

[35]This term for maize beer is thought to be derived from *"teiuinti,"* a general term for intoxicating in those areas where pulque was not used for such drinks (Bruman 2000, p. 78).

[36]It is highly significant that the distribution of cornstalk beverages, and its use as a condiment directly corresponds to the biogeography where maize was originally domesticated, and where the earliest evidence of maize has been recorded archaeologically (see Bruman 2000, Map 8).

and then boiled down into syrup, a process described by various chroniclers (Bruman 2000, p. 57; Smalley and Blake 2003). Hernán Cortés (1963 (1519–1526)) related in his second letter to the Spanish Crown that cornstalk syrup was sold in the great market of Tlaltelolco, as were honey and maguey syrup. The general consensus with regard to the chroniclers is that cornstalk wine was considered to be inferior to maize beer or *tesqüino*, and usually consumed when grain was not readily available or out of season (Bruman 2000, p. 57). Smalley and Blake (2003) have presented evidence to suggest initial exploitation and transportation of teosinte, the progenitor of maize, was for its stalk, which was presumably chewed as a condiment and/or used as an auxiliary catalyst to the fermentation process. In the Tehuacán Valley, two beverages are described using cornstalk juice, pulque and other ingredients, including toasted maize (Bruman 2000, pp. 58–59).There is some evidence to suggest that among the K'iche Maya and the Huaxteca along the Gulf coast near Tampico a drink was consumed in which toasted maize was fermented in cornstalk juice (Bruman 2000, p. 60). A potent drink made of toasted maize, cornstalk juice, and pepper tree (*Schinus molle*) fruit called bone breaker (*quebrantahuesos*) was consumed in Tezcoco and Tacuba (Bruman 2000, p. 59).

The Mesoamerican cultures consumed a whole host of fermented intoxicants, generally referred to as *pulque*; however, these are made from a variety of fruits and succulents, but rarely from maize (Bruman 2000, pp. 61–82). The etymology of the word *pulque* is somewhat contentious. It is suggested to be the Nahua equivalent of *octli* or *uctli* and some accounts state it was brought from Chile and derived from the Arawakan term *púlcu*, which referred to fermented drinks, and maize beer in particular (Clavigero 1844). Bruman (2000, p. 76) suggests that the term has more ancient roots.

What is generally called *pulque agave* refers to fermented intoxicants made from maguey cactus. Maguey is native to the south-central Mexican highlands. The *maguey* worm (*Aegiale hesperiaris*) or *chinicuiles* was a delicacy greatly favored at the Aztec court and is still relished today (Parsons 2006, pp. 113–116; see also Staller 2009, Fig. 2.8). The maguey cactus (*Agave americana*) astonished many of the chroniclers and explorers. Oviedo (1959 (1526)) describes the yellow flower as about the size of a person's hand and resembling a maize cob. Fray Toribio de Benavente Motolinía (1979 (1528), pp. 243–244) called the wine made from maguey very good and healthy, noting that indigenous women would store ground maize flour in its spiky leaves. Maize and beans were commonly cultivated with maguey and such multiple cropping techniques greatly increased the carrying capacity of the central highlands (Parsons 2009, Fig. 4.7).[37] Approximately one million people lived in the Valley of Mexico at the time of the conquest. Maguey

---

[37]Maguey leaves were also used to make thread from which cloth was manufactured, as well as paper "twice as big" as the size of paper produced by the Spaniards (Motolinía 1979 (1528), p. 246; see also Parsons 2006, 2009; Parsons and Parsons 1990). Jeffrey and Mary Parsons (1990, pp. 363–364) have noted that the management and exploitation of maguey and seed crops augmented the carrying capacity of the cold central highland environment (*tierra fria*) almost twofold than if such highland environments would have been solely cultivated with seed crops (see also Parsons 2009).

fiber could potentially have clothed the entire population through cultivation of only five percent (approximately 10,000 hectares) of the total arable land (Parsons and Parsons 1990, pp. 337–338). This same amount of land has the potential for producing about 50–90 million liters of maguey sap or *aguamiel* annually, roughly 6,000 metric tons of cooked maguey flesh, perhaps 8,000–10,000 metric tons of food by multiple cropping maguey with maize or beans (Parsons and Parsons 1990, pp. 337–338). The most widely used kind of *pulque* mentioned by sixteenth century chroniclers was made from agave or maguey. Maguey sap or *aguamiel* is often distilled to make mescal, while a related species blue agave (*Agave tequilana*) is distilled to make tequila;varieties of maguey are made into pulque in different regions of Mexico (Parsons and Parsons 1990, 67–70, Table 3). Jeffrey Parsons (2006, 2009) has provided ethnobotanic and ethnohistoric evidence to suggest that there were dozens of different-named varieties of maguey, in different parts of highland Mexico, distinguished on the basis of their characteristics, for producing fiber, sap, ability to withstand aridity etc. However, the diversity of maguey with respect to specific characteristics and qualities is as yet poorly understood, nor is the extent to which such differences are a product of human selection of specific qualities of the species themselves in terms of their adaptation to different environmental and climatic conditions.

Landa (1975 (1566), p. 49) mentions that the wine or *pulque* made by the Yucatan Maya was made from, "...*honey and water and a certain root of a tree which is cultivated for this purpose, with which they make the wine strong and foul-smelling.*" Landa is apparently referring to *baalche* (*balché*), which is made from by fermenting diluted honey and the bark (rather than roots) of a fruit-bearing Balché tree of the species *Lonchocarpus violaceus*. According to Tozzer (1907, p. 124), balché is milky white in color and has a sour smell, and when first consumed, has a disagreeable taste. *Balché* is the only alcoholic intoxicant of the Yucatan Maya during the time of the Conquest (Bruman 2000, p. 91). Maya societies in Chiapas made what is described as a truly indigenous beverage from sprouted maize kernels and the bark of what is called *mecate colorado*[38] made from what is thought to be either a species of *Hibiscus* or *Heliocarpus* (Bruman 2000, p. 91). With the possible exception of *tezvin* described above, and the ethnographic information from various Maya regions of Guatemala, maize is generally not consumed by these cultures in this manner. One exception, however, is *pulque atole*, which is soured maize gruel that was strained and sweetened. The accounts suggest that Maya beverages in the Yucatan were honey-based, while the highland Mexicans generally fermented maguey or century plant (*Agave* spp.) (Parsons 2009; Parsons and Parsons 1990; Bruman 2000). Maya groups in eastern Yucatan are also known to have managed and cultivated maguey for the purposes of cloth and fermented beverages (Serra and Lazcano 2009, Fig. 5.3).

---

[38]*Mecate* is a Nahuatl term from the word *mecatl* or "cord." The bark is sold in markets in Chiapas in the forms of small skeins composed of long pliable strips used mainly for tying bundles (Bruman 2000, p. 91).

2.5.5   Maize Beer in Ritual and Religion in the Andes

The consumption of maize in the form of beer or *chicha* (*tesqüino*), a term the Spaniards brought from Mesoamerica to refer to alcoholic beverages, appears to have been prevalent throughout Central and South America as well as, the Amazon lowlands at the time of contact. In fact, the cultural geographer, Carl Sauer (1950, p. 494) stated that, *"The Spanish annalists give the impression that more of it was drunk and less eaten ..."* by indigenous populations in these regions. Ethnohistoric accounts of maize beer consumption in the Andes are many, and emphasize the different roles that maize played in traditional forms of reciprocity and redistribution as an alcoholic beverage among indigenous societies under colonial rule. It is apparent that some chroniclers liked the taste of *chicha*. Oviedo (1969 (1526), pp. 136–137) described maize beer as better than the apple cider or wine made and drunk in Biscay, and *"the beer and ale drunk by the English and in Flanders (both of which I have tried and drunk)."* Other explorers, such as Francisco Pizarro, Hernando Pizarro, de Soto, and Diego Trujillo, were more interested in the *aquillas* or *keros*, gold and silver drinking goblets, than the maize beer they contained (Titu Cusi Yupanque 1985 (1570), p. 128; Cobo 1990 (1653), p. 11; Staller 2006b, p. 460). The conquest of the Inca Empire, occurred in 1531, when Francisco Pizarro initially landed with his army on the present-day Esmeraldas coast of Ecuador, some ten and a half years after the total destruction of Tenochtitlan, the conquest of South America brought a new group of explorers and mercenaries. According to the various accounts, these conquistadores appeared at least initially, to be more interested in wealth and power, than converting indigenous populations to Christianity (Xerex 1985 (1534); Cieza de León 1977 (1551), 1998 (1553); Betanzos 1996 (1551); see also Staller 2006b; Traboulay 1994). Regardless of their overt or covert intentions for the conquest of the Inca civilization, it is clear from these accounts, as it was from the earlier chronicles from the Caribbean and Mesoamerica, that there was a clear preference by the Spaniards for certain indigenous food crops. Maize was clearly one of these crops, if not the preeminent food crop desired by Spanish explorers (see Staller 2009).

The large scale consumption of fermented intoxicants particularly *chicha* or maize beer was a critical component of almost every Andean social, political, or religious transaction (MacCormack 2004, p. 107). It is clearly apparent from chronicler accounts that maize beer or *chicha* was central to Andean ritual related to ancestor veneration and to rites associated with agricultural fertility (Arriaga 1968 (1620), p. 56; Staller 2006b, pp. 454–456, Fig. 32.2). During the campaigns to extirpate what the colonial administration considered pagan idolatry, many chroniclers made detailed descriptions of how maize and maize beer was used in Andean ritual practices. For example, Fr. Pablo Jose Arriaga (1968 (1620), p. 41) observed that maize beer or *chicha* was considered to be the principal offering, *"the best and most important part..."* of Inca rituals. It is through the consumption and offering of *chicha* that the religious festivals to *huacas* (sacred places) were initiated.

Spanish chroniclers mention that the drinking powers of the Indians were formidable (Cieza de León 1977 (1551); Cobo 1990 (1653)).

Fr. Bernabe Cobo (1990 (1653) p. 194) referring to what they kept in their houses mentions a preponderance of large earthen jars filled solely with "*their wine or chicha and this does not last them a long time.*" The largest jars held four to six *arrobas* (16–24 gallons) and was suggested to last a man no more than a week (Cobo 1990 (1653), p. 194). These prestigious amounts are probably exaggerated; but clearly emphasize the importance of ritual feasting followed by drinking in Inca festivals and calendar rites. Such ritual practices and customs are still important to Andeans in the present. During the festival of the Sun, Inca rulers would toast the Sun (*Inti*) and at other occasions, *huacas* or sacred places in the landscape, as a form of symbolic alliance, and reciprocity, but it was believed that only the Inca rulers could make the *huacas* "speak" (Vega 1966 (1609), p. 223; Guaman Poma 1980 (1583–1615) p. 220, *fol.* 246, 235, *fol.* 261) (Fig. 2.25). This emphasizes the importance of drinking and toasting maize beer in Andean culture as a recognition of status, as well as hierarchy and rank (MacCormack 2004, p. 107; Staller 2006b, pp. 456–458).

Making *chicha* was under the domain of women in the sierra and men along the coast and in some places they chose girls for this task (Arriaga 1968 (1620), p. 34). Inca administrative centers commonly had *chicheros* or males responsible for making *chicha* from maize they cultivated in the surrounding landscape solely for the purpose of ritual feasting at such centers and to give back in payment to commoner populations for labor carried out for the state (Morris 1993; Morris and Thompson 1985). Along the coast south of Chancay, *chicha* was called *yale* and generally offered with powdered *espingo*, which is a indigenous pepper tree species with a small, dry, round, bitter-tasting fruit, and the drink could be made as strong and thick as desired (Arriaga 1968 (1620), pp. 41, 44). Arriaga (1968 (1653) p. 44) states that this fruit was consumed in powdered form, and that the Andeans paid a high price for it, and would sometimes use it to pay their tribute to the Inca. Sale and consumption of *espingo* or molle (*Schinus molle* L.)[39] was later prohibited by the Church (Archbishop Bracamoros) under the penalty of excommunication (Arriaga (1968 (1653) p. 44). This mixture was poured on the *huaca*[40] (so it can drink), and priests or shaman consumed what was left. Arriaga observed that when religious practitioners consumed it, that it made them act as if they were mad (Arriaga (1968 (1653), p. 41). The Inca used the sweet outer part of ripe fruit to make a drink. Berries were rubbed carefully to avoid mixing with the bitter inner parts, the mix strained and then left for a few days to produce a refreshing and wholesome drink. It was also boiled down for syrup or mixed with maize to make nourishing gruel

---

[39]The term *molle* comes from Quechua word for tree, *molli* (Goldstein and Coleman 2004, p. 523).

[40]*Huaca* is a Quechua term for "sacred or extraordinary". It is an all-encompassing term that can refer to sacred places in the landscape, mountains, certain locations associated with myth and legend, certain objects which were out of the ordinary or unusual in some way, even venerated ancestors (Vega 1966 (1609), p. 73, 76–77; Staller 2008, p. 269–270).

Fig. 2.25 Chapter of the Idols (*Capitulo de Idolos*) Guaman Poma writes in Quechua "*Uaca Billca Incap*" or "the divinity of the Inca." Only the Inca rulers were believed to have the power to make *huacas* "speak." Here he shows the Inca emperor Topa Inca Yupanqui addressing a circle of stone idols. Below Guaman Poma writes that all of these idols and *huacas* "speak" with the Inca. The idols on the summit of the mountain represent mountain spirits, what were referred to as *apu* or *wamani*. (from Guaman Poma 1980 (1583–1615), p. 235 *fol.* 261)

(Coe 1994, pp. 186–187). Garcilaso de la Vega (1966 (1609), p. 182) points out that "...*if mixed with the maize beverage the latter is improved and made more appetizing. If the water is boiled until it thickens, a very pleasant syrup is left. The liquid, if left with something or the other, becomes sour and provides a splendid vinegar*..." There is archaeological evidence to indicate that the fruits of the Peruvian pepper tree *Schinus molle* were used extensively to make *chicha* in the Central Andes between A.D. 550 and 1000 (Goldstein and Coleman 2004).

In the highlands, the beer or *chicha* offered to *huacas* was sometimes made from certain yellow lineages or races, or maize grown specifically for the purpose of ritual offerings. The fields in which maize was grown for this purpose, was often the first to be cultivated, and Andean societies were prohibited by the Inca state from sowing their own fields before they had sown the fields for making *chicha* (Arriaga 1968 (1620), p. 42). The Andean people would also offer or sacrifice round masses or balls of corn meal or porridge mixed with salt called *yanhuar sanco* or *sancu* to their *huacas* (Arriaga 1968 (1620), p. 47; see also Staller 2006b, p. 454). The porridge or *yahuar sanco* was also used in purification rituals during the festival called *Citua* (Cobo 1990 (1653), p. 166; Staller 2006b, p. 454). The rite of *Citua* was held just before the rainy season, the time of the year most people become most susceptible to illness and disease. During purification rituals held during the *Citua* festival, chroniclers indicate that maize was used to purify living space and to effect healing among the sick (Cobo 1990 (1653), p. 166). Ritual purification also extends to what the Spanish chroniclers like Arriaga (1968 (1620), pp. 47–50) interpreted as confession while in the indigenous tradition it related primarily to physical ailments, spiritual cleansing and uncovering social disorder.[41] Before the cleansing rituals were performed during *Citua*, all foreigners (non-Inca) and persons with physical defects were ordered out of Cuzco (Cobo 1990 (1653)). Once they left the imperial capital, all the Inca inhabitants of Cuzco would bathe in the springs and rivers. These natural springs were located along one of the 41 sight lines called *ceques* that radiated from the *Coricancha* in the Temple of the Sun to the various *huacas* in the valley (Zuidema 2002).

The color of maize kernels, as in the case of Mesoamerica, was also important in such rituals (Fig. 2.26). Cobo (1990 (1653), p. 166) observed the Inca "purifying rooms" during the *Citua* festival with maize flour made from both black kernels (which they used first) and then white kernels varieties. The flour from the kernels of these varieties was mixed and then used to scrub the walls and floors of a room. While cleaning the room in this manner, the Incas would also burn some flour as a sacrificial offering. First, maize flour using white and black kernels and then of other colors was mixed with crushed seashells of as many colors as they could obtain and this powdered mix was then put in a sick man's hand. The sick persons are then ordered to chant certain words and then to blow on it as an offering to a *huaca*. In a ritual called *tincuna*, the religious practitioner would place and then rub a pebble they called *pasca* (meaning pardon) on the head and then wash the individual with *yanhuar sanco* and water in a stream where two river channels come together (Cobo (1990 (1653), p. 48). Maize was also used in healing or curing rituals. Andean priests would advise the sick to toss white kernels of maize on the Inca highway, so that the passersby will take away their illness (Arriaga 1968 (1620), p. 77). They also rub the sick with *chicha* and white corn kernels, to take away their illness (Arriaga 1968 (1620), p. 77).

---

[41] Social disorder is generally defined in Inca culture as acts or activities that were detrimental or in some way harmful to the *Sapa* Inca or ruler or to Inca rule.

Fig. 2.26 (a) White *choclo* maize cob cultivated in the sacred Urubamba valley, located northwest of Cuzco where the *Sapa* Inca had an imperial estate in the Inca Empire Such white popcorn varieties were used almost exclusively for ritual purposes and the manufacture of maize beer or *chicha*. (from Staller 2008b, Fig. 9.5). (**b**) Mayan *xmejen-nal* maize variety showing the phenotypic variation maintained within this lineage. The variety is cultivated in North Central Yucatan, Mexico. Many Maya populations continue to use their maize varieties as ethnic markers. These cobs were deliberately cultivated for certain characteristics particularly kernel color and morphology, fast maturation time, and husk coverage (Courtesy of John Tuxill) (Photograph by John Tuxill)

Maize was also central to linking the surrounding Andean landscape to the sacred center of Cuzco and to creating fictive relationships between non-Inca populations to the so-called children of the sun.[42] The Inca referred to themselves as the "children of the sun." They claimed mythological origins from the Island of the Sun in Lake Titicaca and their *panaca* or ethnic group also claimed their original ancestors emerged from a series of caves from the town of Pacariqtambo located south of Cuzco (Urton 1990, 1999; see also Sarmiento 1942 (1572)). The *Capac Hucha* festival involved sacrificing those conquered in battle to the Sun (*Inti*), or selected individuals (Betanzos 1987 (1557); McEwan and van der Guchte 1992, p. 360). The *Capac Hucha* were unblemished children, boys and girls of the empire, who had been promised by non-Inca communities to the Sapa Inca or ruler. The sacrificial victims or *Capac Hucha* were made divine (become *huacas*) by carrying out rituals that first involved rubbing of the dregs of the *chicha* on their bodies (Cobo 1990 (1653)). These children were then led to certain places in the surrounding landscape with great pomp and ceremony, and then ritually interred in carefully selected sites in the four corners of *Tawantinsuyu* (McEwan and van der Guchte 1992, p. 360). Sometimes, they were entombed alive and kept intoxicated by feeding them *chicha* through a tube for 5 days or until they died (McEwan and van der Guchte 1992, p. 360).

The *Capac Hucha* children were buried in specially prepared tombs on mountain peaks and sacred caves and provided ritual offerings and were perceived of as and worshiped as venerated ancestors (Staller 2008b). They symbolized the spiritual power embodied in the natural world, and at the same recognized the political and religious power of the Inca state to transform them into *huacas* (Vega 1966 (1609)). In this important ritual, maize serves as a symbolic medium for spiritual transfer and what all these various rituals demonstrate maize was an ethnic marker used by the Inca to transfer spiritual power to their subjects, and as a form of reciprocity, to maintain balance (*ayni*) and harmony within their empire (Classen 1993, p. 11). Transfer of divinity is through sight, taste, sound, touch, and fluidity (*chicha*), and within a certain prescribed ordered sequence (Cobo 1990 (1653); Classen 1993; and Staller 2008b). These cultural patterns indicate that the Incas did not want to obliterate the boundaries of the senses, but rather order them as a mirror of an underlying duality embodied in the cosmos, e.g., structure/fluidity, male/female, as complementary, yet distinct (Classen 1993, p. 80; see also Staller 2008b, Fig. 9.3). The ethnohistoric accounts clearly emphasize maize played a variety of roles and functions involved with Andean feasting, ceremonial activities, curing rituals and religious rites of passage (Classen 1993, p. 80; see also Staller 2008b). In the Inca examples, maize mostly played not only important social roles but also involved a symbolic reference to elite status, as

---

[42]These mythological claims and legends distinguished them from other Andean *panacas*, and these distinctions were maintained by marriage customs and other cultural practices focused mainly upon veneration of their dynasty (see Rowe 1944; see also Staller 2006b).

well as an ethnic marker central to rituals surrounding the Cult of the Sun and the Dead and in the case Capac Hucha to ancestor veneration. There are numerous iconographic and archaeological examples from various ancient Andean cultures to indicate maize was associated with various religious cults and consumed as *chicha* in ritual ceremonies long before the arrival of the Spaniards (Classen 1993, p. 80; Staller 2003, 2006b, 2008b; Staller and Thompson 2002).

2.6 Maize Ethnohistory: Summary and Conclusions

The discovery of the New World as traced from the early accounts suggest that the cultures, plants and animals of this hemisphere had a profound impact upon western culture and also facilitated the development of modern scientific principles and ways in which to classify and organize the cultural and natural world. Maize, perhaps more than any other plant, introduced from this hemisphere to the rest of the world, has had the most profound impact upon the world's economies. What is telling with the ethnohistoric accounts is that they appear to have conditioned our perceptions of maize to a greater extent than might be initially supposed. The ethnohistoric accounts from the Caribbean, Central and South America suggest that maize was solely a primary economic staple in Mesoamerica, where it was consumed as a grain, and made into bread, tortillas, cakes etc. and all point to its enormous importance to indigenous religious practices and beliefs. Nevertheless, despite these descriptions and what they clearly tell us about maize and its various roles, the fact that it was perceived by early explorers to be a grain, similar in importance to wheat and barley in the Old World, appears to have had an important impact upon western culture and how maize was spread and cultivated subsequently. This impact is also apparent in later archaeological interpretations of its role to New World prehistory.

The perception of maize as a primary cultigen consumed as a grain is evident in the name given the plant by western scholars. Its scientific name, *Zea*, is a Greco-Latin term for "a wheat-like grain," and species *mays*, a Taíno -Arawakan word *mahiz*, or "life-giver." The archaeological evidence and ethnohistoric accounts appear to indicate the Taíno used this term because of its perceived spiritual properties, not because it was an important food crop, a staff of life. However, the discovery of peoples and plants not mentioned in the Bible predisposed European nobility and Spanish explorers to pass off indigenous beliefs as idolatry and superstition. Fanciful depictions of New World cultures, plants and animals continued for many years, even into the 1800s in some circles, as Europeans attempted to connect indigenous cultures to Asia and the Far East or to those societies mentioned in the holy scriptures (see, e.g., Milbrath 1989, pp. 184–185). The fact that the later English and French explorers referred to maize as Indian corn, from the German (*korn*), which the Germans in that time used to refer to barley, oats, rye, and even lentils. Such terms may also have played a role in the general European perception of maize as first and foremost a grain, and subsequently the basis for much of the complex sociocultural development in this hemisphere (Anderson 1947b, p. 3; Cutler and Blake 2001 (1976), p. 2). Perhaps the most important characteristics of maize that made it very important to the survival and sustenance of the Conquistadores are its storabilty and the fact that it can be easily transported (Sauer 1969, p. 242; see also Raymond and DeBoer 2006).

The earliest accounts from Central America (Costa Rica) emphasize the consumption of maize as a fermented intoxicant, despite its apparent economic role in these regions (Sauer 1969, p. 133). Like the Caribbean, Central American societies also consumed maize in the form of cakes. The Italian explorer Girolamo Benzoni recorded the various steps involved in making corn cakes and maize beer (Staller 2006b, Fig. 32-1B). Perhaps because of the enormous scale and wealth of Mesoamerican cultures, and the fact that it was consumed as a grain in these regions, the general emphasis upon its economic role as a grain by Europeans was facilitated and reinforced. It is also clear from the Mesoamerican accounts, that maize had a profound spiritual, ritual and religious significance among these societies. Perhaps more than any other region in the Americas, maize had clear relationship to ethnic identity and to the origins and very existence of the Mesoamerican people (Stross 2006, p. 581). Maize sustained economies of scale in Mesoamerica. Its cultivation involved elaborate and complex forms of irrigation technology, the *chinampas* or floating gardens, and elaborate draining technologies, sustaining urban centers and open-air markets that were comparable and in many cases exceeded in scale those in the Old World. Many chroniclers rationalized the scale of human sacrifice they witnessed among the Aztec and other Mesoamerican cultures as due to an absence of large terrestrial mammals like those of the Old World. Yet, the accounts and the Mesoamerican codices indicate that despite the importance of maize to sustaining these cultures, they consumed a complex variety of maritime resources as well as fresh water fish, fowl, cactus and a complex array of plant foods.

It appears that the overall preference of maize over other indigenous plants by the European conquerors may have played a role in its subsequent cultivation at ever increasingly larger scale in the seventeenth and eighteenth centuries. This pattern is already apparent in the later chronicles with the introduction of the *encomienda* and plantation system. For example, indigenous populations under the jurisdiction of the Franciscan missions in the Gulf Coast were largely involved in the cultivation of maize and beans, which were consumed by the clerics for their own sustenance (Zevallos 2005, p. 89). The accounts suggest that even before the creation of the *encomienda* system, maize was an important trade item, implying that it was primarily grown in areas that were favorable to its cultivation (Thompson 1970, p. 7; Scholes and Roys 1968, p. 39). The Spaniards also emphasize in their accounts those regions or areas with multiple annual harvests. Most of these regions, like the *chinampas* of the Valley of Mexico, and the irrigated fields of coastal Peru produced multiple maize harvests and this is also reflected in the population densities they sustained. The regional variability with regard to maize production no doubt influenced the European perceptions of its role and importance to indigenous economies.

Another perhaps more important factor is the role that maize played in tribute and in maintaining corvée labor for the construction and maintenance of state-sponsored buildings and irrigation fields. The critical role maize played in provisioning warriors and the retribution exacted by indigenous polities through the destruction and devastation of maize fields was no doubt instrumental in how later colonial administrators, and even the early explorers, exacted vengeance upon their indigenous enemies. The role of stored maize in both Mesoamerica and the Andes as a buffer to climatic perturbations, famine and drought most surely impressed its economic importance upon their conquerors and clearly influenced their accounts. The preference of maize as a food source among Europeans is further implied by the continuation of these practices upon indigenous population who resisted their right to rule.

The chroniclers also bring out clear differences in how Mesoamerican and Andean cultures used maize to maintain their authority and power over subject populations. Among Mesoamerican societies, maize appears to have been used economically and symbolically to justify their right to rule, and as a critical consumable in the context of interaction and long-distance exchange. In the Andes, it was a prestige crop used for sealing social and economic alliances through traditional forms of reciprocity and through ethnic identity as recognition of the status and power of the ruling Inca elite (Betanzos 1987 (1551); Morris 1979, 1993; and Rostworowski 1986). In the Inca empire, subsistence cycles were based, in part, upon such labor exchanges that were organized on a rotating basis by *ayllus* while their leaders provided food and beer for the communities as a form of reciprocity several times a month (Vega 1966 (1609); Guaman Poma 1980 (1583–1615)). Among Andean cultures, the economic significance of maize was also closely tied veneration of their patron *huacas* and former rulers and nobles of high rank and status (Valera 1968 (1594); Vega 1966 (1609); and Cobo 1990 (1653)). In contrast, maize among Mesoamerican cultures was symbolically and literally associated with various deities in their pantheons, such as the Maya maize god or the Aztec Maize goddess (see, e.g., Durán 1971 (1588?); Freidel et al. 1993; and Stross 2006).

Perhaps what is most striking about the early chronicles is the extent to which that maize played such an important role in indigenous rituals and religion. The religious importance of maize to Mesoamerican and Andean cultures has been widely documented and described in considerable detail in the accounts. Despite the overwhelming evidence for symbolic and literal association of maize with ancient and colonial religion, most of these descriptions and accounts are unsympathetic to native beliefs, or largely negative in their assessments (e.g., Cortez 1991 (1519–1526); Acosta, 1961 (1581); Durán 1971 (1588; Sahagún 1963 (d. 1590); Vega 1966 (1609); Arriaga 1968 (1621); and Cobo 1990 (1653)). These inherent biases in the accounts, and in some cases, dismissive evaluations of native belief and ritual practice no doubt influenced later interpretations and applications of such information to archaeological reconstructions and interpretations and meaning of their art. The description of syncretism, tortillas and maize gruel, replacing hosts and wine in their mass services among certain native populations must have been disconcerting to colonial proselytizers. The continued melding and syncretism of indigenous and western beliefs among native populations throughout the Neotropics is another indication that such beliefs remain in various forms, and that maize continues to play a central role in such practices (Christenson 2008).

What is perhaps most striking about the role of maize in indigenous ritual and religion throughout the Neotropics is the importance it held to ritual and sacrificial offerings and calendric rites. The differences between Mesoamerica and the Andes is the way in which cobs and kernels were prepared in various ways and then offered to the deities in the former, while maize beer largely played this role in the latter. Indeed, maize beer was central to Andean ritual practice and as a religious offering. Another pattern found in all of these accounts is how maize was critical to religious cults and ethnic identity. These ethnohistoric accounts convey a distinct impression that among native societies of the New World, maize was considered to be a sacred plant as well as an economic staple. The clear preference of maize over other indigenous species appears to have played some role in our current assumptions regarding its economic importance to the development of civilization and its consumption as a grain.

Chapter 3
Scientific, Botanical, and Biological Research on Maize

3.1 Introduction on a History of Science on Maize

This chapter takes an historical approach to maize research. It is focused on early studies in the archaeological and biological sciences: How did these studies indirectly influence the current debates? How were these debates directly influenced by earlier research on plant domestication in the Old World? How are the methodological approaches taken by the New World archaeologists, specialized in domestication and early agriculture, different from those taken by such specialists in the Old World? How do these differences affect the history of research on the origins and spread of maize? Recent groundbreaking results from maize geneticists have indicated that earlier archaeological interpretations of plant domestication and the economic significance of maize need to be reconsidered, yet earlier research and interpretations continue to strongly influence the current research. The term domestication has come to be used in the archaeological and biological literature as referring to a symbiotic relationship among human populations, the local ecology, a mutualism or coevolution that is not necessarily dependent on human involvement, particularly with reference to resource management.[1] In this volume, domestication is defined as the genetic change brought about in a biotic population as a result of interactions with humans, and leads to a dependence relationship (Benz and Staller 2006, p. 665). These definitions on the process of domestication are more in line with those generally published in the biological sciences. Prior to the recent developments in direct dating and molecular biology, archaeologists and historians perceived agricultural practices surrounding primary economic cultigens in terms of a culture history. There appears to have been a general consensus with regard to the

---

[1]Note: this is not the way it has generally been defined in the archaeological literature and to a lesser extent in the biological literature, particularly before the middle of the last century.

economic importance of food plants such as maize in the ancient past, in part because maize was seen as analogous to wheat and barley in the Old World.

The domestication of New World food crops, like maize, has come to be seen as a process of evolutionary change involving the genetics of plant populations. These changes are primarily in response to human influence or deliberate selection for certain favorable traits, conscious selection, or artificial selection (unconscious selection), that is, genetic responses to human modification of the environment or management of plant reproduction. Discussion on the economic importance and the role of maize in prehistoric times would be incomplete without taking into account the early innovative research from the biological sciences, particularly by specialists in genetics, paleoethnobotany, and plant morphology. These data have historically had a profound influence on the American archaeological community and consequently, on research on the pre-historic role of maize in ancient economies. This is particularly true with regard to how the role of maize in pre-historic times was interpreted and what lineages expanded to different areas of the Neotropics and beyond. In fact, many of these pioneering studies continue to play a major role in the methodological approaches that research on maize has taken historically. The following chapter considers how and why the biological and social sciences have framed the current issues – influencing the questions that have been and are still being asked regarding the data. The data presented in this chapter are designed to provide hindsight, and at the same time, an increased sensitivity and awareness of the inner workings of science and research. The primary goal of this chapter is to show how early data and methodologies are historically linked to current debates on the origins of maize, early plant domestication, and the role of economically important plants such as maize in the socio-cultural development.

3.1.1 Comparing and Contrasting Old and New World Approaches

The archaeological, botanical, and biological research on maize in American Archaeology has historically been structured, and incorporated into the research and subsequent literature, in a manner very different from the way in which European scientists have considered the food crops and plant and animal domestication in the Old World (Terrell et al. 2003). Many scholars of early agriculture[2] have taken the position that plants such as maize provided the economic basis for the rise of civilization in the Americas. In fact, maize has generally been seen as the primary catalyst to complex sociocultural development in the Americas from the beginning (Tykot and Staller 2002). Almost a century ago, Herbert J. Spinden (1917) postulated the existence of a "formative" stratum underlying the basis of

---

[2]Agriculture is defined as a symbiotic or mutual interdependence of any food plant and humans (Smith 2001, 2005b; Benz and Staller 2006).

civilization and that the primary constituents of this stratum, which he called "Archaic Culture," included maize agriculture, ceramics, anthropomorphic figurines, and ceremonial mound construction. He also hypothesized that New World cultures developed as a result of the diffusion of maize, beans, and squash out of highland Mexico to other regions of the hemisphere and that agriculture was invented only once (Spinden 1917). The advent of ceramic technology and grinding stones (*metates*) at ancient sites has long been thought to develop simultaneously with maize agriculture in the Americas. Ceramic containers and processing stone were seen as essential for processing this food crop into flour for mass consumption (Lathrap 1970; Lathrap et al. 1975; Staller 2001b). The extent to which pottery and maize agriculture were seen as synonymous is evident by early surveys of formative cultures in the Americas, which were largely based on the comparative analysis of pottery assemblages (e.g., Ford 1969).

Similarly, cereal grains such as wheat, barley, and oats in the Old World have traditionally been reported to be the basis for the early sociocultural development, and ultimately civilization. The appearance of processing tools such as grinding stones and lithic tools for the harvesting of grains was also an indicator of food production in the Old World (Braidwood 1952, 1960; Harris 1989; Bar-Yosef and Belfer-Cohen 1992). Childe (1951b, p. 59) first coined the term "food production" in an attempt to contrast the ancient adaptations involving "food producing" with those of "food gatherer" in analyzing the transition from hunting and gathering to an agriculture economy. Agriculture, the deliberate planting and harvesting of domesticated plants, appeared to many archaeologists working at the turn of the last century and before the development of radiocarbon to be a revolutionary invention. For historical reasons, the integration of cereal grains to ancient economies in the Near East and later Mesopotamia was perceived as an adaptation involving a greater dependence upon certain plants and animals, and generally associated with the beginning of the Neolithic (Childe 1935, 1946, 1951b; Braidwood 1960; Bar-Yosef and Belfer-Cohen 1992). The Neolithic in the Old World is essentially analogous to the Formative in the New World, and thus the transition from food gatherers to food-producing societies was termed the "Neolithic revolution" (Childe 1951b, p. 61, pp. 70–71). The idea of an agricultural revolution was a very popular theory in the 1940–1950s and was first put forward by the archaeologist V. Gordon Childe (1946), who perceived a major economic revolution in prehistoric times brought on during a period of severe drought, and in this vision of prehistoric times, a climatic and environmental stress caused a new relationship to be forced between humans and their natural environment. But, long before Childe's (1951b) book "Man makes Himself" made such theories and ideas popular, the English anthropologist Edward Tylor (1889) observed that "agriculture is not to be looked on as a difficult or out-of-the-way invention, for the rudest savage, skilled as he is in the habits of the food plants he gathers, must know well enough that if seeds or roots are put in a proper place in the ground they will grow." Botanists, anthropologists, and archaeologists later discovered that indigenous farmers, past and present, have considerable knowledge of the food crops they cultivated, much more than that of the deliberately planted seeds that eventually germinate. The gradual interdependence and changes in the adaptation associated with domestication involve significant changes in the archaeological record. One of the primary archaeological patterns associated with the formative was a shift to sedentary, permanent sites associated with rivers and streams and away from the mobile lifestyles of hunting and gathering (Flannery 1972, 1986b, 2002). The epistemological basis of ideas surrounding an agricultural revolution and a formative stratum are in part a product of Western philosophy, particularly the French Enlightenment, and the notion of progress as well as the social Darwinism of Herbert Spencer and ideas about the evolution of civilization and cultural complexity (Spencer 1867, 1897; see also Carneiro 2002). While the criteria used by Old World specialists to divide the "Mesolithic" from the "Neolithic" cultures have changed over the years, the initial assertions that formed the basis for such ideas came out of the late nineteenth and early twentieth century perceptions about the evolution of social stratification, the importance of "food production" as an indicator of progress, and complex sociocultural development (Childe 1951a,b; Harris 1989; Zvelebil 2000; Stiner 2001). The idea of progress was extended to the history of ancient societies and cultures by two of the founders of anthropology, L. H. Morgan and E. B. Tylor, and the Three Age System, a unilinear evolution from Savagery, to Barbarism, and ultimately to Civilization (Willey and Sabloff 1980, pp. 3–4).

American scholars maintained early on that the beginning of agriculture and a dependence upon crops such as maize was a revolution in human history upon which the destinies of cultures, and later, nations would be dependant (Enfield 1866). These perceptions also dominated this period of Western history and were reified by the Industrial Revolution, mechanization of Western agricultural economies, and spread of colonial domination. Consequently, the integration of agricultural economies in the Old World has historically been couched as a product of diffusion, migration, or acculturation in the context of a Neolithic–Mesolithic transition (Barker 1985; Harris 1989; Bogucki 1996; Gebauer and Price 1992; Price 2000; Bar-Yosef and Belfer-Cohen 1992). The transition from food gathering to food production is central to archaeological issues surrounding the development of civilization. Nevertheless, as more ethnobotanic, archaeological, and chronological evidence regarding the transition is reported, there still appears to be considerable disagreement as to precisely when and where to draw the line along this developmental continuum (Terrell et al. 2003; Staller 2004, 2006c).

American anthropology, however, remains somewhat grounded in the four-field approach and more holistic in its breadth and scope, and particularistic and historical in its focus of research than in the Old World (Willey and Sabloff 1980, pp. 130–131). For example, Alfred L. Kroeber (1930) elaborated upon a common agricultural foundation for the formative with identical food plants as those outlined by Spinden (1917) and similar techniques in weaving, metallurgy, and architecture. Such trait lists could then be used by archaeologists to determine the presence or absence of an agricultural economy and the degrees of complexity exhibited by the cultures being studied. Willey and Phillips (1958) presented a historical-developmental interpretation of the formative defined, "by the presence of maize and/or manioc agriculture and the successful socioeconomic integration of such an agriculture into well-established sedentary life" (p. 144). A definition that largely paralleled that of V. Gordon Child (1946, 1951) for the Neolithic, the integration and spread of Old World cereal grains, and the transition from food collection to food production.

Rather than transitions, Ford (1969, p. 5) perceived "centers of domestication" and classified the formative into two parts, the "colonial formative" beginning at about 3000 B.C. and a later "theocratic formative" for the period after about 1,200 B.C. (see also Vavilov 1926). This perspective saw formative lifeways and the associated material constituents radiating to more peripheral centers, initially with the spread of agriculture and later associated with an overarching religious cosmology. This idea was much more closely tied to the European models of "waves of advance" or the assertion that migration and diffusion spread the primary food crops and agricultural technology. In the New World, the early origins of pottery technology, and by extension, maize agriculture in the Americas have also been considered as a result of diffusion (Vavilov 1926; Meggers et al. 1965; Ford 1969; Lathrap 1970). These differences in approach are, in part, related to the fundamental difference in the emphasis on the way in which agriculture is studied in the New versus the Old World (Staller 2006, p. xxi, 2006c; Terrell et al. 2003; Brown 2006). archaeological evidence from the Americas has traditionally been analyzed in terms of how well the data conformed to the definition of a formative way of life, and despite some inherent ambiguity, the presence and absence of such constituents have been consistently applied as a classification scheme (Staller 2006c). Consequently, archaeologists early on were encouraged to integrate different lines of evidence into their interpretations, particularly on the subject of early maize agriculture (Smith 1998, 2001). This is also the case with regard to plant domestication in general, in that investigations were more specifically concerned with the factors and circumstances surrounding plant domestication and also upon the plants themselves (Smith 1986; Hoopes 1994).

In the New World, the development of complex social organization as well as cultural evolutionary approaches has traditionally been perceived as the shift from foraging and a hunting and gathering way of life to a fully developed agricultural economy or formative way of life in terms of a developmental continuum (Spinden 1917; Vavilov 1926). Willey 1964, 1971; Service 1975; Harris 1989; Kelly 1995). However, American archaeologists have historically couched the origins of agriculture in foraging/farming dichotomies that are specific and distinct to different regions of the hemisphere and their associated time periods, rather than an overarching result of migration or acculturation (see, e.g., Flannery 1973, 1986a; Smith 2001). This developmental continuum, however, is the basis upon which there is a dialogue, transitions are a favored approach in both sides of the Atlantic among prehistorians (Harris 1989; Gebauer and Price 1992; Zohary and Hopf 1993). It is the particularistic nature of research from region to region in the Americas versus the emphasis upon acculturation that differentiates research in this case. It should be noted that the predisposition to analyzing the spread of agriculture and primary food crops in terms of transitions is antithetical to Darwinian natural selection, which time and again demonstrates that it is the plant and animal species in landscapes that are consciously and unconsciously being modified in the process of human adaptation (Staller 2004, 2006c). Such "transitions" have occurred in varying degrees for as long as humans have been selecting certain animal and plant species such as maize for their survival (Heiser 1988, p. 78; Rindos 1984; Terrell et al. 2003; Staller 2004, 2006c). The prime mover of culture change is another striking distinction. In the Old World, analyses of the spread of agriculture and cereal grains has been dominated by the human dynamics or "waves of advance" as responsible for the spread of agriculture, largely based upon human DNA, rather than plant genetics, as is the case in this hemisphere (e.g., Ammerman and Cavalli-Sforza 1971, 1973; Diamond 1997; Renfrew 2000; Stiner 2001; Emshwiller 2006).

Later archaeological and chronological evidence in this hemisphere challenged these earlier interpretations and suggested instead that cultural complexity and the associated agricultural economies did not diffuse from the nuclear areas of the Neotropics, but rather were in most cases a result of in situ sociocultural developments (see, e.g., Staller 2001a, 2004, 2006c). However, as more ethnobotanic, archaeological, and later radiocarbon evidence regarding the shift to food production were reported, there still appeared to be considerable debate as to where to draw the line between food gathering and production, and moreover precisely what are the primary constituents that characterize a formative way of life (e.g., Piperno 1999; Staller 2004, 2006c). Subsequently, the focus in American archaeology has been primarily upon culture process (rather than culture history) by considering agricultural transitions rather than look into diffusion and migration for explanations regarding sociocultural development (Staller 2001a, 2006c). Many scholars of early agriculture have taken the position that plants such as maize provided the economic basis for the rise of civilization. Maize (*Zea mays* L.) has been proposed as having played a major if not central role in agricultural subsistence, sedentism, and particularly precocious ceramic innovation throughout the Americas (Staller 2001a, 2006c). In the past three decades, archaeologist have begun to turn their attention away from generalized "Formative" patterns and their presumed associated correlates, and begin to focus upon the dynamics of these processes at a more restricted regional scale (Hoopes 1994).

3.1.2 Research on the Rise of Early Agriculture

The study of agriculture in this hemisphere has placed a major emphasis on the crops themselves: primarily maize and, to a lesser extent, beans and squash, which together formed the basic New World crop assemblage (Flannery 1986a, pp. 5–8). The presence of maize in archaeological sediments and ancient middens has generally been cited as an evidence to what has generally been termed a formative way of life. Maize has historically been an important focus of scientific research – how it originated, what its ancestors were, and how different landraces are related to one another (Flannery 1973, 1986a,b; Smith 1998, 2001; Staller et al. 2006). An emphasis on the waves of advance also has important consequences for issues such as demographic change in prehistoric Europe (Richards 2003), as well as for New World population densities based on an agricultural economy in which crops such as maize can be intensively cultivated. The wave of advance model also implies uniformity in the manner in which agriculture arrived in different parts of Europe, rather than the heterogeneity that has characterized so much of the archaeological research on agriculture and maize in this hemisphere (Thomas 1993, p. 357; Brown 2006, p. 6; Zohary and Hopf 1993, pp. 230–234). In this hemisphere, methodologically driven analyses, designed specifically to uncover precisely where and particularly when food production originally occurred, has generated data sets that focus almost exclusively on the earliest economic plants (particularly, maize) to the exclusion of other wild plants in the paleobotanical inventory and this has generally been the case with cereal grains on the other side of the Atlantic (Terrell et al. 2003; Staller 2006c). The general perception in the archaeological literature is that maize was spread as a food crop from the central highlands of Mesoamerica, the earliest presence recorded in the Tehuacán valley and Guilá Naquitz rock-shelter, north to the American SW, and south along the coast and inland into Andean South America (Fig. 3.1).

In the Old World, the origin and spread of agriculture is rarely looked on as an event in itself, and research on cereal grains is generally undertaken by specialists in the biological sciences, rather than by archaeologists and paleoethnobotanists (Zohary 1996; Zvelebil 2000; Brown 2006; Zohary and Hopf 1993). These distinctions speak of a fundamental difference in what is emphasized in the research surrounding plant domestication and agriculture, and therefore on how food crops like maize and cereal grains in Europe were spread and subsequently related to the development of complex social organization (e.g., Richards 2003). The developmental and evolutionary theories and models we have inherited, in other words, our search for agricultural transitions, centers of domestication, and agricultural revolutions have profoundly affected our understanding of the past (Staller 2006c).

The extent to which such approaches to early agriculture and plant and animal domestication have influenced anthropological and biological science in the USA is

Fig. 3.1 Map showing the spread of maize to different parts of the Old and New World. Maize was brought early on to the Philippines, SE Asia, and southern India by Spanish explorers. The time scales and place of origin for *Zea mays* L. are based upon the evidence from molecular biology and analysis of early 16th century historical documents

also a product of history. Before the end of the World War II, scientific research surrounding maize was largely concerned with creating better hybrids and more productive varieties or landraces (Wallace and Brown 1956, pp. 11–18). After the mid 1940s, European approaches to the origins of agriculture and economic crops, particularly grains, had a slowly increasing influence on the natural and social sciences in this country[3] (Wallace and Brown 1956; Willey and Sabloff 1980). While the initial influences from overseas were minimal, they have gradually increased in the past three decades. After the Second World War, research on maize became more focused on the integration of anthropological, archaeological, and botanical evidence in the pursuit of the origins of domesticated maize, but as with Old World cereal grains, this process was perceived as a product of slow gradual artificial selection (Mangelsdorf 1974, pp. 11–14; Zohary 2004). More recently, the domestication process is perceived as involving the gradual and fortuitous accumulation of genetic mutations that create a form of mutualism that develops between a human population and a target plant or animal population, and has strong selective advantages for both partners (Zeder et al. 2006, p. 139). However, conscious selection for larger grain size and greater productivity could not have occurred in the case of *Zea mays* before certain critical mutations regarding the *tga1* glume architecture occurred (Iltis 2006, pp. 28–29; Dorweiler and Doebley 1997). It was in part through mutation and changes in alleles that teosinte was transformed into maize and became totally dependent upon humans for its reproduction.

The study of maize within American archaeology developed with strong focus on the earliest presence, in the context of the origins of food crops, without a necessary emphasis on locating those origins within the social and cultural context of communities living in the region at a specific timeperiod (Brown 2006, p. 5). Part of this is based on the earlier research on maize morphology and biology, which did not assume a single domestication event related to a specific region – as has long been the case in the Old World with cereal grains (see, e.g., Price 2000; Richards 2003). In the New World, the natural distribution of teosinte now places a geographical limitation on the region of origin (Matsuoka et al. 2002; Freitas et al. 2003). However, earlier research on maize ancestry provided a much wider geographic range to consider the maize origins (e.g., Pearsall 1978, 2002; Piperno 1991; Piperno et al. 1985; Pohl et al. 2007; Zarrillo et al. 2008). The initial biological research emphasized the mutability of maize and the significant morphological and genetic variation displayed by modern specimens, and therefore inferred or assumed the crop was domesticated on multiple occasions in different regions of the Americas (Mangelsdorf and Reeves 1959a,b).

---

[3]There is some reason to believe that this historical shift in theory and methodological approaches in American Anthropology may be related, at least in part, to the passing of Franz Boas in December of 1942. The Father of American Anthropology and a staunch proponent of a holistic, four-field approach to anthropological research, he trained many of the most prominent and influential scholars of their generation (Willey and Sabloff 1980, pp. 84–85, pp. 94–95).

The underlying presumption of a multiple origin model was to have a profound effect on both the archaeological and botanical research focused on the earliest appearance of domesticated maize, and most of this research took place after the discovery of radiocarbon and absolute dating in the early 1950s. These archaeological samples and chronometric results have provided scholars with exciting new approaches to their research on the questions of when, why, and how maize spread out of its Mexican homeland to regions far beyond its region of initial domestication (Smith 2001; Blake 2006; Zeder et al. 2006).

In summary, there historically appears to have been a much more open dialogue and interaction between the social and biological sciences in the Americas, which continues into the present. This is perhaps most clearly apparent when we consider how archaeologists in this hemisphere have used research from the biological sciences, particularly botany, and most recently molecular biology, in their interpretations, and to frame and formulate their research design. Archaeologists have historically exploited paleobotanical results on inflorescence morphology, pollen size, phytolith shape, and abundance, and recently starch grain analysis to document the domestication process (Piperno et al. 2007). These lines of ethnobotanical evidence have played a major role in the Americas in tracing and explaining the origins and the early spread of maize in prehistoric times, but have their limitations with respect to their prehistoric contexts, associated dates, maize identification, and archaeological preservation (Staller 2003; Haslam 2004; Rovner 2004; Reber and Evershed 2004; Thompson 2006). The paleobotanical and biological evidence has been critical in directing archaeological research questions in part because the biological results are presented as hypotheses that can lead to inferences regarding maize culture history (Benz 2006, p. 10). The archaeological evidence is ideally suited to testing whether these biological results and conclusions can be replicated in the field setting. However, this more holistic integration of multiple lines of data continues to be influenced by previous assumptions and methodological approaches in spite of recent molecular evidence from maize DNA that make such interpretations and assumptions no longer tenable. Since the late 1970s, the research coming out of molecular biology has dramatically revised our understanding of the origins and spread of maize throughout the pre-Hispanic Mesoamerica.

3.2 Archaeological, Botanical, and Biological Research on Maize Origins

It was the eminent plant geneticist and maize specialist Paul C. Mangelsdorf who, in the 1960s, encouraged Richard MacNeish to work in a series of caves in Tehuacán, Mexico (MacNeish 1961, 1962, 1967a, 1978; Mangelsdorf et al. 1967). These caves are located in the arid highlands of south central state of Puebla, as it was assumed that macrobotanical remains from such localities would be more preserved than in open-air sites. The historical fact that Mangelsdorf influenced MacNeish to carry out research at the Tehuacán cave sites is an example of how biologists have had a direct influence on where archaeological research on domestication was carried out, although in this instance it was through informal interactions rather than through the published literature.

It has also been the case that scientific research in the botanical and biological sciences have influenced archaeologists in what kinds of questions were being asked of their data, and particularly in the case of maize, what was the origin of this economically important food crop.

Mangelsdorf and Reeves (1931, p. 329, 1939, p. 33) had initially asserted that domesticated maize was the result of a hybridization event between an unknown pre-Columbian "wild" maize, and a species of *Tripsacum*, a related genus. In its final manifestation, the Tripartite Hypothesis later postulated that *Zea mays* L. evolved from a hybridization of *Z. diploperennis* by *Tripsacum dactyloides* (Eubanks 1997, 1999; see also Iltis et al. 1979). The widespread acceptance of the "Tripartite Hypothesis[4]" is an example of how hypothesis testing can influence and promote the focus of archaeological investigations (Mangelsdorf and Reeves 1931, 1938, 1939; Mangelsdorf 1974). Biological scientists working in this hemisphere have moreover been much more attuned to how their research should have direct applications to the field of archaeological research and inquiry (Mangelsdorf and Reeves 1939, pp. 273–302; Anderson 1947b; Weatherwax 1954; Mangelsdorf 1974; McNeish 1985; Eubanks 1995, 1999, 2001). There is some reason to believe that Mangelsdorf and other geneticists who fostered the Tripartite hypothesis may have been influenced to look at such hybridizations as a byproduct of working with archaeologists such as MacNeish and Flannery in their cave and rockshelter excavations.

3.2.1 Early Botanical and Biological Research on the Origins of Maize

The basis for modern botanical and biological research on maize had its inception at the same time that American anthropology was being developed into a distinct field in the social sciences. This was to have a profound effect on the field of archaeology, because it was subsumed into one of the subfields of anthropology, and not an independent field in its own right – as is the case in Europe and much of the rest of the world (Willey and Sabloff 1980, p. 5). Consequently, the holistic approach taken by early anthropologists in this hemisphere had, to varying degrees, an influence on where archaeological research was undertaken and the

---

[4]The role of *Tripsacum* (gama grass) in the origins of maize has been refuted by modern genetic analysis, negating the tripartite hypothesis (DeWet and Harlan 1972, 1976; Matsuoka et al. 2002; Iltis 2006, pp. 43–44).

extent to which archaeologists were predisposed to look to the other sub-fields and to the biological sciences for hypothesis testing and how the questions they would ask of their data should be applied to the archaeological data (Staller 2006a, p. xxi).

During the late nineteenth and twentieth century, the botanical sciences focused considerable attention on the phylogeny, morphology, cytogenetics, and origins of maize. The plant was identified on the basis of its constituent parts in order to carry out systematic scientific research, which would explore its phylogeny and develop ways in which to make maize more productive (Fig. 3.2a). The previous chapter on the sixteenth century ethnohistoric accounts of maize demonstrates that maize played numerous and varied roles among prehistoric and early colonial New World indigenous cultures. One of the most important was its role in ancient religion and by extension ethnic identity (Sandstrom 1991; Raven 2005). Botanists working in the beginning and middle of the last century noticed early on that modern races of maize in the USA and Europe were very different from the maize found in archaeological sites and still being grown in many regions of Latin America (Mangelsdorf and Reeves 1939; Wallace and Brown 1956). The primary reason for these differences has to do in part with the role of maize to indigenous subsistence, which is highly varied from region to region, and its significance to indigenous ethnic identity and traditional religion (Sandstrom 1991; Berlin 1992; Berlin et al. 1974; Raven 2005; Staller 2006b; Christenson 2008).

Wallace and Brown (1956, pp. 39–41) mention that various Amerindian societies understood the function of pollen, and even made metaphorical reference to it in their religions (see also Whorf 1954, pp. 225–251; Christenson 2003). Yet despite their understanding and knowledge of cross-breeding and their devotion to their maize plots, they kept their varieties "pure" apparently over long spans of time (Pernales et al. 2003b). This is directly in contrast to the goals of modern maize research, which at the outset sought, through cross-breeding and hybridization, to create more productive races to supplement the national economy, and to exportation in order to help to sustain the economies of third world nations, particularly in Africa, Europe, and to a lesser extent Asia (Enfield 1866, pp. 11–12; Wallace and Brown 1956, p. 25). One of the most important effects on maize in the process of domestication through genetics is the so-called founder effect of domestication, population bottleneck.[5] This is when a small population experiences a time in its life history where only a small proportion of the genetic diversity is retained, because of random sampling or genetic drift (Emshwiller 2006, pp. 100–101). Maize diversity is in part a reflection of human or conscious or unconscious selection due to changes in the environment brought about by human control of plant reproduction and the result

---

[5]The founder effect is the loss of genetic variation that occurs when a new population is established by a very small number of individuals from a larger population (Mayr 1942, p. 120). The new population may be distinctively different as a result of the loss of genetic variation both genetically and in terms of phenotypic expression (Provine 2004).

was a set of diverse maize varieties or landraces. Questions surrounding the gene flow between domesticated maize and the teosintes such as, what is the genetic basis for morphological change, are some of the research being addressed by the most recent studies (Emshwiller 2006, p. 99; Doebley et al. 1983, 1984, 1987, 1990, 1997). These early researchers were using what was known at that time about maize landraces and attempting through cross-breeding and hybridization to create heartier and more highly productive maize varieties (Fig. 3.2b). Maize

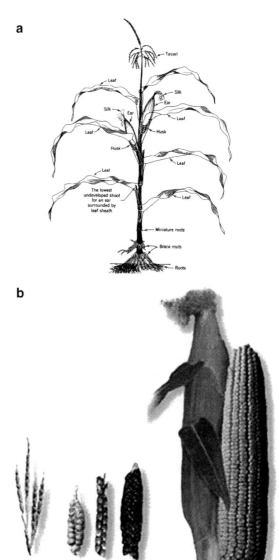

Fig. 3.2 (**a**) Maize and its constituent parts labeled. Maize tassels are distinguished from other wild grasses by their thick and highly condensed terminal spikes (with many spikelets) and their slender and uncondensed lateral branches (from Wallace and Brown 1956, Fig. 1). (**b**) Teosinte spikelet (*left*) compared to three early archaeological maize cobs (*center*) and an ear of modern genetically modified sweet dent corn (*right*). Teosintes are distinguished from maize by their spikelets and small distinctive female inflorescences, which mature to form a two-ranked "ear" of five to ten triangular or trapezoidal, black or brown disarticulating segments, each with one seed enclosed by a hard fruitcase. The fruitcase consists of a cupule or depression in the rachis, and a tough lower glume. There is no archaeological evidence from the various early cave and rockshelter sites to suggest that teosinte fruitcases were exploited for food

farmers in Mexico often grew teosinte in their maize fields (*milpas*) to make them more drought resistant and less prone to insect infestation.

The research goals with regard to maize beginning in the nineteenth century and extending to the middle of the last century was the creation of hybrids, which were more productive, i.e., produced greater yield per acre cultivated. The hybridization of maize lineages or races to further the utilitarian aims of farmers, go to the very beginnings of modern biological research on maize (Enfield 1866; Myrick 1904). This research emphasis sought to produce races that provided the greatest number of kernels, in order to sustain ever-greater populations (Wallace and Brown 1954, pp. 11–13; Fussell 1992, pp. 64–65). The funding and research in the United States that went into developing ever more productive maize varieties also had a profound influence on how prehistorians viewed the role of economic plants to the development of civilization in general, and the role of maize in such developmental processes in particular (e.g., Wilson 1981).

The botanical research on cross-breeding to develop varieties which were more drought resistant, more productive, and resistant to predation, began at around 1878 (Wallace and Brown 1956, p. 14). They are an indirect outgrowth of inventions, such as the microscope, and a byproduct of the religious wars that more or less ended at the beginning of the eighteenth century in the United States (Wallace and Brown 1956, p. 44). Cotton Mather, a puritan minister,[6] became interested in the maize reproduction, when he noticed in 1716 that his neighbor's garden had varieties of maize with multiple color kernels. Mather thought that the row of blue and row of red kernel varieties in the center of this garden were affecting the kernel colors particularly on the windward side of the garden, where the most multicolored rows of maize were standing (Wallace and Brown 1956, pp. 44–45). A member of the Royal Society wrote on maize pollination and hybridization, with pollen being the male element in the reproductive process (Fig. 3.3a, b). It was such insights that fostered later research in cross-breeding and hybridization.

One of the first discoveries made by geneticists at the early part of the last century was that genes appeared to control some of the particular traits that are affected by domestication. Beadle (1980) believed that it was one trait for each gene (Doebley 2001). These early maize studies considered how domestication has been manifested genetically in the crop, but in order to carry out such studies maize must be compared to its wild progenitor, and it is at this point where research on the origin of maize begins. Most of this early research on maize origins was focused on the identification of an unknown "wild" maize from which *Zea mays* L. was domesticated (Mangelsdorf et al. 1967, Fig. 96). Many of these early cross-breeding

---

[6]Cotton Mather was a Quaker, a puritan divine who condoned the persecution of "witches" but not the extreme methods of execution used by their prosecutors. Mather was aware of the effect of corn pollen from one variety falling on the silks of another and reported on this in various publications. Paul Dudley, a wealthy aristocrat, whose family were bitter enemies of the Mather clan, reported on the same phenomenon 8 years later without citing Mather's published work (Wallace and Brown 1956, pp. 45–47). Dudley referred to pollination as "wonderful copulation" (Wallace and Brown 1956, p. 47).

Fig. 3.3 (**a**) Teosinte spike (*right*) and maize cob (*left*): while maize seed dispersal is dependent upon humans, teosinte fruit cases are not. Maize is highly mutagenic and kernel color and ear morphology are highly affected by wind pollen from different maize varieties grown in surrounding fields. This led scholars to deduce that particular phenotypic characteristics and traits in *Zea mays* were genetically controlled. (**b**) The silk of the maize plant is a female element, while pollen is male. The silks are fertilized by wind pollen from tassels of other maize plants. (Courtesy of John Tuxill) (Photograph by John Tuxill)

studies were initially interested in understanding *Zea* phylogeny, and searching for a wild form, which would have given rise to the many races found throughout the Americas. Botanists at the Botanical Garden received a seed from Mexico, which was said to be from a plant known in that country as *maíz de coyote* and assuming it was a new species, classified the plant as *Zea canina* (Watson 1891). Subsequent research of this species indicated that it was not a distinct species, but rather that it could potentially be the progenitor to domesticated maize (Bailey 1892). John Harshberger (1893) compiled a treatise on the various races of maize and assumed that *Zea canina* was the ancestral form. Ultimately, all of these conclusions had to be reconsidered when it was disclosed from Mexico that the plant in question was a hybrid of maize and teosinte (Harshberger 1896a; cf. Weatherwax 1956, p. 142). The teosintes are a group of large grasses of the genus *Zea* found in Mexico, Guatemala, and Nicaragua. Despite the importance of agriculture to scientific research in the social sciences, their contributions to our understanding of agricultural origins have been minimal (Ford 1985a, p. 13). Basically, cultural geography and anthropology as well as archaeology and linguistics have made important discoveries and promoted ground breaking research. However, most of the literature from these fields was to some degree derived from a close symbiosis with the biological sciences in general and botany in particular. Early studies in the social sciences on the origins of plant domestication and early agriculture were more closely aligned with psychological arguments that appealed to ideas that the developmental processes or advances through time were the product of inventions of an individual genius, or part of human nature, and show a tendency to assume either explicitly or implicitly an idea of progress and cultural superiority.

Ascherson (1875) initially discovered and reported on teosinte, and in the process, provided the first evidence of the possible existence of "wild maize." Teosinte[7] was known to grow spontaneously in various regions of Mexico and Central America, and closely resembled maize in a number of its characteristics, particularly the silks of the ears – to the point where it was often mistakenly identified as a variety of maize (Collins 1921). Hybrids of teosinte and maize were long known to be able to be self-pollinating, to cross-pollinate with one another, and even to backcross with either parent, thus resulting in many intermediate forms (see, e.g., Mangelsdorf and Reeves 1938; Mangelsdorf 1974; Benz 2001). Some of the hybrids morphologically resembled maize very closely, yet had enough teosinte germ plasm to maintain a wild state for a considerable period of time. Various scholars in the biological sciences presented evidence of the genetic similarities of maize and teosinte in that both had ten pairs of chromosomes, and readily hybridizes with maize (Beadle 1939, 1980). Hybrids such as these were referred to early on as possible candidates for wild maize (Collins 1917, pp. 395–396; Weatherwax 1954, p. 142; Mangelsdorf and Reeves 1939, pp. 29–32). Research on hybridization was focused on exploring the origins of maize and its phylogenetic relationships to other wild grasses (Beadle 1939; Mangelsdorf and Reeves 1939). Teosintes were generally not considered by botanists and plant morphologists to be related to *Zea mays* L. because they were seen dramatically divergent in morphology, and it was on the basis of morphological similarity that wild taxon may be supported by the genetic and morphological data as a probable progenitor, or as a closely related taxon that was nevertheless not the progenitor (Emshwiller 2006, p. 100). Once the progenitor taxon is identified, the crop and the progenitor are usually considered to belong to a single species, because they are not only capable of continued interbreeding, but usually are indeed continuing to exchange genes (see Zohary and Hopf 1993; Smith 2006).

Research into the origins of maize in the biological sciences continued to be the primary focus for the rest of the twentieth century (Mangelsdorf 1948, 1958, 1974, 1986; Iltis 1971, 2000; DeWet and Harlan 1976; Beadle 1972, 1978, 1980). The early research in the biological sciences on maize was concerned with making maize varieties more productive, and increasing yield as evidenced by the various studies just outlined. During and after the 1930s, the scientific research on *Zea* became more directly focused on its phylogeny and particularly the origins of maize. The shift in emphasis was related in part to scientific advances in the molecular sciences, particularly on research on microfossils and plant DNA (Benz 2006, pp. 10–11). The ground breaking genetic research by George W. Beadle was primarily concerned with how genes work within cells, and as a result

---

[7]Teosinte takes its name from the Nahuat *teocintli*, which means good or evil grain and is used to widely refer to seven taxa of wild grasses closely related to maize (Benz 2006, p. 9).

he was awarded the Nobel Prize in Physiology in 1958 for his insight that "genes act by regulating definite chemical events" (Fussell 1992, p. 77). He observed early on that although maize was the most productive New World grain, it was at the same time the least able to reproduce itself. He revived the research of Ascherson and others that the ancestor of maize could be found in one of its relatives in the grass family and focused on annual teosinte (*Euchlaena mexicana*), because he observed that when its seeds were heated, they produced popcorn identical to that of popped maize (Beadle 1939, p. 247, Fig. 8, 1972, p. 10, 1980). Beadle (1939, pp. 246–247) suggested that since there was not any clear evidence to infer that Mexican or Guatemalan farmers consumed teosinte as a source of food, that perhaps it may have initially been exploited and consumed as popcorn, thus providing an incentive for its cultivation. Research in the biological sciences are more concerned with answering "how" anatomical changes and speciation occurred, while the social sciences emphasize "why" questions. This results in a focus on the complexity of human behavior and a search for causation in human history, particularly with regard to culture change (Ford 1985a, p. 13).

3.2.2 Historical Interface of Biological and Archaeological Maize Research

Beadle (1972, p. 10) noted in passing that the sugary maize stalks could have been chewed as a condiment, and still is by Mexican nationals (see also Bruman 2000, pp. 57–61; Iltis 2006, p. 26, Fig. 3-2A). Smalley and Blake (2003) recently expanded along these lines to propose that maize was initially exploited for the sugary pith and other edible parts rather than as a grain. The consumption of maize for the sugary stalk provides another possible explanation as to why teosinte was exploited by archaic and preceramic populations[8] (see also Mangelsdorf et al. 1967, p. 194; Bruman 2000; Iltis 2004). Excavations at the Tehuacán caves and at Guilá Naquítz produced quids of plant stalks of various species including maize (Mangelsdorf et al. 1967, pp. 194–197, Figs. 117–118; Smith 1986, Fig. 19.4). However, such quids were found in the later levels of these sites, suggesting that preceramic and archaic peoples occupying such localities in their annual cycles were consuming other grasses (Fig. 3.4a, b). Flannery (1973, p. 297, 1986b, p. 25) reports that teosinte and other small-seed plants particularly foxtail millet (*Setaria* sp.) were consumed during the Late Archaic Period (see also Callen 1967, Fig. 170). Other plants such as prickly pear, roasted agave, pochote root, mesquite, chiles, maguey, acorns and pinyon nuts, hackberry, and wild avocado were also present in

---

[8]Mangelsdorf (1974, pp. 72–73) asserted sugarcane of the genus *Saccharum* can cross with *Zea*, producing infertile hybrids, but others consider this claim unproven (Clayton and Renvoize 1986, p. 331). However, both sugarcane and maize stalks produce sweet juice that can be easily extracted and once the sugar is concentrated, consumed in making syrup and particularly alcoholic beverages or *pulque* (Smalley and Blake 2003, p. 675).

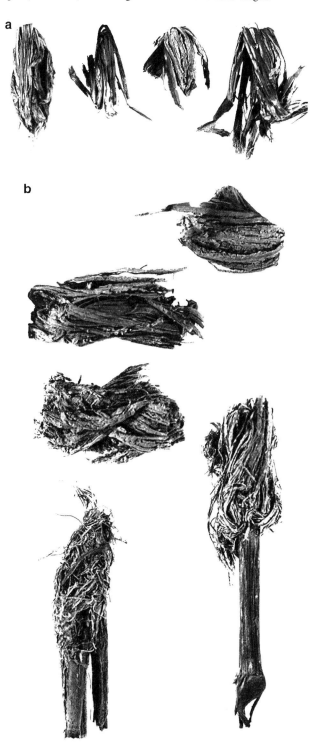

Fig. 3.4 (a) Quids in various stages of maceration from San Marcos Cave, Palo Blanco Phase (A.D. 0–1000) actual size (from Mangelsdorf et al. 1967, Fig. 117). (b) Quids of mostly chewed

archaeological deposits (Callen 1967, pp. 269, 271, 279). However, the archaeo-logical evidence appears to indicate that teosinte seeds, because of their bitter flavor and hard fruit cases, were largely absent in Early Archaic Period sites (Coe 1994a, p. 33; Iltis 2000, pp. 23–24; Smalley and Blake 2003, p. 677). Teosinte seeds are unpalatable and for this reason were shunned by archaic populations (Coe 1994a, p. 33). If consumed at all, it would have been during periods of resource scarcity, i.e., "starvation of food" (Flannery 1973, p. 290). Iltis (2000, pp. 23–24) states that utilization of the teosinte grain is out of the question, since it is encased in a hard fruit case. Teosinte has, according to Flannery (1973, pp. 290, 296–297) the following characteristics; it is difficult to harvest efficiently. It has a high percent-age of roughage – up to 53%. When mature, the seed generally shatter from their casings when bumped or disturbed. The evidence from archaeological sites for early and intense exploitation of teosinte, however, has been lacking, which left open the possibility for a hypothetical wild progenitor (Harlan and deWet 1973).

Early research on domestication and the evolution of food crops such as maize primarily involved analysis of macrobotanical remains for botanical identification (Ford 1985a, p. 8). Some ethnobotanists, however, suggested that such ancient samples could provide evidence for maize origins and the analysis of undigested seeds could potentially provide evidence on its importance to the ancient diet (Harshberger 1896b). Mangelsdorf (1974, p. 156) mentions in this regard that a number of coprolite specimens excavated at La Perra Cave in Tamaulipas included teosinte fruit cases that were still "hard" and "bony," essentially unchanged (see also MacNeish 1947, 1958). Callen (1967, pp. 266–267) observed that of the wild grasses, foxtail millet (*Setaria* spp.) appears in coprolites from all five of the archaic cave sites in Tehuacán, and was in fact the "principal plant" consumed in the El Riego and Coxcatlan Phases at Coxcatlan Cave (Fig. 3.5). With regard to *Zea mays*, Mangelsdorf et al. (1967, pp. 180–181) state that on the basis of their analysis of the earliest cob samples from San Marcos and Coxcatlan Caves that the separation of the staminate and regular pairing of the pistillate spikelets, among other morpho-logical features, provide convincing evidence that an extinct pre-Columbian species of "wild" corn, a hypothetical polystichous species, rather than teosinte or *Tripsa-cum* was the progenitor of cultivated maize (Fig. 3.6a, b).

The shift in focus from biological research involved with maize productivity in the late nineteenth and early part of the twentieth century, to a concern with phylogeny and particularly earliest occurrence or origin was related in part to archaeological discoveries with large macrobotanical inventories that included obviously ancient and archaic looking maize cobs. Mangelsdorf et al. (1967, p. 180) suggested that the earliest maize cobs were identified in excavations from San Marcos and Coxcatlan Caves, and in fact, since these pioneering studies, most of the ancient macrobotanical samples were recovered from rockshelters and caves rather than open air sites in the Americas. The botanists and geneticists working with the archaeological macrobo-tanical samples recreated a hypothetical wild maize on the basis of morphological

Fig. 3.4 (continued) maize stalks from San Marcos Cave Palo Blanco phase (A.D. 0–1000) actual size (from Mangelsdorf et al. 1967, p. 194, Fig. 117)

Fig. 3.5 Coxcatlan Cave, Tehuacán Valley, where prehistoric deposits extended from the right edge of the cave mouth to the left hand margin of the image from MacNeish (1962)

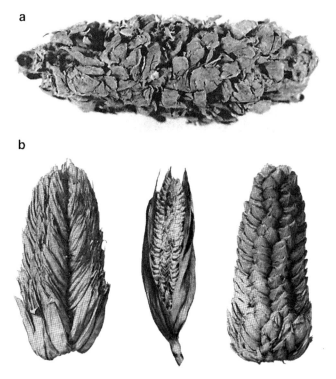

Fig. 3.6 (**a**) Intact cob of what was believed to be "wild" corn. The authors emphasize that the long soft glumes gave the general impression of pod corn. This early maize specimen is from excavations at San Marcos Cave in the Tehuacán Valley and was subsequently determined to represent early domesticated Zea mays L. (from Mangelsdorf et al. 1967, Fig. 97). (**b**) Examples of different varieties of pod corn (from Weatherwax 1954, Fig. 49)

information gathered from these samples (Fig. 3.7). These early studies also provide a clear indication of the extent to which the biological sciences influenced archaeological research on the origins of maize, and reveal that despite evidence for other edible parts of the maize plant, the primary focus was almost exclusively on the seeds or kernels, rather than the stalks (Doebley 1994).

Despite Beadle's (1939) observation that teosinte seeds could be made into popcorn that was edible, ethnobotanical evidence and analysis of ancient coprolites indirectly supported the assertion by many in the biological sciences that an unknown wild maize was the progenitor. Beadle (1972) later suggested that maize may also have been exploited for its sugary stalk, but the scientific community largely ignored this insight until many years later. This is because many of the biologists, botanists, and archaeologist during the early and middle part of the twentieth century were focused almost exclusively on the ear morphology and

Fig. 3.7 Wild maize reconstructed on the basis of an ancient maize cob from San Marcos Cave. The authors assert the husks enclosed the young ears, but opened when the cob was in a mature state, presumably permitting the natural dispersal of the kernels (from Mangelsdorf et al. 1967, Fig. 124)

kernel shape and size, when attempting to understand the origins of domesticated maize. Moreover, the hard fruit cases, and bitter flavor of teosinte seeds continued to be considered as an unlikely candidate progenitor by most scientists outside of plant genetics. It was genetic research on maize that created the possibility for considering teosinte as the progenitor just as it was that such research provided the basis for considering these wild grasses as the closest relative of *Zea mays* (Langham 1940; Kellogg and Birchler 1993).

3.2.3   Teosinte and the Search for the Origin of Maize

In the same year that Beadle published evidence of teosinte representing the proposed ancestor of domesticated maize, Mangelsdorf and Reeves (1939) proposed that there was a species of "wild" maize, that is, an undomesticated form of the plant, and that the most likely candidate for the origin of cultivated maize was pod corn (Fig. 3.8). Moreover, the hypotheses they set forth provided no genetic basis for the close similarities of maize and teosinte.

Beadle (1939) pointed out early on that any hypothesis that proposed an extinct or wild prototype must remain unsatisfactory until tangible evidence for its existence was forthcoming. This is brought out in his statement on the question of the genetic relationship between maize and the teosintes that;

> ... would account for the close relationship between teosinte and maize by assuming that teosinte arose through contamination of cultivated maize by the addition of segments of chromosomes of Tripsacum. While this view that teosinte is secondarily derived has points

Fig. 3.8 Cobs of "wild" maize associated with the Coxcatlan Phase occupations at San Marcos Cave, Tehuacán Valley. The authors emphasized uniformity and the relatively long glumes and fragile rachises (from Mangelsdorf et al. 1967, Fig. 96)

in its favor, the hypothesis still assumes an extinct or undiscovered wild prototype of maize... The sequence of operations employed by Mangelsdorf and Reeves' to obtain the F_1 hybrids artificially is such that the natural occurrence of these hybrids would appear to be precluded. In addition, the transference of the particular four of five necessary chromosome segments through rare crossing over involves another sequence of events *a priori* [original italics] improbable. Finally, admitting the close relationship between teosinte and maize, it appears to be quite unnecessary to postulate an hypothetical wild prototype of maize which through hybridization with Tripsacum gave rise to teosinte (Beadle 1939 p. 246).

Beadle also hypothesized that each morphological trait that separated maize from its teosinte ancestor was regulated by one gene, however, later research would provide evidence of a more complex genetic profile (Dorweiler 1996; Dorweiler et al. 1993; Dorweiler and Doebley 1997; Doebley 2001; Doebley and Stec 1991; Doebley et al. 1993, 1997). What made teosinte an unlikely progenitor was morphological; the most recognizable feature of *Zea mays* L., the massive husked ear, is absent in the teosintes (Beadle 1939, 1972). It was the massive husked ear that inspired Beadle to refer to domesticated maize as a "biological monstrosity." The inflorescences and pronounced difference in the ears of domesticated maize to other related wild grasses, was what was so perplexing to researchers. Recent research on maize DNA has established that the ancestor of maize was a variety of teosinte because this species is the only large-seeded, wild, annual grass in the tropical Americas (Doebley and Iltis 1980). Teosinte lacks a cob; instead, its seeds are contained in a spike of hard fruit cases (Weatherwax 1954). Its seeds are dispersed through shattering, but in other respects annual teosinte varieties are very similar to maize and they produce fully fertile offspring when interbred with the domesticated plant (Beadle 1972; Iltis 2006).

Some members of the related genus *Tripsacum* are locally referred to as "teosinte," plants that have compound leaves, a Cycad. The term "teosinte" accompanied the first Guatemalan accession and does not appear in Mexico. There are six annual and perennial species of teosinte (Iltis 2006). *Zea diploperennis* and *Z. perennis* are perennial, while all other taxa are annual[9] (Iltis 2006, Fig. 3.5). The two annual teosintes are most similar to domesticated maize, but one of them had the specific proteins from one of these annual teosinte subspecies (*Zea mays* spp. *parviglumis*) which were indistinguishable from maize (Matsuoka et al. 2002). Teosinte and maize have similar tassels and DNA, as well as amino acid and nutritional compositions (Matsuoka et al. 2002). Even examination with a SEM cannot distinguish the pollen of annual types of teosinte from small-seeded varieties of corn such as early domesticated maize (Horn 2006; Pohl et al. 2007). The geographic distribution of the *parviglumis* subspecies occurs at an elevation of

---

[9]The most puzzling teosinte is *Z. mays* ssp. *huehuetenangensis*, which combines morphology rather like *Z. mays* ssp. *parviglumis* with many terminal chromosome knobs and an isozyme position between the two sections. Phenotypically, the most distinctive and the most threatened teosinte is *Z. nicaraguensis*, which thrives in flooded conditions along 200 meters of a coastal estuarine river in the northwest Nicaragua (Iltis 2006). There may be some questions regarding the Nicaraguan species, as geneticists do not mention it (see e.g., Matsuoka et al. 2002; Vigouroux et al. 2003).

400–1,200 m (1,300–3,900 ft) along the slopes of the central Balsas river drainage, 250 km (155 mi) west of the Tehuacán Valley (Piperno et al. 2007). Teosinte often grows in the same fields with maize, beans, and squash. Teosinte is a weedy pioneer that thrives in disturbed areas such as seasonally wet streambeds and abandoned campsites. Because of their hard fruit cases, the teosintes are not easily ground into flour, and their seeds have a bitter taste (Mangelsdorf et al. 1967). Teosinte, however, has a sugary stalk and the stalk may occasionally have been chewed for its sugary taste or used as an auxiliary catalyst for fermentation (Beadle 1972, 1978). It appears to have been the morphological differences between the teosintes and domesticated maize that fostered such intense research into the possible existence of a "wild" maize progenitor.

Mangelsdorf was also a geneticist, and he was particularly interested in cross-breeding and plant taxonomy (Fussell 1992, p. 78). He believed that domesticated maize was the progenitor of teosinte; Mangelsdorf and Reeves (1939) concluded on the basis of their research that the teosinte grasses arose as a result of contamination of domesticated maize by the addition of chromosomes from *Tripsacum*, and that maize was derived from an unknown, possibly extinct, "wild" maize (Beadle 1939, p. 246). Mangelsdorf (1974) carried out extensive field survey in various regions of Mexico, Guatemala, and Peru searching for maize landraces or varieties that might provide clues to this hypothetically unknown, presumably extinct, wild ancestor. It was this field experience that reinforced his interdisciplinary approach to maize origins, because it became quickly apparent that such an ancestor could only be identified in archaeological investigations in the Neotropics. This was to some extent realized by the research directed by Richard S. MacNeish in the Tehuacán valley. However, as they were discovered before the advent of AMS radiocarbon dating, there was no way to directly determine the age of these macrobotanical remains (Smith 2005a). The earliest occurrences and antiquity of maize was largely a matter of estimated chronology by associated artifacts and conventional ^{14}C dates. Later developments in absolute dating exposed problems with the contextual reliability of associated conventional dates and the morphological viability of extinct "wild" maize progenitors (Fig. 3.9).

Excavations in the Tehuacán Valley spanned over 40 caves, five of which San Marcos, Coxcatlan, Purron, Tecorral, and El Riego had enormous quantities of maize macrobotanical remains. More than half of these macrobotanical samples are whole cobs. Excavations at five Tehuacán caves uncovered over 24,000 specimens of maize cobs from various timeperiods, and more than half were whole or almost intact and initially ^{14}C dated by association to a time span of a little over 7,000–5,000 years ago (Mangelsdorf et al. 1967, p. 179). The earliest ears were small (about the size of an index finger) and contained no more than four to eight rows of kernels. Mangelsdorf et al. (1967, p. 180) provided six *morphological* and *archaeological* reasons to suggest that 27 of the earliest cobs from San Marcos Cave and 44 from the Coxcatlan Cave may be considered as examples of "wild" maize (see Fig. 3.6b). Later genetic evidence proved that this was not the case. The conventional ^{14}C dates for maize at Tehuacán appear in cave deposits dating to the end of the sixth millennium B.C. (Benz and Long 2000;

Fig. 3.9 *Tripsacum* or gamma grass was thought to be one of the progenitors of domesticated maize, and a central part of the Tripartite Hypothesis. *Tripsacum dactyloides* grows in the Midwest and has a much greater natural range than the teosintes, which are confined to present day Mexico, Guatemala, and Nicaragua

Benz et al. 2006). The initial estimates of the earliest occurrence of *Zea mays* L. based on conventional radiocarbon assays were later considerably revised with the advent of direct dating by Accelerator Mass Spectrometry (AMS) dating.[10] Nevertheless, the associated dates from the Tehuacán caves had a profound influence in the field archaeology since it provided an incentive for those studying the origins of maize to look for Archaic sites dated to about 7,000 years ago (Long and Fritz 2001, p. 87). Since the mid-1980s, AMS radiocarbon dating, with its small sample size requirement, has allowed seeds and other small plant parts to be directly dated (Long et al. 1989). This dating technique has ended the reliance upon age estimates obtained by conventional, large sampling dating of charcoal or other organic material found in close proximity to botanical samples. Directly dated seeds and other plant parts that exhibit morphological changes, consistent with domestication, now constitute the primary class of evidence for plant domestication in the Americas (Benz et al. 2006). The ability of researchers to later date minute maize cob samples has provided a basis for establishing a reliable, absolute chronology for its initial appearance and eventual dispersal (Long and Fritz 2001, p. 87; Blake 2006, p. 55; see also Smith 1997a, b, 2000). The early AMS dates based on direct ^{14}C AMS dates of the early maize cobs

---

[10]Subsequently, direct AMS dates on some of the earliest of the San Marcos and Coxcatlan cobs have shown that they were more recent in time, raising considerable controversy and debate in the field of archaeology and palaeoethnobotany (Long and Fritz 2001; see also Smith 2005a).

themselves produced results to the mid-third millennium B.C., several 1,000 years younger than previously thought (Table 3.1). The new AMS dates, therefore, provide a new timetable for the arrival of maize into the Tehuacán Valley (Benz and Long 2000; Benz et al. 2006).

Some anthropologists took the early botanical results by Asherton and others to correctly surmise that teosinte was ancestral to domesticated maize, and that the two plants were taxonomically related to one another (Thompson 1932). However, Mangelsdorf and his associates still sought the elusive extinct wild species and spent the rest of their career maintaining that pod corn represents wild maize, and the key was to find the so-called extinct or unknown progenitor connecting pod corn to *Zea mays*[11] (Mangelsdorf 1974). Because of the research by Mangelsdorf and others, many anthropologists, archaeologists, botanists, and some plant geneticists remained skeptical and subsequently continued to pursue other plants as ancestral to *Zea mays* L. and this history of research is to some degree reflected in the taxonomies which were initially established for the various grass taxon. The phylogenetic relationships of the various maize varieties, was of central concern to research interests in the origins of maize. Mangelsdorf was extremely influential in the research and scholarship surrounding maize. From his laboratory at Bussey Institution at Harvard University, his pursuit of internally consistent interdisciplinary evidence provided him a basis for working and collaborating with botanists, plant geneticists, and particularly archaeologists and anthropologists to look for answers to the origins of maize (see, e.g., Mangelsdorf et al. 1967). He long believed that domesticated corn was the result of a hybridization of *Tripsacum* with another unknown wild grass related to teosinte, the species called *Manisuris* (*Euchlaena mexicana*[12]), which pertains to the same family as maize and teosinte (Mangelsdorf 1974, p. 13; see also Ascherson 1875; Bailey 1892). Mangelsdorf and Reeves (1939, pp. 220–221) dismissed, for morphological reasons, the possibility of teosinte as the progenitor of maize early on, and instead asserted it was a descendent, the result of a hybridization of an unknown wild maize and *Tripsacum*. Later research indicated that all the cobs

---

[11]Mangelsdorf et al. (1967, pp. 179–180, Fig. 97) believed that they had identified a specimen of pod corn derived from San Marcos Cave, Zones F and E, which were dated by association to 7,000 years ago (see Fig. 3.5a). Later he joked that his initials stood for "Pod Corn Maize." Mangelsdorf continued to declare until his passing in 1989 at the age of 90, that pod corn represented the wild ancestor of maize (Fussell 1992, p. 78).

[12]Manisuris is teosinte now classified as *Zea mays* ssp. *mexicana*. Yoshihiro Matsuoka and his collaborators (2002:6083) have emphasized the importance of gene flow from *Z. mays* ssp. *mexicana* as contributing to the maize diversity. This subspecies grows as a weed in many highland maize fields (above 1,800–2,500 m) where it frequently hybridizes with maize, whereas *Z. mays* ssp. *parviglumis* often grows at lower elevations (below 1,800 m) and rarely hybridizes with maize (Wilkes 1967, 1977). *Z. mays* ssp *mexicana* is also adapted to lower rainfall (500–1,000 mm) than subspecies *parviglumis* (Pearsall and Piperno 1998, p. 161). These various factors appear to play a role in this subspecies' role to maize diversity.

Table 3.1 Direct AMS dates on maize macrobotanical samples

Country/Region	Site	Dated material	^{14}C method	^{14}C years B.P.	Sample ID No.	Reference
SW United States						
New Mexico	Bat Cave	Maize	AMS	3010 ± 150		Wills (1988)
New Mexico	Fresnal Shelter	Maize	AMS	2945 ± 55	AA-6402	Tagg (1996, p. 317)
Arizona	Milagro	Maize	AMS	2930 ± 45		Huckell et al. (1995)
Arizona	Three Fir Shelter	Maize	AMS	2880 ± 140		Smiley (1994)
Arizona	Fairbank	Maize	AMS	2815 ± 80		Huckell (1990)
Arizona	Cortaro Fan	Maize	AMS	2790 ± 60		Roth (1984)
Arizona	West End	Maize	AMS	2735 ± 75		Huckell (1990)
New Mexico	LA18091	Maize	AMS	2720 ± 265	UGa-4179	Simmons (1986)
New Mexico	Jemez	Maize	AMS	2410 ± 360		Adams (1994, p. 79)
New Mexico	Sheep Camp Shelter	Maize	AMS	2290 ± 210	A-3396	Simmons (1986, p. 79)
New Mexico	Tularosa Cave	Maize cob	AMS	1920 ± 40	β-166755	Jaenicke Després et al. (2003, p. 1207)
Mexico						
Oaxaca	Guilá Naquitz	Maize cob	AMS	5420 ± 60	β-132511	Piperno and Flannery (2001, p. 2102)
Puebla	San Marcos Cave	Maize cob	AMS	4700 ± 110	AA-3311	Long et al. (1989, p. 1037)
Puebla	El Riego Cave	Maize cob	AMS	3850 ± 240	AA-52684	Benz (2006, Table 5.1)
Puebla	Coxcatlan Cave	Maize cob	AMS	3740 ± 60	AA-3313	Smith (2005a, Table 1)
Tamaulipas	Romero's Cave	Maize cob	AMS	3930 ± 50	β-85431	Smith (1997b, p. 373)
Tamaulipas	Valenzuela's Cave	Maize cob	AMS	3890 ± 60	β-85433	Smith (1997b, p. 374)
Chiapas	San Carlos	Kernel	AMS	3365 ± 55	β-62911	Clark (1994), Blake et al. (1995, p. 164)
Chihuahua	Cerro Juanaquena	Maize cob	AMS	2980 ± 50	INS-3983	Hard and Roney (1998, p. 1664)
Tabasco	San Andres	Maize cob	AMS	2565 ± 45	AA-33923	Pope et al. (2001, p. 1372)
Sonora	La Playa	Maize	AMS	1885 ± 50	n/a	Matson (2003, Table 1)
Central and South America						
Honduras/inland	El Gigante	Maize cob	AMS	2280 ± 40	β-159055	Scheffler (2002)
Ecuador/coast	Loma Alta	Kernels	AMS	3500	n/a	Pearsall (2003, p. 223)

Ecuador/coast	Loma Alta	Carbonized kernels	AMS	2490 ± 40	β-103315	Zarillo et al. (2008, p. 5007)
Peru/highlands	Pancan	Maize cob	Conventional	1500	n/a	Smith (1998), Rodriquez and Aschero (2007, p. 261)
Brazil/Minas Gerais	Peruaçu Valley	Maize cob	AMS?	990 ± 60	?	Freitas (2001)
Argentina/Mendoza	Gruta del Indio	Maize	Conventional	2065 ± 40	GrN-5396	Gil (2003, p. 297)
Catamarca/highlands	Punta de la Pena 4	Kernel	AMS	740 ± 40	UGA-15089	Rodriquez and Aschero (2007, pp. 259, 268) date given: 560 ± 50
Chile/North Coast	Ramaditas	Maize cob	AMS	2210 ± 55	GX-21725	Rivera (2006, Table 29-4)
	Guatacondo	Maize cob	AMS	1865	UCLA-1698c	Rivera (2006, Table 29-4)
	Rixhasca	Kernel	AMS	1025	GX-21748	Rivera (2006, Table 29-4)
	Tiliviche 1-b	Maize cob	AMS	920 ± 32	AA-56416	Rivera (2006, Table 29-3)

Only the earliest direct dates on macrobotanicals are included in this table analyzed from this site were already fully domesticated *Zea mays* L. (Benz et al. 2006, Tables 5-2 and 5-3). However, Mangelsdorf and Reeves were convinced that the progenitor of maize was a non-extinct pre-Columbian wild grass, and for many years pursued this line of thinking in seeking the origins of this plant (Mangelsdorf and Reeves 1933, 1939). Those grasses were maize, pod corn, teosinte, and *Tripsacum* (Mangelsdorf 1974, p. 13). They carried out numerous cross-breeding experiments over many years in the pursuit of the origins of domesticated maize. Since they were concerned with several species of wild grass, they could pursue the quest for wild maize in terms of multiple domestication events in geographically different regions of the Neotropics (Mangelsdorf and Reeves 1959a,b). They state early on that, since they dismissed teosinte as a possible progenitor, "there remains no necessity for seeking the center of origin either in Mexico or Central America" (Mangelsdorf and Reeves 1939, p. 241). They go on to state that it was paradoxical that their conclusion on the origin of teosinte as a hybrid of Zea and *Tripsacum* suggested to them that, "we seek the center of origin of maize or at least the primary center of domestication in a region where its relatives do not occur" (Mangelsdorf and Reeves 1939, p. 241).

The broader implications of these conclusions had profound consequences for archaeological and later ethnobotanic research on early domestication in general and the origins of maize in particular. The archaeological research became focused early on upon where to look for such domestication events, and how early in time maize was originally domesticated. Archaeologists in this hemisphere were, therefore, predisposed to look for earliest occurrence, and to seek out Archaic sites in Mexico, Central and northern South America. Moreover, such research in this hemisphere has generally interpreted the factors responsible for the adoption of agriculture and plant domestication at a particular site as complex and multifaceted, that is, very different from one site or region to another (Brown 2006, p. 3; see also Smith 2001). Although there has historically been a strong focus on earliest occurrences, the spread of maize and its integration into the agricultural economy has in recent years been generally interpreted as distinct from region to region (Hoopes 1994; Smith 2000, 2001, 2006). Teosinte was also dismissed as a possible progenitor by many archaeologists because there is no prehistoric evidence from any of the Tehuacán cave sites to suggest it was consumed as flour (Mangelsdorf et al. 1967). Scientists have been exploring questions surrounding the origin of maize since research on the crop began. However, important data on how the domestication process affected the transformation from teosinte to maize have only recently been addressed (Emshwiller 2006). These include how the genetics of maize populations or varieties are affected by conscious selection, how the extent of hybridization and gene flow between domesticated and wild grass populations influences the genetic basis of the morphological and other changes that occur during domestication. These new areas of inquiry were made possible by recent DNA research which provided the molecular tools now applied to systematics, population genetics, and genetic mapping (see, e.g., Matsuoka et al. 2002; Doebley et al. 2006). Had such tools in genetic research been available in the last century, the search for the wild progenitor may have been quite different.

3.2.4 Approaches to Finding Wild Maize

The search for wild corn took two approaches; one was the quest to find wild corn plants still growing. This approach was inspired in part by study or an eighteenth century document by the Italian explorer Lorenzo Boturini (1746, p. 21) who stated that he found wild maize in the jungle forests of Mexico, and that their kernels were small and had an excellent flavor (cf. Weatherwax 1954, p. 141). The search for information of a wild variety of maize began by an examination of the legends, myths, folktales, and early chronicler accounts. Boturni's account combined both of these goals in that he couched his descriptions of "wild" maize in historical speculations involving ages of history. The first age being of course occupied by the gods who on burning the forests of Mexico found parched grains of maize and other seeds, which they tasted and found good to eat. Seeds, which were not damaged by the fire, were presumably planted in the same soil as the mother of plants, which eventually produced good harvests (Weatherwax 1954). Studies of indigenous folklore and mythology revealed that the appearance of maize was generally through some supernatural agents or mythological ancestors.

Later scholars had a different way of seeing the natural world, replacing the medieval way of reading the world's diversity as symbolically as historical ages in God's poem, into a textual analysis of nature's prose (Fussell 1992, p. 60). These more systematic scholars also applied human analogs to plants, as Indian myth makers had done, but for very different ends. Where Amerindians imbued maize with concepts of gender and sacred mythology,[13] Linnaeus and others created taxonomies based on phylogenetic relationships to make all God's creatures one (Godfray 2007).

It was also in the eighteenth century that some Europeans began to use also human analogies to separate the plant world from the human one and to reconstitute plants into groups of autonomous interrelated families. New tools for observation, such as the microscope, led plant scientists to presume they were discovering "the thing in itself" and around these observations there developed a new language, which systematized their "discoveries" (Fussell 1992).

During the colonization of the New World, clerics and priests systematically learned and wrote down many indigenous languages for the purpose of religious conversion. Later linguists would use such vocabularies and create new ones to trace when and how terms for the plant were introduced into various New World indigenous populations (Swadesh 1951; Berlin 1992). The anthropological research involving this methodological approach has largely been dominated by historical

---

[13]For example, this Osage myth; "For the fourth time Buffalo threw himself upon the earth, And the speckled corn, Together with the speckled squash, He tossed into the air, Then spake, saying: "What living creature is there that has no mate?" And thus he wedded together the speckled corn, a male, to the speckled squash, a female. He continued: "The little ones shall use this plant for food as they travel the path of life. Thus they shall make for themselves to be free from all the causes of death as they travel the path of life." (Rankin 2006, p. 563).

linguistics (Swadesh 1959a; Kaufman 1994; Campbell 1997; Dixon 1997; Hill 2001, 2006; Brown 2006). Folk taxonomies provide anthropologists and archaeological a sense of how various indigenous groups thought about maize and exploring the vocabularies surrounding the cultivation and preparation of maize (and other cultigens), provided a linguistic basis for understanding how it spread and was consumed (Swadesh 1959b; Campbell 1997; Hill 2001, 2006; Rankin 2006; Hopkins 2006). In the case of historical linguistics, there has been relative success in identifying linguistic cognates specific to maize being introduced and borrowed by neighboring populations, and also in tracing the beginnings of maize by an indirect glottochronology (Swadesh 1951, 1959a; Berlin 1992; Dixon 1997; Brown 1999, 2006; Rankin 2006). Such linguistic data were, therefore, critical to scholars attempting to understand the antiquity and spread of maize since it was believed at the time that it could have been domesticated in numerous regions of the Neotropics. For example, Mangelsdorf and Reeves (1939) initially believed that maize originated in the Amazonian lowlands because of historical references to pod corn in that part of the Americas. The possibility of an Amazonian origin influenced later research on maize cultivation in this area of the Neotropics (Lathrap 1970; Bush et al. 1989).

3.2.5   Pod Corn as Wild Maize

Pod corn is a grass in which the kernels are enclosed in floral bracts, similar to the majority of grasses and other cereal grains (Mangelsdorf 1948, pp. 377–378; Weatherwax 1954, Figs. 49–50). The premise that pod corn was the progenitor of maize was appealing because their principal morphological characteristic, enclosed kernel in floral bracts, is nearly a universal characteristic of wild grasses (Mangelsdorf 1974, p. 13). Nineteenth century botanists attempted to make a compelling case for pod corn as wild maize classified it within the genus (*Zea tunicata*). Pod corn because of its morphology also intrigued later botanical and biological scholars; it has ears known to resemble maize except that six small husks cover each grain or kernel (see Fig. 3.8). Pod corn was known to be more resistant than other varieties to insect predation. Its morphological characteristics were ideally suited to what many scholars had associated with a wild variety of *Zea* (Kellerman 1895). Although these traits made it a compelling candidate for wild or undomesticated maize, pod corn had little to recommend it as an economic staple (Weatherwax 1954, p. 143). Nevertheless, it is for these morphological reasons that it was, even after being rejected as a possible candidate for wild maize, later resurrected again in association with an elaborate array of evidence (Mangelsdorf 1948, 1974).

Like other early searches for wild maize, scholars studied ethnohistoric accounts and herbals. Since Mangelsdorf and Reeves (1939, p. 248) saw the Andes as the primary center of domestication of maize. Moreover, they believed that they identified representations of pod corn in prehistoric Peruvian pottery. Although pod corn is unknown in Peru they began to explore its presence in other regions of tropical South America. In looking through sixteenth and seventeenth century herbals they noted that, "pod corn first appeared in the herbal published by Gaspard Bauhin in 1623" (Mangelsdorf and Reeves 1939, pp. 222, 226). Early South American accounts of wild maize and maize varieties stated that pod corn was mentioned by the Guaicuru Indians, and the Guarani in present day southern Bolivia and the state of Entre Rios in northern Paraguay where it was said to be indigenous to this region and believed to grow there in the humid forests (Mangelsdorf and Reeves 1939, p. 226). The 1829 reference to pod corn by Saint Hilaire to the French Academy of Sciences, refers this species as *Zea Mais* var. *tunicata* (Mangelsdorf and Reeves 1939, p. 226). Mangelsdorf and Reeves considered the ethnographic evidence from Paraguay among the Guarani and Guaicuru to be so important that he concluded that maize originated in northern Paraguay and that in its natural state the grains were covered with glumes, and these were presumably lost following cultivation (Mangelsdorf and Reeves 1939). Recent ethnographic research among the Guarani in southern Bolivia has demonstrated that maize (*Zea mays* L.) is an important economic staple and closely associated with their rituals and Native religion (Ortiz et al. 2008, pp. 179–180). However, these ethnobotanical and ethnographic data provide no comfort to pod corn as important economically or symbolically to present day Guarani culture.

The ethnohistoric literature and early herbals provided a compelling case for pod corn as the progenitor of maize. These data also explain why Mangelsdorf and others were predisposed to multiple domestication events over a vast geographic area, and why such thinking was later taken as a given by archaeologists and ethnobotanists searching for earliest occurrences. What perhaps made a great impression was the first known illustration of pod corn in Bonafous monograph,[14] as it suggested to them that this was a heterozygous form of the plant, and the ethnohistoric accounts indicated that different varieties of it could have been consumed, and therefore cultivated by this culture (see Fig. 3.10). Bonafous (1836) reported that this plant species was referred to as *pinsingallo*. The term was seen as a corruption or modification of the word *bisingallo*, which Dobrizhoffer (1822 (1784)) used in connection with a kind of corn grown by the Guarani. Dobrizhoffer, a Jesuit missionary in Paraguay from 1749 to 1767, states: "The Guaranis sow various kinds of it, ... the abati hata [is] composed of very hard grains, the abati moroti, which consists of very soft and white ones, the abati mid, which ripens in 1 month, but has very small dwarfish grains and bisingallo the most famous of all, the grains of which are angular and pointed" (cf. Mangelsdorf and Reeves 1939, p. 226). The use of the expression "most famous of all" as well as "angular and pointed" suggested to these scholars a type of maize that was different

---

[14]Bonafous' classical monograph "*Histoire naturelle, agricole et economique du Mais*", published in 1836, makes reference to pod corn which he called *Zea cryptosperma*. Pod corn is currently classified as *Zea mays var. tunicata* Larrañaga ex A. St. Hilaire.

Fig. 3.10 Earliest known illustration of the heterozygous form of pod corn, ca. 1836, reproduced in the Bonafous botanical, *Histoire naturelle, agricole et economique du Mais* (from Mangelsdorf and Reeves 1939, Fig. 87)

from the other varieties, and they were intrigued by his description of pointed grains of pod corn with the glumes attached (Mangelsdorf and Reeves 1939, pp. 226–227). Furthermore, the terms, *bisingallo* and *pinsingallo*, initially reported by Bonafous with reference to pod corn, and this additional linguistic evidence led them to conclude that the missionary was describing pod corn (Mangelsdorf and Reeves 1939, p. 226). They point out that this description was earlier than that of Azara (1809). Azara (1809) was the Spanish commissioner and commandant of Paraguay from 1771 to 1801 and made early reference to its natural history (cf. Mangelsdorf and Reeves 1939, p. 224). Varieties of pod corn were also identified in the Far East, China, and India presumably taken there by Portuguese and Spanish traders in the early sixteenth century (Mangelsdorf and Reeves 1939).

The first American publication to report on pod corn was by Teschemacher (1842) of a botanical sample from Texas, which he concluded did not represent a distinct species. A report to the Horticultural Society of London by Lindley (1846) stated that this pod corn was from the Rocky Mountains as Native Indian corn. Lindley who published the first comparative data on pod corn and maize pointed out that the cob or rachis of maize was much larger, "as if the deterioration of the latter [glumes] had caused the enlargement of the former [cob]" (Mangelsdorf and Reeves 1939, p. 228). Moreover, Salisbury (1848) subsequently reported that pod corn was composed of several subspecies with white, yellow, red, or purple kernels. This was particularly compelling to archaeological and botanical scholars studying the origins of the plant because it was identical to the variability known for maize varieties in Mexico and the American Southwest, where Native American taxonomies generally consider kernel color, shape, size, and row number as key morphological traits (Berlin 1992; Berlin et al. 1974; Perales et al. 2005; Huckell 2006; Benz et al. 2007; Tuxill et al. 2009). Such phenotypic and morphological traits are also used in folk taxonomies in other parts of the New World to distinguish different varieties or landraces.

An illustration of pod corn appeared in the Annual Report of the Commissioner of Patents (1852) accompanied by the statement of an anonymous writer that this type of corn is found in a wild state in the Rocky Mountains and in the humid forests of Paraguay (Mangelsdorf and Reeves 1939, p. 228). Candiole (1855) was one of the earliest scholars to remain skeptical. After discussing pod corn morphology at length he concluded that there was insufficient evidence to consider this plant as a progenitor to *Zea mays* L. Similarly, Charles Darwin (1868) observed that pod corn reverted to its normal state upon cultivation and therefore ruled out this type as a progenitor, but nevertheless conceded that its morphology was compelling, surmising that the original form "would have had its grains thus protected" (cf. Mangelsdorf and Reeves 1939, 228–229; see also Kellerman 1895). Later Galinat (1954) considered pod corn as a "false" progenitor of maize and saw this plant as exhibiting morphological traits associated with a remote ancestor of the family *Maydeae*. Since the progenitor of maize was not definitively established until after the twenty-first century, a number of different wild grasses continued to be plausible as progenitors for different races of maize.

3.2.6 Teosinte as a Progenitor of Maize

In the New World, the natural distribution of teosinte in Mexico and Central America places a geographical limitation on the region where maize agriculture could have first appeared archaeologically, but the initial interpretation, based on the substantial morphological and genetic variation displayed by modern maize, was that it was domesticated on multiple occasions (Mangelsdorf 1974, pp. 14, 163; Galinat 1988, pp. 95–97). It was in fact the discovery of fossil pod corn in Mexico that reinforced the possibility of multiple domestication events, i.e., that maize evolved into its present form in multiple locations in Mexico and particularly South America (Mangelsdorf 1974, p. 14, 1983; Mangelsdorf and Reeves 1959a, b). Mangelsdorf (1974) reported that wild maize differentiated into at least six races and that these landraces then spread to North, Central, and South America, while teosinte was confined to Mexico, Guatemala, Nicaragua, and Honduras. This hypothesis presumed that teosinte evolved from maize, rather than being an ancestral progenitor. Since *Tripsacum* underwent extensive divergence and differentiation in its glume architecture and number of chromosomes during its speciation process, it was more diverse morphologically and genetically, thus, able to adapt to a wider range of New World environments (Galinat 1976, p. 94).

Mangelsdorf and Reeves (1939) noted early on that maize farmers often planted teosinte around maize fields in order to acquire cross-pollination and disease resistance. Maize and teosinte hybrids are highly fertile indicating that there was still gene flow between the species. It is in fact the mutability of corn that inspired much of the research into its origins, and it is this characteristic that also has made it such an important staple (Iltis 1983, 2000). Hybrid varieties of maize were developed by cross-breeding, that is putting the pollen or male element from *one* variety on the silks or female element of *another* variety (Wallace and Brown 1956, pp. 14–15 [original emphasis]; see also Mangelsdorf and Reeves 1939, pp. 77–83; Beadle 1939; Langham 1940).

The first approach of looking for a wild ancestor of maize occupied much of the scholarly research on the topic of maize origins in the biological sciences for the first part of the twentieth century. The reason that pod corn played such an important role early on in the quest for maize origins is that many scholars in the botanical and biological sciences were to varying degrees looking to Old World grains and approaches to agricultural origins. Thus, many became fixated upon finding a homozygous, true breeding, earless form of *Zea* with seeds in the terminal inflorescence, in other words, a subspecies which morphologically approximated maize phenotype, but which was at the same time "primitive" in appearance. Secondly, they were focused on the kernels, and did not consider the possibility that the young sugary stalk may have also been an attraction to early archaic peoples as a condiment and/or used in the fermentation process (Beadle 1972, p. 10; Smalley and Blake 2003; Iltis 2004). Moreover, there was no archaeological evidence to suggest that teosinte was exploited by humans for food or that its seeds were consumed in prehistoric times, although the stalk was very high in sugar (Smith 1986, pp. 273–274; Staller 2003; Smalley and Blake 2003). A review of the botanical and biological research on maize suggests that "pre-domestication cultivation" during which a crop was partially managed by a distinct population, but not completely isolated in a reproductive sense from its wild relatives, genetic bottlenecks, and disruptions in gene flow was of primary interest (Eyre-Walker et al. 1998, pp. 4441; Brown 2006, p. 5; Benz 2006, pp. 12–15; Benz et al. 2006, pp. 78–79).

All the evidence from comparative morphology of that time suggested that cultivated maize had in its ancestry a "perfect flowered plant having covered seeds on a terminal inflorescence, with brittle branches" (Mangelsdorf and Reeves

1939, p. 231). Precisely, what would be expected if such a potential wild maize did in fact exist. One of the reasons botanists and biologists looked beyond teosinte for so long as a possible progenitor is that the glume architecture and the caryopsis or seeds were morphologically distinct (Benz 2001, Fig. 1, 2006, Fig. 2.2; Iltis 2006, Fig. 3.4). These early studies did not take into consideration artificial selection, asserting that what was referred to as a "naked-seeded corn" (as opposed to a pod corn with covered seeds) would not have survived long in nature without human management, and despite the fact that it had been established that teosinte was the closest living relative of *Zea mays* among the wild grasses (Beadle 1939; Langham 1940; Wilkes 1967).

These early studies not only provided them with a vast geography in which to look for maize origins, but also emphasized that the domestication of maize could have been the result of multiple events. The archaeological evidence of an intermediate species of maize, which shared morphological characteristics of both pod corn and domesticated maize was not forthcoming. Mangelsdorf (1974, p. 13) later dismissed pod corn as a possible progenitor because it was inconsistent with the archaeological evidence, but then resurrected the tripartite hypothesis with the discovery of the teosinte *Zea diploperennis* (Iltis et al. 1979).

3.3   Maize: Morphological, Biological, Genetic, and Taxonomic Approaches

The second approach to the discovery of wild maize was deductive and applied by many ethnobotanists and biologists in the late nineteenth and throughout the twentieth century. The basic approach involved attaining an intimate knowledge of the plant morphology and biology, particularly in depth knowledge of the plant phylogeny (Kellogg and Birchler 1994). This research was heavily grounded in methodologies surrounding plant morphology and genetics, particularly cross-breeding and comparative analysis of what were believed to be related plant species (Weatherwax 1954, p. 140; Galinat 1983; Iltis 1983; Eyre-Walker et al. 1998). Despite attempts to cross-breed maize and/or presumably related species, no self-sustaining maize plants were ever produced using such scientific approaches (Bennetzen et al. 2001). Part of the reason for this had to do with the fact that the initial assumptions and associations, both theoretical and comparative, were flawed, and the predominant theory and methodological approach to the evolution and origins of maize was the tripartite hypothesis.

Mangelsdorf and Reeves (1939, p. 52) carried out numerous cytogenetic experiments to obtain true hybrids of *Tripsacum*, with *Zea mays* L. and *Euchlaena*, and also a trigeneric hybrid involving all three genera. They did this by repeatedly backcrossing from teosinte, translocation segments originally from *Tripsacum* that distinguish it from maize. In order to create a naked type of maize with many "wild" genes, not the same genes that characterized the original wild corn in its genetic complex, perhaps, but genes presumably similar in their effects (Mangelsdorf and Reeves 1939, p. 241). By superimposing genes for poddedness through repeated backcrossing, they believed they would create a plant quite similar in its essential features to the primitive wild maize. They also believed that such a plant would be capable of perpetuating itself in nature, at least in those regions where teosinte now grows in the wild (Mangelsdorf and Reeves 1939). Their results provided evidence suggesting that phylogenetically teosinte was more closely related to maize than *Tripsacum*, but the hybrids produced by these experiments produced no fertile maize specimens (Wet and Harlan 1978, p. 135). This was for many years dismissed as relating to genetic variability related to geography and adaptation to distinct environmental settings (Fig. 3.11).

Fig. 3.11 Teosinte (*Zea mays* ssp. *mexicana*) from the central highlands of Mexico (Weatherwax 1954, Fig 46)

Galinat (1964) had suggested that *Tripsacum* is a hybrid of wild maize and *Manisuris*. Later, when considering the cytogenetics, he found that teosinte was more variable than either maize or *Tripsacum,* and therefore, correctly concluded that it probably evolved earlier in time – since both teosinte and maize had ten pairs of chromosomes that they were related[15] (Galinat 1976, p. 94, 1983; Doebley 2001). Weatherwax (1954) considered the tripartite hypothesis to be overly complex and laden with assumptions that would fall like a house of cards if any of the premises were found to be inaccurate.

During the 1970s and 1980s, the long held hypotheses proposed many years before by Beadle (1939) and Mangelsdorf and Reeves (1939) as well as many others regarding the progenitor of domesticated maize were slowly being generated by archaeologists, botanists, and plant genetics. These data, with few exceptions, seemed to point to maize having its earliest presence in ancient Mesoamerica, and its progenitor being one of the species of teosintes from this region. These data, presented in a series of published reports, for the most part, directly challenged the tripartite hypothesis, and the role of *Tripsacum* in the origins of *Zea mays* L. Mangelsdorf (1974) and others (Randolph 1975) nevertheless continued to hold fast to the idea of wild maize and genetic divergence. Research from ethnobotany on pollen, plant morphology, and genetic studies of chromosome structure consistently ruled out *Tripsacum* as a possible ancestor of maize (DeWet and Harlan 1978, pp. 137–138; Galinat 1985; Mangelsdorf 1986). Mangelsdorf (1986) conceded at least in part that domesticated maize originated from the hybridization of perennial teosinte, and a primitive "pod-popcorn." Hugh Iltis (1983) proposed the idea that the hybridization from teosinte to corn was a result of epistasis, multiple genes affecting a single trait, which in this case resulted in a sudden evolutionary transformation involving the male tassel spike at the lateral branch of teosinte suddenly transforming into the female ear of corn attached to a central stem. He later conceded that rather than being a sexual transmutation that the origins of maize more likely had to do with sexual translocation of its genetic makeup (Iltis 2000). Cytogenetic research by Walton Galinat (1970) established that the progenitors of *Tripsacum* were a result of wide-cross hybrid of *Manisuris* from the family *Andropogoneae* and teosinte – a result of an alloployploid followed by a doubling of the chromosomes. Galinat (1983, 1985), in contrast to Iltis, perceived the process of domestication of *Zea mays* to be a gradual one, involving the paired kernels gradually transmuting without sex change into the soft kernels. In other words, the kernels liberated themselves outward from their hard teosinte spikes (Galinat 1985).

---

[15] All teosinte species are diploid ($n = 10$) except *Z. perennis*, which is tetraploid ($n = 20$). The different species and subspecies can be readily distinguished based on morphological, cytogenetic, protein and DNA differences and on geographic origin, although the two perennials are sympatric and very similar.

3.3.1 Perennial Teosinte and a Reconsideration of the Tripartite Hypothesis

Field research in 1978 by the Mexican botanist Rafeal Guzman in the state of Jalisco provided evidence of a new species of teosinte *Zea mays* ssp. *diploperennis* or perennial teosinte which could easily produce offspring when cross-bred with maize (Fussell 1992, pp. 81, 82). Subsequent research at the Herbarium at the University of Wisconsin indicated that this perennial species of teosinte had the same number of chromosomes as maize ($n = 10$), and given its reproductive capacity in backcrosses with *Zea mays* L., it was perceived as a potential progenitor to maize (Iltis et al. 1979). Despite the seemingly indisputable evidence that *Tripsacum* was not involved in the domestication of maize, and thus the tripartite hypothesis no longer tenable, Eubanks (1995) continued to pursue such lines of research. Eubanks (1995, 1997) attempted to replicate the tripartite hypothesis for the origin of maize by postulating that its descent resulted from a hybrid cross between an extinct maize and a wild ancestor. She crossed and backcrossed *Tripsacum* with the diploid perennial teosinte, *Z. diploperennis* (Eubanks 1995, 1997, 2001a–c). Eubanks (1995, 1997) had previously demonstrated that crossing diploid perennial teosinte with *Tripsacum dactyloides* will produce fully fertile hybrids, some of which closely resemble the archaeological specimens of maize found in the Tehuacán Valley. There was a strong impetuous to maintain the tripartite hypothesis by some scholars because it appealed to previous research largely predicated upon multiple domestication events, and the natural distribution of grass species related to this hypothesis[16] (Mangelsdorf and Reeves 1939, 1959a, b; see also MacNeish and Eubanks 2000; Piperno 1984; Pearsall and Piperno 1990). Galinat (1978, p. 94) found teosinte to be cytologically more variable than either maize or *Tripsacum*. It was ultimately the research from geneticists and molecular biologists, which would have the most dramatic effects on the viability of *Tripsacum*, *Z. diploperennis* or teosinte being the progenitor of domesticated maize (DeWet and Harlan 1972, p. 137; Doebley et al. 1984, 1987, 1990; Dennis and Peacock 1984; Buckler and Holtsford 1996).

Maize was discovered to have the least cytological variability on the *Maydeae*, this despite the fact that more than 200 races had been identified morphologically (Galinat 1978; see also Goodman 1968, 1973, 1978). Identification was first and foremost on the basis of ear morphology, kernel shape, color and size, number of rows, etc. Such classifications were initially believed to pertain to different lineages and reinforced the belief that extant wild maize still existed and that the origins of maize predated that of teosinte (Weatherwax 1954, pp. 139–149). Goodman and McKBird (1977) identified 219 races for South America alone, using multivariate statistical analysis of morphological traits. Galinat (1978, p. 106) believed that

---

[16]Collins (1931) noted early on that the region that includes Peru, Bolivia, and Ecuador has a greater diversity of maize varieties than the whole North American continent, presumably making it a center of origin (Mangelsdorf and Reeves 1939, pp. 242–243).

when ancient maize was taken out of landscapes that included teosinte, the previous systems of obligatory epistasis between blockers were broken down resulting in greater morphological variability due to a recombination of genes. This was particularly apparent in the ears of maize varieties in Peru and Guatemala (Galinat 1978). In a final rebuttal of the tripartite hypothesis and the later tripsacum-diploperennis hypothesis, a number of maize genetics and evolutionary biologists synthesized the evidence from the biological sciences and molecular biology (Bennetzen et al. 2001). These scholars concluded that the evidence supporting teosinte as the progenitor of maize was overwhelming and that although it may be possible to produce crosses with *Tripsacum-diploperennis* hybrids the research did not demonstrate they were successfully hybridized (Bennetzen et al. 2001, p. 85). Previous hybrids of teosinte and maize were found to be highly fertile and produced hybrids that approximated those found at Archaic sites (Wilkes 1967). The hybrids produced from *Tripsacum* – teostine hybrids were found to be sterile (Talbert et al. 1990; cf. Bennetzen et al. 2001, p. 85). Moreover, the molecular data presented by Eubanks (1995, 1997) do not appear to establish that the hybrids are real (Bennetzen et al. 2001, p. 85). These scholars also pointed out that despite the years of research with crossing and backcrossing *Tripsacum* there was no direct evidence ever produced for its contribution to maize ancestry (Bennetzen et al. 2001).

3.3.2   Maize Antiquity and ^{14}C and AMS Chronologies

The various articles on maize and the chronological spread of the crop reported in 2001 in the first issue of the journal *Latin American Antiquity* represented a crossroads of sorts for archaeologists investigating the origins of agriculture and maize in the Americas. The article publications in this journal also unleashed a number of controversies that extended beyond the antiquity and phylogeny of maize, to issues surrounding its biogeography based on ethnobotanical remains. It also extended to published research on maize origins based on plant microfossils as well as more general theoretical questions involving plant domestication. Despite the overwhelming evidence for teosinte as the ancestor of maize, and a more recent chronology for its original domestication and spread, many in the archaeological and biological science continued to maintain the possibility of multiple domestication events, and maintain the original chronological data reported for its domestication and spread to different regions of the Americas (e.g., MacNeish 2001a, b; Eubanks 2001). Reasons for continued archaeological research on the earliest presence and the possibility of multiple domestication events had to do with a paucity of macrobotanical remains recovered from Archaic sites in Mexico. The well-known early maize cobs were derived from dry caves or rockshelters in three separate regions Tamaulipas, Tehuacán, and Oaxaca (Smith 2001, p. 1325; Blake 2006, p. 56). The first directly dated maize remains using AMS radiocarbon dating from Archaic contexts are restricted to 12 samples from the Tehuacán Caves, six from Coxcatlan Cave and six from San Marcos (Long et al. 1989), six from the Ocampo Caves in Tamaulipas (Smith 1997b, pp. 373–374; Jaenicke-Despres et al. 2003), and the two cobs from Guilá Naquitz in Oaxaca (Piperno and Flannery 2001). The dendrocalibrated AMS dates of the two of the earliest cobs at Guilá Naquitz dated to ca. 4200 and 4300 CAL B.C. (Piperno and Flannery 2001, Table 1). The original chronology for the earliest "wild" cobs at San Marcos Cave from associated charcoal and organic remains were reported to extend back to about 7,000 years ago (Johnson and MacNeish 1972, p. 17). The dates ranged from ca. 5000 to 3500 B.C. for the Coxcatlan Phase and 3350 to 2000 uncal. B.C. for Abejas Phase. The later AMS dates from Coxcatlan Caves produced assays of 3560–2850 CAL B.C. and moreover, all the cobs analyzed thus far were found to be already fully domesticated (Benz and Iltis 1990; Benz and Long 2000, Table 1; Benz et al. 2006, Table 5.1, Figs. 5.2, 5.3). The AMS dates on several early Abejas Phase cobs from Coxcatlan Cave reported by Bruce Smith (2005a, Table 1) suggest a chronological range of between 2850 and 2040 CAL B.C. for the earliest samples. This indicates that when the major landmark articles from the biological and molecular sciences appeared at the beginning of this decade, there was little direct evidence on the distribution of early maize (see Table 3.1). Subsequently, maize chronology was dramatically revised for its early presence throughout the Americas by direct AMS dates (Fritz 1994; Smith 2000, 2005a).

In the article by MacNeish and Eubanks (2000), they not only attempted to resurrect *Tripsacum* as a possible ancestor of maize, but they also rejected the direct AMS dates and revised chronologies from the ancient cobs found in the Tehuacán Caves (see also MacNeish 2001b). Long and Fritz (2001) mention the previous assertion that the dates produced by AMS dating were false because they had been contaminated.[17] Long et al. (1989) addressed possible questions surrounding the contamination in some detail in their initial report, and moreover, MacNeish selected all the cob samples they AMS dated (Long and Fritz 2001, p. 88). Ultimately, the results indicated that the earliest AMS dates were derived from Guilá Naquitz Cave in Oaxaca were 5410 ± 40 and 5420 ± 60 B.P. [β 132511] and from San Marcos Cave in the Tehuacán Valley, dated to 4700 ± 110 B.P. [AA-3311] (see Table 3.1) (Long et al. 1989, Table 1; Piperno and Flannery 2001, Table 1). Maize appears to have radiated slowly northward as indicated by the 3930 ± 50 B.P. [β-85431], date from Romero Cave in Tamaulipas (Blake 2006, p. 57). In the north coast of Tabasco, only two AMS dates have been reported 2565 ± 45 B.P. [AA-33923], and another from the El Gigante rockshelter in Honduras, 2280 ± 40 B.P. [β-159055] (Blake 2006, p. 57). Along the Pacific

---

[17]MacNeish and Eubanks (2000, p. 15) claimed that the Tehuacán cob samples AMS dated by Long et al. (1989) were contaminated by Bedacryl, a polymethyl acrylate used to preserve fragile botanical materials. The Mexican curators, however, stated emphatically that they never treated the macrobotanical remains since they arrived at the national museum in Mexico City in the early 1970s (Long and Fritz 2001, p. 88). Moreover, when Benz conducted his morphological analysis of the remains he commented to Long and Fritz that they were covered with soil and ash from the original excavation (Long and Fritz 2001).

coast, the earliest dates from Soconusco range from 3365 ± 55 B.P. [β-62911] to 3000 ± 65 B.P. [β-62920] (Clark 1994). Blake and his associates report finding dozens of carbonized maize kernels and cob fragments in Early Formative period household trash deposits from sites on the Pacific coast of southeastern Mesoamerica, which were dated by conventional methods on associated charcoal and ceramic diagnostics as well as direct AMS dates (Clark 1994; Blake et al. 1992, 1995). Their results indicate that six of seven AMS dates on maize macro remains fell within their expected phase time ranges ca. 3700–2800 B.P. (at the one sigma level) and the seventh did so at the two-sigma level (cf. Blake 2006, p. 60; Blake et al. 1992, 1995). These results indicate that maize has a very early presence in the coastal lowlands of Soconusco, but is of secondary importance to the formative diet (Chisham and Blake 2006, pp. 166–167; Blake et al. 1992). The overall revised chronology from AMS dates from the regions of central and southeastern Mesoamerica suggest that domesticated maize spread slowly, over a period of 2200 years from its earliest appearance in Oaxaca and the Tehuacán valley, to the north and south coastline of Mexico and neighboring Honduras. This chronological range is in stark contrast to the earlier estimates based on conventional ^{14}C dates from the various cave site in Tehuacán and are consistent with early dates of other domesticated food remains from those sites based on direct AMS dates (Smith 1997b, p. 342; Kaplan and Lynch 1999, p. 264; Long and Fritz 2001, p. 89).

Maize does not appear to have spread to the American SW until later ca. 2000–3000 B.P. (Table 3.1). Jaenicke-Després and her associates (2003) reported that with the spread of maize into the American SW at 1920 ± 40 B.P. [β-166755] the maize cobs they analyzed from Tularosa Cave, New Mexico still maintained an allele of the starch gene *su1* common to teosinte. The *sugary 1* (*su1*) gene encodes a starch debranching enzyme expressed in kernels (Rahman et al. 1998). Together with the branching enzymes, this enzyme determines the structure of the amylopectin, the chain length of which as well as the ratio of amylose to amylopectin is critical to the gelatinization properties of starch (Whitt et al. 2002; Rahman et al. 1998). Since the *su1* allele affects the gelatinization properties of starch it also affects the textural properties of tortillas (Jane and Chen 1992; Jane et al. 1999). These data indicate that early maize in the American SW could not have been consumed as flour (Jaenicke-Després et al. 2003, p. 1206; Jaenicke-Després and Smith 2006).

The three genes that they considered in their study were involved in the plant architecture, storage protein synthesis, and starch production and they analyzed archaeological cob samples from both Mesoamerica and the American SW (Jaenicke-Després et al. 2003; Jaenicke-Després and Smith 2006). Their results indicate that Mesoamerican maize had allelic selection typical of contemporary maize by 2,200 B.C., but that one of the genes (*su1*) was as yet incomplete at 2000 B.P.[18]

---

[18]The fact that 2,000 B.P. maize cobs from Tularosa carried an *su1* allele which occurs today in teosinte but is very rare in modern maize, suggests that the selection process at *su1* was incomplete at that time making such maize unsuitable for the manufacture of tortillas. This implies that the selection for *su1* starch properties occurred long after the initial domestication of maize in central Mexico.

These data indicate that early varieties of maize brought into the American SW lacked the high quality starch necessary for making tortillas found in the later and modern maize varieties found at this and other sites in this region (Fig. 3.12). The missing alleles, however, do not appear to have affected the cob morphology. This genetic discovery emphasizes the importance of how analysis of ancient DNA can be used to integrate archaeological and genetic approaches, to understand the early role of maize to pre-Columbian subsistence economies and explore several different sets of questions on the early history of maize (Jaenicke-Després and Smith 2006, p. 83).

The other major proposed center for the domestication of maize, northern South America and the Andes, have few directly dated cob or kernel samples. Exceptions in this regard are one of five cobs from the site of Ramaditas in the Atacama Desert of northern Chile with date of 2210 ± 55 B.P. [TO-4810] (Rivera 2006, Fig. 29.3, Table 29-4). This date pertains to one of the late occupations and the site compound is associated with a vast formative period irrigation network (Staller 2005). Most of the directly dated maize from Andes and other regions of South America are microfossils from carbon residues in ancient pottery (Staller 2003; Staller and Thompson 2000, 2002; Thompson and Staller 2001). The earliest dates from coastal Ecuador are approximately 2,000 years earlier than those from Ramaditas in northern Chile. These dates and the data from plant microfossils, both pollen and phytoliths, are discussed below with regard to the spread of maize from the Mesoamerican heartland. Research is being undertaken by Michael Blake and Bruce Benz to directly AMS date Latin American macrobotanical remains and

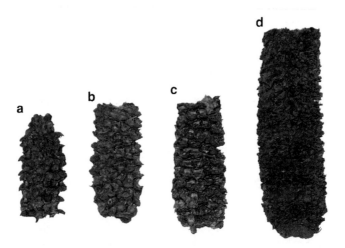

Fig. 3.12 Maize cob fragments maintaining the *sul* allele for starch, common to teosinte. Cutler (1952, pp. 464, 465) excavated all these cob fragments from Unit 3R2, Level 11 [catalogue # 315267] at Tularosa Cave, New Mexico. Right to left; (**a**) 10 rowed, 7 cm. long, 2.0 cm. diameter; (**b**) 10 rowed, 7.3 cm. long, 2.2 cm. diameter; (**c**) 12 rowed, 9.5 cm. long, 2.5 cm. diameter (AMS date: 1920 ± 40 Cal B.P. [β-166755]); (**d**) 10 rowed, 11.3 cm. long, 2.7 cm. diameter (Note that these genetic distinctions do not appear to effect the phenotypic characteristics of the archaeological maize (Courtesy of Field Museum Collections)) (Photograph by John E. Staller)

study their morphology and microbiology. These data will provide a much clearer understanding of the biogeography and spread of maize into the Neotropics in the near future – beyond the early AMS dates for maize from Mesoamerica mentioned in this chapter. As more dates are being reported, many in the archaeological community are becoming increasingly aware, that the direct AMS dates on early cobs from Mesoamerica document a more recent domestication event than reported previously (Fritz 1994; Staller and Benz 2006). Consequently, some archaeologists and paleobotanists have had to considerably revise their thinking about associated processes and timing of agricultural transitions.

The revised chronology and genetic data have established that teosinte is the ancestor of domesticated maize and that the so-called wild maize long sought by various scholars does not exist. The early spread of macrobotanical maize to different regions of the Americas is synthesized by Michael Blake (2006, Table 4.1, Fig. 4.1) and supports the most recent data from genetics and the revised chronologies provided by various archaeologists with direct AMS dates (see Table 3.1). The preliminary evidence appears to indicate that both conscious and unconscious selection have resulted in an extensive mosaic of maize landraces (Heiser 1988). Since all of these landraces were initially identified before the advent of AMS dating, and many are clearly distinct in terms of their morphological and phenotypic characteristics, their evolution and spread over different regions of the Americas have been based on the assumption of multiple domestication events.

The recent genetic and chronological research brought on by AMS dating has also generated data resulting in some significant revisions of maize antiquity and taxonomy. Given the paucity of certain kinds of data from different regions of the Neotropics, focus is given to the kinds of modifications that appear to have occurred with the earliest cobs due to conscious selection. Morphological analysis of maize cobs from the Tehuacán Valley suggests that prior to 2500 B.C. (ca. 4450 B.P.) there was a conscious selection for ears with more kernels and in the later periods for larger kernel size (Benz and Long 2000, p. 463). After 2500 B.C., the rate of change in ear morphology stabilizes and human selection may have been more concerned with increasing ear numbers per plant. Genetic analysis of maize from the Ocampo caves indicates that people were selecting for increased protein and starch quality and that some specimens were similar to modern maize by 4450 B.P. (ca. 2500 B.C.) (Jaenicke-Despres 2003, p. 1207). A comparison of the Ocampo cob morphology with the earlier cobs from Guilá Naquitz suggests that there was a continuous selection by early farmers for increased cob size and the presence of *pbf* and *su1* involved in protein and starch quality, respectively, in cobs dated to 4,400 years ago. This suggests that kernel quality was, along with the cob size, selected early on by ancient farmers. Just how early this artificial selection process began is at this point unknown, but it may be sometime around 4,400–4,300 years ago. The absence of alleles necessary for starch production among early cobs found in the American SW has broader implications for the role and spread of maize landraces in this part of the Americas and in the North American continent in general. Such data speak directly of the importance of genetic markers in providing a basis by which to study the evolution and biogeography of maize, and to incorporating these data with morphological information and phenetic traits that characterize different landraces to understanding the interrelation of the various maize varieties and their spread to different regions of this hemisphere.

3.3.3 Phylogenetic Considerations

The phylogeny of maize landraces has generally been based first and foremost on the morphological characteristics of the ear traits, rather than on the anthropogenic neutral morphological traits. Although such traits are of considerable importance to botanists in the classification of such varieties, they are not necessarily the best traits for understanding evolutionary relationships (Doebley 1994, p. 102; see also Doebley 1990a, b; Doebley and Stec 1991; Kellogg and Birchler 1994). Doebley (1994, pp. 102–103) emphasized that phenetic traits, and their overall similarity, can be useful indicators when assessing similarities and differences among maize varieties. However, the most effective manner to analyze the biogeography and evolutionary relationships through phenetic traits is at the molecular level of their chromosome constituents (McClintock 1978, p. 159; Vigouroux et al. 2002, 2003; Doebley et al. 2006).

Phylogenetic inferences from phenetic analysis predisposes analysts to linearity, that is to require one to identify which traits are primitive and which are more "advanced" states for each morphological trait (Doebley 1994, p. 104). Many archaeologists and botanists have come to assume that such lineages are useful for tracing the spread and origin of maize through distinct domestication events (Wellhausen et al. 1952, 1957; Grobman et al. 1962; Timothy et al. 1961, 1963). Subsequent ethnographic research has shown that distinct varieties were closely associated with ethnic identity and largely a product of conscious selection for specific traits (Sandstrom 1991; Perales et al. 2003a, b; Wright et al. 2005; Benz et al. 2007). The recent research in molecular biology has had profound consequences for plant taxonomy and phylogenetics in general and for maize phylogeny in particular. Analysis at the level of DNA is incontrovertible and speaks directly of the kinds of questions and relationships that plant taxonomists and geneticists have been asking for centuries.

The remarkable diversity of maize landraces has long been held to be consistent with multiple domestications; however, a comprehensive phylogenetic analysis for maize and teosinte has found that such diversity is equally consistent with a single domestication and subsequent diversification (Matsuoka et al. 2002, p. 6080; Freitas et al. 2003; Doebley and Wang 1997; Doebley et al. 2006). The microsatellite study by Matsuoka et al. (2002) was the first to reconsider maize phylogeny on the basis of 99 microsatellite loci that provide broad coverage of the maize genome and a sample of 264 maize and teosinte plants. The implications of these molecular data have been profound for ethnobotanical and archaeological research on the origins of maize. These researchers concluded that maize is monophyletic and arose from a single domestication in southern Mexico about 9,000 years ago (Matsuoka et al. 2002, pp. 6083–6084; Piperno et al. 2007, Fig. 2). Rather than using genetic distance markers, Matsuoka et al. (2002, p. 6080) used the proportion of shared allele distance with multilocus microsatellite data to construct phylogenetic trees. Their results indicate that *Zea mays* L. is a direct domestication of a Mexican annual teosinte *Zea mays* ssp. *parviglumis*, native of the Balsas River drainage[19] (Matsuoka et al. 2002, p. 6083, Fig. 2a). Research has shown that modifications involving conscious and unconscious as well as natural selection for the so-called *tga1* mutation for soft glume architecture resulted in a mutation in which teosinte (*Zea mays* ssp *parviglumis*) began to approximate some, but not all, of the morphological characteristics of maize, representing a significant departure from teosinte morphology (Benz 2001, 2006). The morphological changes from the stiff hard fruit case of teosinte to the soft glume architecture of maize are among the first mutations associated with the domestication of the plant (Benz 2001, pp. 2104–2105, 2006). Three genes of critical importance to the transformation of wild teosinte into maize some 6,000 years ago were – teosinte branched-1 (*tb1*), which affects the overall plant architecture; the prolamin box-binding factor (*pbf*), which regulates the expression of protein storage in the kernels; and sugary 1 (*Su1*) is involved in the starch biosynthesis pathway (Jaenicke-Després and Smith 2006, p. 85). Research into these various genes have not only uncovered the evolutionary relationship between teosinte and maize, but have also played a major role in how the plant taxon is currently classified.

Plant taxonomists classify maize (*Zea mays* L., *sensu lato*) as a member of the grass family Poaceae, which is divided into the family *Andropogoneae* (maize, sugar cane, sorghum, and teosinte), two genus of the subfamily *Maydeae* (*Tripsacum* and *Zea*) and finally into the genus *Zea*, which contains six distinct taxa classified into four species (Doebley and Iltis 1980, p. 982). The genus *Zea* includes *Z. mays* ssp. *mays* (cultivated maize) and the teosintes represent the various subspecies (Anderson and Cutler 1942, pp. 69, 70; Iltis 1972, pp. 248–249; Doebley and Iltis 1980, p. 982; Doebley et al. 1984, 1987, 1990; Matsuoka et al. 2002, Fig. 2a–b). The fruit of *Poaceae* is a caryopsis, that is, it has the appearance of a seed. Flowers among the *Poaceae* are usually arranged in spikelets, each has one or more florets further grouped into panicles or spikes (Weatherwax 1954, pp. 150–151; Mangelsdorf 1974, p. 71). All taxa of *Zea* have a central spike or terminal branch, the continuation of the central inflorescence axis (Doebley and Iltis 1980, p. 986, *f.*; Iltis and Doebley 1980; but see Wilkes 1967, p. 103). Male inflorescences (tassels) have generally 12 or more branches (except for maize), a central terminal spike that is stiffer, stronger, and more densely beset with lateral spikelets that are highly exaggerated than those of others in the *Zea* taxon (Doebley and Iltis 1980, p. 988). Spikelets consist of two (or sometimes fewer) bracts at the base, called *glumes*, followed by one or more florets (Fig. 3.13a, b). A floret consists of a flower surrounded by two bracts; the external is

---

[19]Maize DNA studies have shown that up to 12% of its genetic material was obtained from *Zea mays* ssp. *mexicana* through introgression or gene flow, more specifically, by backcrossing an interspecific hybrid with one of the parent species (Matsuoka et al. 2002, p. 6083).

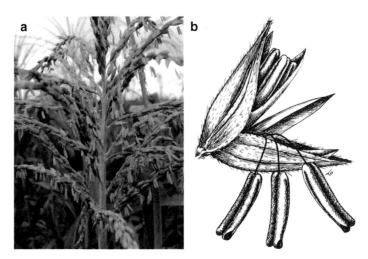

Fig. 3.13 (**a**) Close up of a tassel from the *xmejen-nal* maize variety of Yucatanm, Mexico. Maize spikelets occur at the terminal portion of the tassel (Courtesy of John Tuxill) (Photograph by John Tuxill). (**b**) A single maize spikelet (greatly enlarged) showing three anthers ready to shed pollen. The male spikelets occur in the terminal tassels and they drop their pollen on the maize silks (from Wallace and Brown 1956, Fig. 2)

called the *lemma*, while the internal *palea*. The grasses are usually hermaphroditic. Maize is an exception in this regard, it is monoecious, that is, each flower has a stamen or a pistil, but not both (Weatherwax 1954, p. 150; see also Iltis 2006), and pollination is always wind borne or anemophilous. Mexican farmers are apparently aware of this and sometimes plant teosinte in their *fincas* to make them more resistant to fungi and other natural parasites (Tykot and Staller 2002). The success of grasses is directly related to their morphology and growth processes, as well as their physiological diversity (Anderson and Cutler 1942).

The teosintes contain four species of large grasses classified under the genus *Zea*, with wild stands in Mexico, Guatemala, and perhaps Nicaragua. There are five recognized species of teosinte: (1) *Zea diploperennis* (Iltis, Doebley, and Guzman), (2) *Zea perennis* (Hitchc.), (3) *Zea luxurians* (Durieu and Asherson), (4) *Zea nicaraguensis,* and (5) *Zea mays*. The last species is cultivated maize or corn, the only domesticated taxon in the genus *Zea*. The teosintes are further divided into four subspecies: *Zea mays* ssp. *huehuetenangensis*, *Z. mays* ssp. *mexicana*, *Z. mays* ssp. *Parviglumis,* and *Z. mays* ssp. *mays*. Teosinte varieties in Guatemala are classified as *Z. mays* ssp. *huehuetenangensis*. The two perennials teosintes, *Z. diploperennis*, and *Z. perennis*, and the annual *Z. luxurians* are more distantly related and considered unrelated to domesticated maize (Matsuoka et al. 2002, p. 6080).

The subfamily *Maydeae* is further divided into three main groups, and one of these groups includes the genus, *Zea* (maize and teosinte), and *Tripsacum* – all are native to the Americas (Anderson and Cutler 1942. pp. 69, 70; Weatherwax 1954, pp. 150, 151; DeWet and Harlan 1976, pp. 130–134). Both *Tripsacum* and teosinte are known to occur in the wild state in the Americas; but maize is the only food crop of this group, and as has just been discussed in detail, much research went into attempting to identify this species in its wild state; no such species has been identified. Cross-breeding experiments encouraged botanists and later ethnobotanists and archaeologists to search for a "wild" that is, undomesticated form of *Zea mays* L. The relationships of maize to its relatives in this subfamily were initially obscure (Mangelsdorf and Reeves 1934, p. 33).

Wild grasses can be divided physiologically into two groups, on the basis of photosynthetic pathways for carbon fixation, that is, they are either C_3 or C_4. Maize is a C_4 grass and its photosynthetic pathway is linked to specialized Kranz leaf anatomy that particularly adapts this species and other C_4 grasses to hot climates and an atmosphere low in carbon dioxide (Sage et al. 1999). The C_4 photosynthetic pathway leaves traces in human bone chemistry and can therefore be used as a quantitative measure of its dietary importance (Tykot 2006, pp. 136–137).

Grasses generally spread out from a parent plant. Growth habit refers to the type of shoot growth present in particular grass plants and is directly related to their ability to spread out from the parent plant and ultimately form a clonal colony. There are three general classifications of growth habit present in grasses: bunch-type, stoloniferous, and rhizomatous (Weatherwax 1954, pp. 154–156). The genus *Zea* includes all of the various landraces of maize (McClintock 1978). Although maize comes in a great variety of forms and botanists attempted for many years to divide the genus into varieties or subspecies (landraces), which could be classified into a binomial system, they are all now classified taxonomically into a single genus (McClintock 1978, p. 156). It was these phenotypic and systemic characteristics that motivated scientists in the botanical and biological sciences to identify a related species which would provide a basis for the origins of domesticated maize in the form of a still extant wild ancestor, but it was ultimately the genetic and molecular research which provided the answers to this scientific mystery (Beadle 1972, 1980).

3.3.4   Early Research on Maize Landraces and their Classification

The search for more productive maize varieties, which had its inception at the end of the nineteenth century, continued throughout the next century and into the present century. However, this scientific research soon became the domain of the industrial sector of the first world, and was an anathema to much of the third world where maize had its origins and where it continued to play an important role to ethnic identity beyond its importance as a food crop (Raven 2005; Pernales et al. 2003a, b). The identification of various landraces of maize initially involved an attempt to gain an understanding of the diversity of those maize varieties, which were of commercial importance – created through hybridization and crossing (Fussell 1992, p. 86). Classifications were initially based on the composition of the endosperm of the kernel. The maize classification by E. Lewis Sturtevant (1899) was published into a monograph entitled "Varieties of Corn" and still seen by botantists as "comprehensive" almost a century later (Mangelsdorf 1974, p. 102). Sturtevant classified the variability of maize landraces into six main groups, five of which were based on the endosperm composition of the kernel and this classification was used by biological and social scientists almost without modification for a period of 50 years (Mangelsdorf 1974). Differentiating flint varieties from dents, and popcorns, from sweet corns and those with lesser or high levels of starch (flour varieties) was done first and foremost through economic incentives,[20] and secondarily for scientific reasons, i.e., to develop classifications, trace the spread and biogeography of different landraces, or how such races related to different domestication events (Fig. 3.14). Classifications such as these were primarily concerned with the morphological characteristics of the kernels and the cobs, as significant identifiers for the uses and forms of consumption of the various maize varieties.

Such classifications have an inherent problem in that they are only indirectly related to taxonomy and phylogeny, yet many scholars in the social and biological sciences have historically treated them as if they represent a phylogeny of maize ancestry and evolution, as was the case with Strurtevani's sixth group, pod corn. Anderson and Cutler (1942) first sensed a problem with this classification and noted that it was almost entirely based on the characteristics of the endosperm. They reported that several of these traits in floury and sweet varieties were known by geneticists as primarily dependent upon single loci on a single chromosome for their expression. In their opinion, a "natural" classification, should take into account the entire genetic constitution of the plant.

Anderson and Cutler (1942) did emphasize that the original classification by Strurtevani made an important contribution in showing that the maize tassel, whose central spike has long been recognized as the homolog of the ear, is valuable to classification of maize varieties in that it was one of the more easily measured characteristics of the plant (see also Anderson 1944, 1952). Mangelsdorf (1974, p. 102) later disagreed stating in defense of an earlier botanical study that domesticated species vary primarily in those parts of the plant which are consumed, and the other characteristics are either unmodified or present trifling alterations. Doebley and Iltis (1980, p. 985) later observed that the tassels are critical to maize classification and clearly differentiated from those associated with the teosintes (see Fig. 3.13a). They state that the central spike or terminal branch, i.e., the continuation of the central inflorescence axis, is present in all species of the *Zea* taxon (Doebley and Iltis (1980, p. 986*f*).

In summary, the initial pattern with regard to the classification of maize landraces has been for botanists and plant taxonomists to focus on those parts of the plant which are related to conscious selection, i.e., the part of the plant that is consumed and then compare them with related species in the wild state (e.g., Einfield 1866; Strurtevant 1899; Harshberger 1893, 1896b; see also Wallace and Brown 1956; Mangelsdorf et al. 1967). It was John Harshberger (1896b) who

---

[20]His sixth category, pod corn, was at the time a botanical curiosity, but one, which in later years, had a profound effect on research on maize origins and phylogeny.

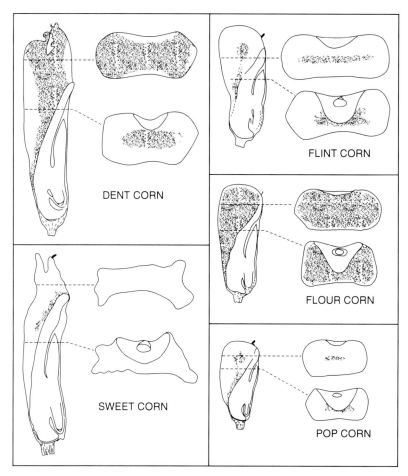

Fig. 3.14 Different kernels of maize from the principal varieties or landraces to show the arrangement of flinty and floury parts (from Weatherwax 1954, Fig. 63)

first proposed the idea of a symbiosis between botany and archaeology to studies of plant domestication. He recognized the necessity of botanically trained specialists who understood plant morphology and who could analyze plant fragments often left in layers at archaeological sites. Hershberger also recognized the importance of such specialists to study plant phylogeny and taxonomic identification. Consequently, research on the origins and domestication of plants in the United States has formed the scientific basis for research on prehistoric plant production (Ford, 1989a, p. 13). The phenetic differences are generally assumed to be a measure of how the plant had been changed through time and space by different human populations and by artificial and natural selection (Fig. 3.15). An interest in a more systematic consideration of maize taxonomy and phylogeny was not forthcoming until after World War II, and this was related in part to the increasing

Fig. 3.15 Variation within a single maize landrace (*xmejen-nal*) showing the effects of human selection for kernel shape as well as ear size and shape. Note that the kernel shape is maintained throughout, yet kernel color varies, effected by wind pollen from nearby male inflorescences. This landrace is cultivated and maintained by a single Yucatec Mayan farmer, Esteban Cuxin. Such phenotypic divergence within and among landraces is related to the founder effect and bottleneck phenomenon. Indigenous farmers have for centuries selected certain characteristics or traits, which are often tied to cultural norms related to ethnic identity or how different varieties are processed and consumed (Courtesy of John Tuxill; Photograph by John Tuxill)

research by plant morphologists and geneticists focused on the origins of maize (e.g., Wallace and Brown 1956; Harlan 1975; Galinat 1970, 1976, 1983; Hastorf 1994; deWet and Harlan 1972, 1976; Harlan et al. 1973; McClintock 1976; Iltis and Doebley 1980; Doebley and Iltis 1980; Doebley and Stec 1991; Doebley et al. 1983, 1990).

3.3.5   Maize Landraces in the Americas

The classification of domesticated species or landraces has historically lagged behind the creation of plant phylogenies in general. Edgar Anderson (1952) observed this as an issue with domestic plants in general, and with maize in particular – what Doebley and Iltis (1980, p. 983) referred to as the "Great Ethnobotanical Paradox." Phenetic or morphological analysis had long been a mainstay of ethnobotanic research on cladistics and these analyses have only recently begun to have a profound influence on the archaeological reconstructions for the spread of maize lineages (Doebley 1994; Doebley et al. 1983, 1990, 2006; Hill 2006; Rainey and Spielmann 2006; Vierra and Ford 2006). The beginning of the pursuit of identifying and classifying landraces has like studies into the origins of maize, undertaken during and just after World War II (Mangelsdorf 1974, p. 101). The search for finding landraces or maize varieties had its inception with

Mangelsdorf and the Rockefeller Foundation in association with an agricultural program with the Mexican Ministry of Agriculture (Mangelsdorf 1974). The stated intent was the initiation of a program of "practical maize improvement" and in order to carry this out, a comprehensive inventory of what maize varieties were being cultivated and available to plant breeders was undertaken (Mangelsdorf 1974). When Mangelsdorf and his collaborators began their initial classification of the Mexican maize varieties in 1948, they focused almost exclusively on the ear morphology as the basis for distinguishing different landraces (see e.g., Wellhausen et al. 1952, 1957; Grobman et al. 1961; Timothy et al. 1961, 1963). The reasoning was that since this is the part for which maize is cultivated, it is the most highly specialized organ of the plant and it is the ear structure that distinguishes *Zea mays* from all other species of grasses. The initial presumption was that the ear and not the tassel would offer those diagnostic characters most useful in classification and differentiating maize landraces (Mangelsdorf 1974, p. 102).

A systematic survey was undertaken and varieties from all parts of Mexico were assembled at Harvard University in Cambridge, Massachusetts for controlled experiments (see, e.g., Wellhausen et al. 1952, 1957; Brown 1960; Goodman 1976; Goodman and McKBird 1977). This provided Mangelsdorf with the necessary assemblages to test his ideas surrounding the origins of maize and to identify its possible progenitor. In the context of the project, these varieties were tested in order to assess their potential productivity, disease resistance, and other characteristics of general importance.

The creation of natural species taxonomies grew out of the realization of the extraordinary diversity of the various maize landraces of Mexico. Out of this diversity various cytological, genetic, and particularly botanical studies emerged (Mangelsdorf 1974, p. 101). Botanical surveys were specifically geared to finding little known varieties in the remote corners of Mexico (Wellhausen et al. 1952; MacNeish 1961, 1962, 1967b). Subsequently, such classification studies and associated research was expanded to different regions of Latin America (e.g., Anderson 1947a; Anderson and Cutler 1947; Wellhausen et al. 1957; Grobman et al. 1961; Timothy et al. 1961, 1963). The Mexican results produced a total of 25 "more or less" distinct races and were then further divided into four major groups on the basis of their presumed chronological origins: Ancient Indigenous, Pre-Columbian Exotic, Prehistoric Mestizos, and Modern Incipient (Mangelsdorf 1974, p. 101).

The numerous publications that arose from this early research are still having a major influence on how archaeologists reconstruct the archaeological record with regard to the movements of ancient populations, and their influence on the sociocultural development in different regions of the Americas. Wellhausen and his associates (1952) presented very explicit hypotheses with regard to the evolutionary relationships of the various Mexican landraces, based on the phenetic analyses of morphological traits. They concluded that most Mexican landraces arose from the hybrids of other races, i.e., were a product of artificial selection, whether deliberate or not (Fig. 3.16). Given that later research showed maize to be monophyletic, and the only characteristic being considered in the classification was ear morphology,

Fig. 3.16 The distinctive
Mexican Popcorn Palomero
Tluqueño; a landrace
distinguished by long, pointed
"dingy" pericarp colored
kernels, and small cone
shaped cob as it appeared in
the 1950s. Popcorn varieties
are found throughout the
Americas and are associated
with the early spread of maize
into central and Andean South
America from Wellhausen
et al. (1952)

this conclusion would appear in retrospect to be a self-fulfilling prophesy. Studies
by ethnographers and linguists among indigenous Mexican populations had shown
that ear morphology, kernel color, and shape were deliberately being selected by
different groups as a means of differentiating themselves and their maize from those
of linguistic and culturally distinct surrounding indigenous communities. These
early botanical studies were, however, methodologically encouraged by the fact
that many of these similar varieties were directly related to one another from a
developmental standpoint. In other words, one landrace was consciously modified
from a variety brought into the communities in the past, and the introduction of
different landraces could be traced using traditional plant taxonomies and project-
ing them back in time with historical linguistics (Swadesh 1951, 1959b; Kaufmann
1994; Campbell 1997; Hill 2001, 2006; Brown 2006).

The indigenous societies that grew these popcorn varieties did so even though
more productive races could be locally obtained. They emphasize that their studies
appeared to indicate that the four different Mexico races identified on the basis of
ear morphology had been cultivated since pre-Columbian times and maintained
their phenetic characteristics throughout the centuries to the present (Grobman et al.
1952; Mangelsdorf 1974). However, the hybridization of certain landraces by
indigenous populations has not been necessarily related to greater productivity –
particularly, in the later periods where societies were closely associating maize
varieties (on the basis of certain phenotypic characteristics, kernel color, cob shape,
etc.,) with their ethnic identity. We may infer from these data that the classification
of maize varieties under Mangelsdorf's direction saw characteristics as leading to
greater productivity as being more "advanced" morphologically. Such linearity was

also involved in the search for wild maize, and in the identification of ancient maize cobs and ultimately with the spread of landraces through different regions of the Neotropics. The San Marcos Cave maize cobs from the Coxcatlan Phase were interpreted as "wild." Researchers interpreted these "wild" maize macrobotanicals as resembling pod corn (see also Fig. 3.6a, b) while the rest of the archaeological cobs were fully domesticated (Mangelsdorf et al. 1967, Fig. 122). It was in these archaeological examples that the conscious selection for more productive varieties is most apparent, although in this case it was clearly for adaptive reasons associated with domestication, agriculture, and a greater interdependence upon hunters and gatherers and certain food crops and other plant species (see, e.g., Benz 1999, 2001; Benz and Long 2000).

3.3.6   Maize Landraces and Colonial Bioprospecting

The multiyear research on the identification of maize landraces involved considerable ethnographic data derived from interactions with indigenous and mestizo maize farmers in Latin America. The classification of maize landraces using primarily anthropogenic morphological characteristics has for these reasons tended to take on a life of its own, particularly since many archaeologists studying maize biogeography perceived such classifications in terms of phylogeny and evolution through time and space. Moreover, the tripartite hypothesis and search for an apparently extinct species of "wild" maize, created initial conditions by which scholars in the biological and social sciences could consider the possibility of multiple domestication events over a wide geographic area. When reviewing this literature, it is immediately apparent that the evolutionary relationships and their biogeography with regard to the phenotypic similarities and differences of various landraces was what was particularly emphasized – this despite the overall tendency for indigenous populations to maintain "pure" landraces for cultural reasons for long periods of time. This is in stark contrast to what has occurred with regard to maize breeding and cultivation in the colonial and post-colonial world. With the advent of the European governments of New Spain and Peru, and their later appearance in various regions of the North and South American continent, government officials and colonial administrators generally encouraged indigenous societies to produce ever more productive varieties (Cutler 2001; Rainey and Spielman 2006, Tables 34-5-34-8; Huckell 2006, Fig. 7-2, Table 7-2). This is not to infer that various landraces are necessarily unrelated to one another, in some cases they have directly led to the appearance of recognized landraces, but rather, that they have had a direct effect on how archaeologists have in some cases come to understand the origin of maize and its biogeography in different regions of the Americas.

Botanists and archaeologists have historically tended to use morphological characteristics of domesticated plants on the one hand, and earliest appearances on the other, as directly reflecting the evolutionary relationships and then using these relationships to trace the early spread of plants such as maize on the basis of supposed primitive and more advanced traits[21] (Goodman 1968, 1988). In both cases, analysts generally did not take into account how different environmental settings and human selection have affected and modified maize landraces in different regions of the Americas (Darwin 1868; see also Gould 2002). Darwin's observations on the great variation of cultivated species infers that such variability is primarily a reflection of human or conscious selection to increasingly broader phenotypic variation. This is particularly apparent in North America where the original maize varieties were dramatically changed early on in the pursuit for greater productivity and ability to tolerate dramatic climatic and environmental variability (Enfield 1866; Myrick 1904; Wallace and Brown 1956). This was also apparent in later periods in Mexico, in association with the so-called pre-Columbian Exotic varieties. The collaborators concluded that four other races of maize were introduced from further south (Andean South America) and these four, called Cacahuacintle, Harinoso de Ocho, Oloton, and Maiz Dulce, hybridized with the Ancient Indigenous races and with teosinte to produce an incredible diversification of Mexican landraces resulting in the creation of 13 new races classified as Prehistoric Mestizos (Wellhausen et al. 1952). The Mexican Ministry of Agriculture published the classification and descriptions resulting from these intensive studies in 1951. The Bussey Institution of Harvard University published another English edition in 1952 (Wellhausen et al. 1952). Both Spanish and English editions of the monograph by Wellhausen and his collaborators were distributed throughout the hemisphere and have subsequently had a major impact among corn breeders, geneticists, botanists, and archaeologists. Later studies of maize varieties were funded for Latin America in Ecuador, Peru, Bolivia, Chile, Venezuela, and the West Indies (Brown 1960; Wellhausen et al. 1957; Grobman et al. 1961; Ramirez et al. 1960; Timothy et al. 1961, 1963; Goodman 1976; Goodman and McKBird 1977). These publications along with the original work on the landraces of Mexico, represented a comprehensive inventory of maize landraces of this hemisphere, and the germ plasm collected from these surveys was made available to corn breeders. Maize varieties were also classified, that is given distinct names on the basis of how they were classified by indigenous and mestizo farmers and these folk taxonomies are based in part on the phenotypic characteristics, some non-adaptive, such as kernel color, or adaptive such as ear size, but also by characteristics or traits that are in response to both human and natural selective pressures (Benz et al. 2007). In this way, the different language groups who have worked with the maize researchers are also contributing to maize diversity through their distinct linguistic terminologies. Later studies on maize genetics has uncovered evidence to indicate that some biochemical traits, such as the presence of certain alleles such as those

---

[21]The general tendency in the archaeological and biological literature has been to assume an overall pattern of selection for more productive (larger kernels and/or more rows) Pre-Columbian landraces over time.

involved in the storage protein synthesis, and starch production are not apparent by ancient cob morphology. Thus, it is logical that different varieties are at times classified not solely on the basis of phenotypic variability, but also on how and with what they are consumed (Jaenicke-Després et al. 2003). This suggests that such folk classifications would have been different among native groups involved in breeding and cultivating the plant than in those given among consumers (Benz et al. 2007, p. 290; see also Tuxill et al. 2009).

Fifty years later, various government-sponsored programs to introduce more productive varieties appear to have had only limited success in Mexico. Similarly, indigenous and mestizo populations in several Andean nations have resisted the introduction of more productive and now genetically modified varieties, and this is related for the most part to cultural rationales. The maintenance of particular landraces to certain cultures and social groups has ancient roots (Perales et al. 2003a, b; Tuxill et al. 2009). Despite more recent attempts by corporations and first world governments to introduce genetically modified species, certain maize lineages have been steadfastly maintained, by indigenous and mestizo societies (Raven 2005; Thomson 2007).

In some regions, these studies went hand in hand with linguistic analysis involving the movement of prehistoric populations and the spread of certain terms and terminologies (Swadesh 1951, 1959a; Berlin et al. 1974). However, Mangelsdorf (1974, pp. 103–104) had little confidence in linguistic folk taxonomies or historical reconstructions – this, due in part to the fact that what was interpreted as distinct landraces were given the same or similar names (see, e.g., Pernales et al. 2003a; Tuxill et al. 2009).

The comparative analysis of maize phenotypes involved taking the landraces of Mexico or another Latin American country and looking for phenotypic similarities between these and the varieties from all other countries (Mangelsdorf 1974, p. 104). Other criteria considered included adaptation to altitude, earliness, kernel color, and hardness, tiltering, pilosity, basically characteristics associated with the cob portion of the plant. The conclusions drawn from the diversity of maize landraces was that Mexico and Peru were the major centers for the origin, evolution, and the later diversification of maize. In other words, the basic premise regarding all these publications is based on the assumption of a multiple-origins hypothesis, the assertion that maize was a product of an extinct pre-Columbian wild maize. In the context of these initial premises, a conclusion was reached on the basis of the high frequency of so-called endemic varieties unique to these countries. In light of the recent genetic and phylogenetic revisions of the taxon, these patterns regarding a high incidence of endemism among landraces in Peru and Mexico may be explained in a number of ways. With regard to Mexico, the role of the teosintes, particularly *Zea mays* ssp. *mexicana* in the hybridization and variation of maize landraces is no doubt a critical factor (Goodman and Brown 1988; Matsuoka et al. 2002). Other factors would no doubt be related to the movement and exchange among indigenous populations in Pre-Columbian Mesoamerica, conscious selection for certain characteristics or traits, and of course, the later imposition of colonial governments and their demands for more productive maize for tribute and taxes (Rainey and Spielman 2006). With regard to the perceived endemism in Peru, more recent genetic research has determined that such varieties are a product of conscious selection of particular traits in combination with adaptation to the extreme environmental conditions that characterize this region of the world (Sevilla 1994, pp. 233–237). Variation as a result of adaptation to extreme conditions in the Andes is apparent in the Copacabana Peninsula where an endemic variety of maize grows in altitudes over 3,100 masl (Chávez and Thompson 2006, pp. 416–417). The systematic identification of maize varieties in the context of the research funded by the Rockefeller Foundation and others, as well as the more recent research in various parts of Latin America has provided clear evidence of the incredible mutability of *Zea mays* ssp. *mays* and its ability to adapt to a whole host of environmental and climatic conditions.

The systematic identification of maize varieties in the context of the research funded by the Rockefeller Foundation and others, as well as the more recent research in various parts of Latin America has provided clear evidence of the incredible mutability of *Zea mays* ssp. *mays* and its ability to adapt to a whole host of environmental and climatic conditions. Mangelsdorf and his associates believed that their research represented a complete inventory of the maize landraces of this hemisphere, and that their collections in storage represented virtually all the germ plasm available to farmers and breeders. There was no doubt on the part of the various researchers that the majority of the maize landraces were classified and described and reported by their 11 subsequent publications (Mangelsdorf 1974, p. 103). However, research on maize and teosinte genetics and a reconsideration of the *Zea* taxon would ultimately revise much of this earlier research, but not before it became widely accepted and spread through the scientific communities of both Latin and North America. What the early studies of maize landraces provided is a comprehensive data on the biogeography of maize as well as a great deal of botanical data on maize morphology from different regions of Latin America. The classifications of these various landraces also provided a basis for recon-structing lineages and evolutionary relationships by later scientists and scholars of prehistoric times.

3.3.7 Morphological versus Genetic Maize Landraces

Research on maize landraces appeared to have documented different morphological changes over time, primarily as a product of conscious selection, and secondarily related to selective pressures as maize varieties adjusted to distinct environmental and climatic circumstances. The importance of maize landraces dealing with regions of the Neotropics appears to be derived from the close ethnic association with certain places and to particular ethnic groups within those regions (Huckell 2006; Newsom 2006). The various survey and analyses of maize landraces resulted in the classification of more than 300 landraces, which Mangelsdorf (1974, p. 113) and others attempted to categorize, on the basis of genetic and morphological evidence, into a smaller number of what they referred to as "super races." Their attempts were, however, largely unsuccessful. What was discovered upon closer analysis was that the majority of the landraces could be assigned to a limited number of lineages; that they were descendant from a common progenitor (Mangelsdorf 1974). The concept of descent from a common ancestor was by no means new, but had never before been applied to landraces of maize, and at the time, the idea was seen as somewhat revolutionary, and largely met with enthusiasm from a phylogenetic standpoint, but later disproved by the molecular evidence (deWet and Harlan 1972, 1976; Doebley 1990a,b; Doebley and Wang 1997). The revolutionary aspect of this idea of "superraces" of maize was that it postulated at least six different landraces of "wild" maize from which all present-day maize varieties had descended (Mangelsdorf and Reeves 1959b; McClintock 1960; Mangelsdorf and Galinat 1964; Mangelsdorf 1974). In other words, they were attempting to create phylogenies of landraces, and it was at this level where the complications, derived from their initial assumptions of multiple domestication events, and origins based on an extinct pre-Columbian wild maize landraces, first arose and then later took on a life of their own (see, e.g., Grobman et al. 1961). Recent cautionary research on maize landraces is somewhat justified, particularly the avoidance of race names to refer to archaeological maize, excepting perhaps the Northern Flints (Benz and Staller 2006). Nevertheless, tracing morphological variation that is anagenic versus cladogenetic[22] evolutionary processes through time and across space can lead to a greater understanding of the dispersal of landraces and/or human migration and markets, and given the recent genetic research on maize DNA suggests that the future potential for such research may be particularly enlightening with regard to maize biogeographic selective pressure (Benz and Staller 2006; Hard et al. 2006; Jaenicke-Déspres and Smith 2006).

One important point for the early spread of maize, which was uncovered and appears to have been verified by later research, is that pointed popcorn lineages, under various names spread from Mexico to Guatemala, Colombia, Ecuador, Peru, Bolivia, Chile, Argentina, Venezuela, and Brazil as well as the United States (see Figs. 1.26a, b, 3.15, 3.16). The apparent absence of pointed popcorns in the West Indies may be related to the relatively late arrival of maize in that part of the Neotropics, and explain in part the secondary role of maize in the pre-Columbian subsistence economy of that subregion (Freitas et al. 2003; Newsom 2006; Newsom and Deagan 1994; Blake 2006; Bonzani and Oyuela-Caycedo 2006).

The pointed Mexican landrace referred to as Palomero Toluqueño, one of the popcorn races described by Wellhausen and categorized as an ancient indigenous landrace, was asserted to be the source of the ancient and wide spread of maize to

---

[22]Anagenesis refers to the persistence of one or a suite of biological traits that over time leads to varietal divergence. Cladogenesis refers to the development of evolutionary novelty through the extinction of pre-existing forms (Benz and Staller 2006, p. 665).

other regions of the Neotropics (Mangelsdorf 1974, p. 103; see also Wellhausen et al. 1952). Although this has yet to be demonstrated on the basis of molecular and ethnobotanical data, some researchers have reported that some of the earliest maize cobs in regions outside of the Mexican heartland are associated with pointed popcorns, called Canguil and Pisankalla in Ecuador and Argentina, respectively (Thompson 2006, pp. 89, 90, Table 7.4; see also Dorweiler and Doebley 1997). However, the phylogenies that they created and their assessments and interpretations of the spread and origins of the various wild landraces of maize were substantially revised by research from plant taxonomists and particularly geneticists. The introduction and the extensive movement of genotypes through artificial and natural selection during the post-Columbian period have complicated research regarding the origin of South American maize (see, e.g., Pearsall and Piperno 1990). The diversity of maize landraces can only be understood through a comparative genetic analysis of modern varieties with those of primitive landraces and preserved maize remains (e.g., Freitas et al. 2003).

The research by Mangelsdorf, Wellhausen, Grobman, Grant, Timothy, and their various associates can be summarized as an attempt to use phenetic characteristics of the ear morphology in order to create a phylogeny of maize landraces (Sevilla 1994; Bonzani and Oyuela-Caycedo 2006). What resulted was a history of relationships (phylogeny), based on the assumption of a linear progression of traits, i.e., that eight row varieties would lead, through conscious selection to ten and ultimately to varieties with more rows and larger kernels. The general consensus on the basis of the early maize from the various cave and rockshelter sites in Mexico was that archaic and preceramic societies consciously selected for ever more productive landraces. These scholars sought to identify extant races in prehistoric contexts and therefore emphasized anagenetic change as the principal mechanism by which such maize races evolved (Benz and Staller 2006, p. 672; see also Benz and Staller 2009). Persistence over time of maize phenotypes characterized by one or a suite of traits implies that varietal divergence would have occurred through anagenesis (Benz and Staller 2006, pp. 672–673). This is particularly the case with archaeological collections in that one can infer that if two distinct forms are present in one stratigraphic layer and one landrace is present in lower earlier strata, then the subsequent landraces are often asserted to be the result of human and natural selection having given rise to new varieties. Punctuated equilibrium is the evolutionary model used to describe this phenomenon while species selection is the mechanism (Gould 2002). Cladogeneis on the other hand, emphasizes the development of evolutionary novelty as a result of the extinction of the ancestor or pre-existing maize variety (Benz and Staller 2006, pp. 672, 673). In this scenario, new landraces are generally seen as the product of technological innovation or social reorganization and new varieties as primarily a product of cultural selection pressure. The pioneering research on extant landraces in the 1950s and 1960s emphasized cultural selective pressure as a primary basis for phenotypic divergence, yet at the same time referred to archaeological samples of maize macrobotanicals to connect the various landraces to one another, and by extension provide a basis for early biogeography.

3.3.8 Genetic Research and Paradigm Shifts

The previous evidence indicates that maize exhibits a complex array of phenotypic and genetic diversity and that it is distinguished by having the broadest cultivation range among all cultivated plants. Geneticists and archaeologists have to varying degrees worked together and independently on questions surrounding the origins of maize and the diversity of landraces. Archaeologists were initially limited to interpreting the phenotypic characteristics of ancient cobs. Geneticists interpreted the genetic consequences of conscious and natural selection by analyzing modern maize landraces and more recently the teosintes. Recent innovative approaches in molecular biology and genetics have incrementally increased our understanding of the maize genome and provided a possibility for a more comprehensive understanding of the domestication process. The long and complex history of genetic modification outlined in this chapter has been shown by recent direct AMS dates to span more than six millennia and encompass a wide range of morphological and genetic traits. The transformation of teosinte into maize was a result of conscious selection for certain characteristics such as row number and kernel size as well as unconscious selection, that is, an adaptation by the crop to the different kind of selective pressures associated with planting, harvesting, and ways of preparation for consumption (Jaenicke-Després and Smith 2006, p. 84).

The initial exploitation of teosinte appears to have been related to its sugary pith and other edible parts as evident by the quids identified in some ancient sites (Smalley and Blake 2003). Research in molecular biology undertaken in the 1990s (Doebley 1990a,b, 1994; Doebley et al. 1990; Doebley and Stec 1991, 1993) revealed that only few major quantitative trait loci (QTL) were involved in the morphological differences between maize and teosintes (Camus-Kulandaivelu et al. 2008, p. 1108). The *tb1* gene is involved in plant architecture and has been shown to be responsible for the reduced tillering of maize compared to teosintes (Dorweiler and Doebley 1997; Doebley et al. 1997). Recent analysis of the patterns of *tb1* nucleotide variation among maize and teosintes (*Z. mays* ssp. *parviglumis*) indicate a high recombination rate and the possibility of recurrent crosses with wild individuals during the domestication process[23] (Camus-Kulandaivelu et al. 2008; see also Wang et al. 1999). Clark et al. (2004) demonstrated that the selective sweep on *tb1* 59-non-coding region encompasses a 60–90-kb region later shown to include a regulatory region in this gene that plays a central role in the realization of the cultivated phenotype (Clark et al. 2006).

Recent genetic studies have also demonstrated that although the morphological characteristics, particularly ear morphology were useful for distinguishing the various landraces, they were not indicators of phylogenetic relationships (Doebley 1994, pp. 106–107, Fig. 8.2). Despite these problems, many in the field continue to use such classifications to identify the biogeography and phylogeny of landraces. The problem with phylogenetic inferences from phenetic analysis is that it creates a predisposition to linearity, i.e., a tendency to perceive certain traits as "primitive" and others as more "advanced" morphologically (Doebley 1994, p. 104). The concepts surrounding what was primitive and advanced in maize morphology was conditioned on the one hand by the archaeological maize recovered in archaic and preceramic sites, and on the other by a preconception of a "wild" maize resembling to varying degrees pod corn. For example, the four Mexican landraces: Palomero Toluqueño, Nal-Tel Chapalole, and Arrocillo Amarillo, were categorized as Ancient Indigenous, and were seen as primitive in that all were popcorns (Wellhausen et al. 1952; Mangelsdorf 1974). Although popcorn landraces have been shown to have an ancient distribution out of the Mexican heartland, the phenetic characteristics as described by the early studies made a set of inaccurate assumptions regarding how conscious selection for particular traits would manifest themselves among different landraces. The hybridization hypothesis proposed by Wellhausen et al. (1972, p. 153) also maintained that if one race was intermediate between two other races, it must be a hybrid derived from those races. It was these assertions which Mangelsdorf (1974) later summarized in this volume on maize evolution and improvement and since most of these earlier publications were readily available in Latin America and in some cases translated into Spanish, they became integrated into archaeological interpretations and formed the basis for which much of the later ethnobotanical research with plant microfossils was carried out.

The recent genetic evidence has shown that such evolutionary changes due to unconscious selection are initiated by the changes in the relationships between human populations and a target species (Smith 2006; Doebley et al. 2006). The selective pressures can be seen in terms of a causal relationship involving behavioral change toward a maize landrace, selecting for certain traits which induce a genetic response, and ultimately morphological change (Smith 2006, p. 16). While it is true that in the case of maize, that the ear and kernels provide more reliable traits for the classification of landraces than either vegetative (leaves) or tassels, such traits do not necessarily provide reliable information about the phylogenetic (historical) relationships (Doebley 1994, p. 102).

The early researchers were encouraged to use such morphological traits for analysis because it was these parts of the maize plant which was most often found preserved in archaeological sites (see, e.g., Mangelsdorf et al. 1967). Moreover, these traits were known to have considerable genetic to environmental variance over time and space and therefore enabled the various researchers involved in the initial classification of maize diversity into discrete landraces to recognize so many different landraces (Mangelsdorf 1974, pp. 113–115). Taxonomic research revealed that traits that have large genetic and small environmental variances, like those selected by the researchers of maize races, were found to provide reliable and consistent phenetic classifications (Stuessey 1990). Their classifications, however, were to encounter considerable difficulty in trying to derive historical or phylogenetic relationships among the various landraces, and this is because phenetic analyses can also be misleading if used to infer evolutionary relationships among maize races (Doebley 1994, p. 103; Doebley et al. 1994; Doebley and Wang 1997; Wang et al. 1999; Camus-Kulandaivelu et al. 2008). John Doebley (1994) used the eight-kernel row variety as a hypothetical example.

He noted that varieties with 10 and 12 rows all had a common ancestor with eight rows of kernels. The hypothetical phylogeny would assume that the switch from 8 to 10 rows occurred independently on two different occasions. While subsequent switches to 12 rows also occurred twice independently, which would suggest that the varieties with 12 rows of kernels were not necessarily more closely related to one another than they would be to other races of maize (John Doebley 1994, pp. 102–105, Fig. 8.1). In other words, the assertion that row number accurately reflects a phylogeny among two or more maize races can lead to considerable misinterpretations regarding the phylogenetic relationships among maize races, because such varieties may have evolved independently and not necessarily through conscious selection (John Doebley 1994, p. 104). One might also erroneously conclude that the occurrence of 12-rowed maize in two different regions represents the transfer of 12-rowed maize from one region to the other. These early efforts to attain a clearer understanding of the early evolution and diversification of maize landraces were initially limited to general comparative morphological studies, focused on modern maize. Recent genetic research has shown that morphological traits associated with different parts of the maize plant are limited to the extent to which they can trace phylogenetic relationships across time and space.

One of the most effective approaches in molecular biology involves research at the genetic level that allows botanists and archaeologists to understand and directly document how genes affect the phylogeny of maize landraces (Matsuoka et al. 2002; Vigouroux et al. 2002, 2003; Camus-Kulandaivelu et al. 2008). It is in fact understanding the phenotypic variation expressed by the different landraces through anagenesis and cladogenesis which is the future for further exploring the bottleneck phenomenon and exploring why certain landraces are maintained while others disappear (Smith 2006; Pernales et al 2003a,b, 2005; Benz et al. 2007).

Geneticists demonstrated a relationship between the cupule and chaff morphology through a genetic study of the related grass teosinte (Dorweiler and Doebley 1997; Dorweiler et al. 1993). Some researchers emphasize that molecular data in general provides the most abundant evidence regarding the evolutionary relationships among maize varieties (Goodman 1978, 1994; McClintock 1978; Doebley 1994; Goodman and Stuber 1983; Doebley et al. 1986, 2006). This is particularly the case with allozymes, as they are different allelic forms of the same enzyme and are believed to be neutral in that they are essentially unaffected by either human or natural selection (Doebley 1994). Their mutation rate was considered to be a constant, and thereby provided a basis for considering time and allozymic variation among maize varieties diverging from a common ancestor – precisely, what was being considered with regard to the spread of maize varieties (see Kimura 1986). However, morphological traits, particularly the ear morphology, are under strong conscious selection, and thus misleading when considering phylogenetic relationships (Doebley 1994, p. 106; Doebley et al. 1997). Doebley and his various collaborators demonstrated how previous hypotheses regarding the evolutionary relationships of maize lineages were inconsistent with allozymic evidence and that such molecular data provided a more reliable basis for considering spread and the interrelationships of maize varieties. Genetic studies were also critical to providing direct evidence documenting that *Tripsacum* was unrelated to maize evolution and origins, that maize was monophyletic and originated from the wild grass teosinte, as Beadle (1939, 1980) had suggested years earlier. Some of the maize research coming out of molecular biology involved morphological traits directly affected by human selection. These researchers revealed a single Mendelian locus, *teosinte chaff architecture* (*tga1*), that controlled several aspects of chaff morphology, both macroscopic and microscopic including the deposition of silica such as phytoliths and epidermal cells (Dorweiler and Doebley 1997, p. 1314). Their research indicates that the previously mentioned genetic locus, called *tga1*, controls the induration, orientation, length, and shape of the chaff and that the pleiotropic effects suggested that *tga1* may represent a regulatory locus (Dorweiler 1996; Dorweiler and Doebley 1997). Benz (2110) later noted in his analysis of the earliest cobs from Guilá Naquitz that it was changes in the glume architecture which is one of the first distinguishing characteristics which separated fully domesticated maize from its wild progenitor teosinte (see also Staller 2003). These differences were dramatic and essentially based on the effect of a single genetic locus. Researchers investigating the genetic changes involved in the evolution of the maize ear from the ear of teosinte provided the evidence for which later microsatellite research, or simple sequence repeat (SSR) of genes was based (Vigouroux et al. 2002, 2003; Matsuoka et al. 2002). These researchers concluded that maize (*Zea mays* L.), the principal domesticated crop of the New World, originates from one or more varieties of teosinte (*Zea mays* ssp. *parviglumis* and *Zea mays* ssp. *mexicana*) (Matsuoka et al. 2002, Fig. 2).

The diverse mosaic of morphology displayed by maize landraces imply that there is high genetic diversity among races, suggested to be related to a founder effect, resulting from a population bottleneck followed by intense conscious selection for specific traits (Smith 2006, pp. 15, 16; Tanksley and McCouch 1997, pp. 1063, 1064; Freitas et al. 2003, p. 906). However, the precision of the genetic data of maize conflicts with this hypothesis (Eyre-Walker et al. 1998). Isozyme studies have instead indicated that the high maize diversity may be related to a rapid dispersal and evolution, but this conflicted with DNA evidence from ancient South American maize cobs dating from 400 to 4500 B.P. (Goloubinoff et al. 1993). The genetic diversity of the alcohol dehydrogenase 2 (*Adh2*) allele sequences obtained from South American cobs by Freitas et al. (2003) suggest that maize diversity remained more or less stable in the past 4,500 years, contradicting the premise of a rapid evolution rate for maize landraces. However, Frietas et al. (2003, p. 906) indicate that ancient maize races spreading from Mexico had a nucleotide substitution rate 130–135 times higher than other grasses (Frietas et al. 2003). Thus, the assertion that maize landraces evolved from multiple domestication events among several wild ancestral populations, or domestication by a variable teosinte population by repeated cross-hybridization between the crop and wild teosinte are unnecessary. *Adh2* sequences reaffirm microsatellite data initially published by Matsuoka and his associates (2002) for a single origin in highland Mexico and a subsequent spread to various regions of South America in two distinct expansions (Freitas et al. 2003, pp. 904–905). The first, associated with a highland culture that spread from

Central America through highland Panama into the Andes and the western coast of South America, and a second lowland expansion from coastal Panama along the northeast coast of South America (Freitas et al. 2003). There are very few direct AMS dates associated with the earliest maize in Latin America, and using associated dates they suggest an initial expansion between 7000 and 4500 B.P., and a subsequent lowland expansion at around 2000 B.P., this corresponds with the spread of maize landraces to the north into the American SW (Staller and Thompson 2002, Table 9a; Jaenicke-Després and Smith 2006; Jaenicke-Després et al. 2003).

The compelling and detailed evidence derived from the genetic research on maize in the past two decades and particularly in the last few years is shattering the previous paradigms regarding its origin, spread, diversification, and even role of maize in complex sociocultural development. Other scientific breakthroughs, beyond direct chronometric dating techniques, such as stable carbon and strontium isotopic studies, are also greatly revising our understanding of the role of maize in the pre-Columbian diet, and its movements during prehistoric times in different regions of the Americas.

Chapter 4
Ethnobotanic, Interdisciplinary and Multidisciplinary Methodologies

4.1 Methodological and Technological Breakthroughs

This chapter is primarily focused on the methodological approaches and technological innovations used by the archaeologists and ethnobotanists to answer the larger questions on plant domestication, early agriculture, and human adaptation. The research on maize has generally evoked the broader, more theoretical questions surrounding early agriculture and its role in complex socio-cultural development. However, the early archaeological research on plant domestication in the New World was primarily focused on the origins of maize and the various roles of maize in such developmental and evolutionary processes. Particular emphasis is laid on ethnobotany, plant macrobotanical remains (kernels, cobs, etc., recovered primarily from archaeological sites), microfossils (pollen and phytoliths, taken from lakes and swamps as well as archaeological contexts), and paleodiet through bone chemistry involving carbon and strontium isotope analysis. The scientific literature comprising research from these disciplines has significantly influenced the archaeological reconstruction of the roles of maize in the ancient New World economies in the past three decades. Such data have provided an ever-increasing detail on the contextual associations, and the economic importance of maize throughout prehistory. Most of the archaeological studies have been focused on issues of plant domestication and early agriculture in the Americas. Therefore, discussion on maize must necessarily be within the context of such multidisciplinary research. In fact, few topics have generated as much theoretical speculation as the origin of agriculture and the role of maize in such adaptive and developmental changes.

4.1.1 Comparing Research on Old and New World Ancient Economies

The early archaeological and biological research surrounding early plant domestication in the Old World have generally provided evidence that suggests an adaptive shift to agricultural economies, arising from a single prehistoric event in which the domesticated strains of cereal grains were developed in a single localized area (Flannery 1972a, 2002; Hillman and Davies 1990, 1992; Zohary and Hopf 1993; Zohary 2004; Brown 1999, 2006). The existing evidence suggests the possibility that small-scale cereal cultivation could have occurred independently in a broader region on the Near East during ca. 12–10000 BP, when major environmental changes associated with glacial retreat, rise in sea level, and geomorphology were occurring, but the general tendency is to see such developmental processes as occurring from a particular region or "center" (Braidwood and Reed 1957; Bar-Yosef and Cohen 1992; Hillman et al. 1989). Some ethnobotanists have suggested that only a few foraging societies in Fertile Crescent combined the harvesting techniques, selective planting, and plant tending that resulted in more productive domesticated strains, totally dependent upon humans for their successful reproduction (Hillman and Davies 1992; Hillman 1996; Zohary 2004). These plants and the associated cultivation practices were then seen as radiating from the centers of domestication to other regions of the Indo-European continent (Harris 1989; Bellwood 2005; Brown 2006). A general focus on the independent centers of agricultural origin continues to influence the methodology and theory on domestication in New World prehistory as well (Smith and Yarnell 2009). However, human and natural selection associated with plant cultivation in the New World is generally believed to have begun simultaneously in various regions. Consequently, there has long been a strong focus on the earliest presence by ethnobotanists working in both temperate regions and the Neotropics. This methodological and research focus is, in part, related to providing direct evidence of a chronological base line for the shift to food production, that is commonly known an agricultural or formative way of life (Staller 2006c).

Almost a century ago, the archaeologist Herbert J. Spinden (1917) postulated the existence of a "formative" stratum underlying the basis of civilization in the Americas, and an important component was "maize agriculture." Willey and Phillips (1958) presented a historical-developmental interpretation of the formative stratum defined as, "by the presence of maize and/or manioc agriculture and the successful socioeconomic integration of such an agriculture into well established sedentary life" (p. 144). This definition largely parallels V. Gordon Child's (1951b) definition for cereal grains and the Old World Neolithic shift from food collection to food production. Although the rates of domestication in many New World crop plants are currently unknown or at best somewhat imprecise (see, e.g., Smith 1997a, b, 2000; Matsuoka et al. 2001), the general tendency has been to assume, in contrast to Old World domestication, that multiple domestication events occurred in different regions of the Neotropics (Mangelsdorf 1959b, 1974; Mangelsdorf et al. 1978; Sauer 1950, 1952). Many scholars believe that such evolutionary and developmental processes appear to have occurred simultaneously over large areas of Mesoamerica and South America (Piperno and Pearsall 1998, p. 30). However, as more ethnobotanic, archaeological, and chronological evidence regarding the shift to food production are reported, there appeared to be an ever-increasing lack of consensus as to where and when to draw the line between food gathering, management and tending of wild plants, and food production (Ford 1981, 1985b; Terrell et al. 2003; Staller 2006c).

The subsequent focus in New World archaeology has primarily been on agricultural transitions rather than diffusion and migration as plausible explanations for the integration of an agricultural economy over a vast geographic area (Brown 2006). The most current research suggests that these processes generally occurred in distinct localized regions, but that hybridization with related species and/or human selection leading to morphological change subsequent to their initial dispersals occurred over a longer span of time in some food crops than in others (Wilkes 1985; Smith 1997a,b, 2005a, 2006; Smith and Yarnell 2009). Methodologically driven analyses, designed specifically to uncover precisely where and particularly when the food production originally occurred, has generated data sets that focus almost exclusively on the earliest economic plants (particularly, maize) to the exclusion of other wild plants in the paleobotanical inventory[1] (Staller 2006c). Future studies must incorporate all plants in the paleobotanic inventory, as was done in the Tehuacán Valley and Oaxacan research if we are to understand the adaptive processes associated with early plant domestication that eventually led to a dependence upon certain food crops, economic and medicinal plants by ancient societies in different regions of the world (Staller 2006c; Terrell et al. 2003). On the other hand, the introduction of new approaches and methodologies in the recovery of organic matter have also created an enormous potential for ethnobotanical research in the humid lowlands of the Neotropics and provide data for our understanding of the early food production. The most recent data suggest that broader distinctions need to be made between the beginning of plant cultivation and the appearance of certain plant domesticates (see, e.g., Smith 1997a; Smith and Yarnell 2009).

The transitions between hunting and gathering and farming are for the most part gradual over time and space in the New World when compared with the Middle East or Fertile Crescent (Braidwood 1952, 1969; Braidwood and Reed 1957). One of the major factors associated with this difference is the generally small body size of herbivores and ungulates species in the Neotropics (Binford 2001). Amerindian populations continued to derive most of their meat protein from wild resources rather than major meat producing herd animals (Flannery 1972; Binford 1980, 2001). An exception in this regard was the societies of the high Andes of Peru and Bolivia, where llamas, alpacas, and guinea pigs were domesticated. Dogs and turkeys were domesticated in Middle America, but were not widespread meat staples (Bellwood 2005). Another major difference between the ancient Old and New World agricultural economies is a general lack of "centricity" in the

---

[1]Important exceptions in this regard are the published reports by the Tehuacán and Oaxaca Valley research (e.g., Mangelsdorf et al. 1967; MacNeish 1967b,c, 1992; Flannery 1973, 1986c). The macrobotanical inventory from these projects still represents the most detailed record in archaeology on the beginnings of agriculture (Benz and Staller 2006, 2009).

Neotropics. Rather than geographically circumscribed regions in which there is an extensive macrofloral and faunal remains to indicate evidence for early agricultural societies, evidence of plant domestication appears in rockshelters and caves in semi-arid regions peripheral to rich river valleys. Scholarly debate along these lines has somewhat clouded the conceptual distinction between where agriculture first appeared as related to food production, as opposed to regions where individual crop species were first domesticated (Staller 2006a, 2006c). Consequently, there has also been an emphasis on early presence of food crops such as maize among social and biological scientists working in this hemisphere. There is documented evidence in certain regions where food crops were domesticated early on, that food production based on a suite of domesticated food crops did not appear until much later such as the late Woodland Period (AD 400–1000) or only after the contact period (Rose 2008, pp. 427–433). This is particularly true for the Great Plains and Eastern Woodlands of North America where large terrestrial mammals and birds provided a significant proportion of protein to the ancient diet. More recent evidence, however, suggests that the initial formation of a crop complex of as many as five domesticated seed-bearing plants formed a coherent complex of low-level food production as early as 3,800 years ago in the Eastern North America[2] (Smith and Yarnell 2009, Table 1).

Plant domestication and food production were not as centralized in the New World. Evidence of distinct plant complexes in different regions of the Americas suggests that food production developed independently and that agricultural economies and their associated food crops were later superseded by the introduction of maize, beans, and squash from the Neotropics (Fig. 4.1a, b). Until about 3000 BC, most Native societies were hunter/gatherers – with a possible but disputed earlier investment in horticultural activity claimed for some tropical regions. Dated macroscopic plant remains for domesticated staple food plants, as opposed to snack foods, ceramic containers, and condiments, occur only after 4000 BC and are, for the most part, a great deal later in time (Bellwood 2005, p. 149; see also Smith 1998, 2001; Benz 2001; Piperno and Flannery 2001). archaeologists have long maintained that the middle Mississippi drainage basin and its tributary valleys represent a subregion of seemingly independent development of plant domestication (Fowler 1971; Ford 1985a). The analytical predisposition to the search for centers of domestication is related, in part, to archaeological research in the Old World, where the Fertile Crescent and Middle East were seen as the regions where the domestication of the

---

[2]AMS dates and reanalysis of macrobotanical assemblages from a brief occupation at the Riverton Site in Illinois documents the cultivation of domesticated bottle gourd (*Lagenaria siceraria*), marshelder (*Iva annua* var. *macrocarpa*), sunflower (*Helianthus annuus* var. *macrocarpus*), and two cultivated varieties of chenopod (*Chenopodium berlandieri*), as well as the possible cultivation of squash *Cucurbita pepo* and the perennial grass, little or foxtail barley (*Hordeum pusillum*) (Smith and Yarnell 2009, pp. 6561–6563).

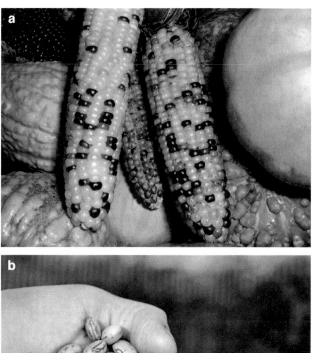

Fig. 4.1 (**a**) Maize (*Zea mays* L.) and varieties of squash (*Curcurbita* spp.) (**b**) Many varieties of beans (*P. vulgaris* L.) are seeds with a hard outer seed coat to protect the embryo during dispersal. Thus, beans require overnight soaking in water to soften the seed coat. Beans along with maize and squash formed the primary food crops of ancient Mesoamerica. (Photograph by John Tuxill)

basic food crops that would later constitute the economic basis for sedentary village agriculture would occur (Braidwood 1952, 1957; MacNeish 1992).

The fact that Native American agricultural economies did not include draft animals, plows, or wheeled transport is another factor affecting the spread of early agricultural economies in the New versus the Old World. archaeologists and prehistorians have long emphasized how such technological innovations were critical to understanding the seemingly different trajectories that early agriculture took in the New World versus the Old World Neolithic and Bronze/Iron

Ages[3] (Braidwood 1960; Bellwood 2005). At European contact, about half (probably more) of the land area of the New World was occupied by hunters and gatherers rather than by agricultural societies (MacNeish 1992; Bellwood 2005). The only highly productive New World food crop was maize, and the earliest domesticates were mainly root crops, vegetables, fruits, and industrial plants (e.g., manioc, beans, avocado, chili pepper, gourd, cotton) rather than productive food staples (Iltis 2000, p. 37). It is for these reasons that scholars in the biological and social sciences generally perceive widespread agriculture as appearing later in the New World (Bellwood 2005, p. 146). The current consensus among archaeologists is that plant and animal domestication began in some parts of the world at about the same time, and that the changeover from hunting and gathering to food production was a long process involving a complicated set of interacting variables (Binford, 1965, 2001; Ford 1981, 1985a; Smith 1998, 2006; Terrell et al. 2003). These interpretations were drawn from earlier theoretical model building, which sought to get away from cultural historical reconstruction to view the development of complexity, early agriculture, and the roles of food crops such as maize in ancient economies as a cultural process (Binford 1964, 1965, 1968, 1989; Smith 1977; Flannery 1972a,b, 1973, 1986b–d). The most recent attempts to understand these processes evoked such factors as population pressure and adaptive changes to how societies practicing some cultivation, particularly, maize, beans, and squash, adjusted to their environments over time (Flannery 1973, 2002; Smith 1998, 2001). These explanations usually involve placing newly domesticated plants such as maize in their ecological and cultural context. The chronological and contextual evidence from these later studies have had a profound influence on subsequent approaches, in which data were increasingly quantified on the basis of statistical analyses and was based on a more interdisciplinary methodological framework (see e.g., Binford, 1989: 55–58, 2001).

Scholars in the biological sciences as well as prehistorians have long assumed that a dependence upon agriculture initially arose in the Neotropics, particularly in Mesoamerica and NW South America (Sauer 1950, 1952; Mangelsdorf 1974; Mangelsdorf and Reeves 1939, 1959b; Mangelsdorf et al. 1967; Lathrap 1970). The most recent archaeological and ethnobotanical evidence suggests that regions such as the Valley of Mexico and lowlands of the Peten were not necessarily where the earliest evidence of domestication occurred. Moreover, the extent to which such adaptive patterns gave rise to large-scale agricultural societies has been the focus of much debate and archaeological research. With the advent of new technologies and methods for uncovering the biological and technological traces of such activity in the archaeological record, focus has in the past two decades shifted to the tropical lowlands and mid-altitude regions of Middle America and northwestern South

---

[3]The expansion of agricultural economies in some parts of the Old World is interpreted as a migration or wave of advance of fully agropastoral food producing cultures extensively replacing hunting and gathering societies as for instance in Neolithic Europe, and Iron Age central and southern Africa (Childe 1950; Bellwood 2005; Brown 2006). The idea of an agricultural revolution was supported to some extent from linguistic evidence, which showed that in some cases such economies were associated with certain language groups (see, e.g., Childe 1950).

America. In some of these regions, seasonal foragers continued to exist alongside the agricultural economies up until the conquest period[4] (Staller 2006c). These regions are also characterized by complex biodiversity and broad undulating rivers, where the primary suite of food crops, particularly maize, beans, and squashes, appear to have been identified (Sauer 1952; Lathrap 1970; Pearsall and Piperno 1990; Piperno 1991, 1999; Piperno et al. 1985; Pope et al. 2001). The cultural geographer Carl Sauer (1952) challenged the most conventional theories regarding the origin of agriculture, instead proposed that it was associated with the tropical lowlands, and arose from the necessity to procure seasonally restricted plant resources by sedentary fishing societies (see also Lathrap 1970).

4.1.2 Paleoethnobotany: Methodological Approaches to Domestication

Much of the research on early agriculture in the New World was inspired by methodological and scientific approaches taken by researchers in the biological sciences, particularly botany, and in the context of such studies originated the field of ethnobotany. The decade of the 1980s was a productive period in the history of ethnobotany. Methodological tools such as the scanning electron microscope (SEM) and AMS dating and macrobotanical data from an increased number of well-preserved prehistoric plant fragments provided researchers with the ability to identify the presence of plants through microfossil analysis. The result was an incremental increase in the scientific knowledge on the variety of domesticated and cultivated plants and their history in ancient times. Considerations regarding the preservation of macrobotanical remains in the archaeological record and the more recent use of infrared spectroscopy[5] in the identification of charred remains have also provided valuable tools for the presence and identification of maize varieties in archaeological contexts (Staller and Thompson 2002; Thompson 2006). Palynological studies of lake cores have been more widely applied in recent years because, in contrast to macrobotanical remains, pollen survives well in humid sediments and in lower elevations (e.g., Bush et al. 1989; Dull 2006; Pohl et al. 1996, 2007; Pope et al. 2001; Piperno et al. 2007). Phytoliths, microscopic pieces of plant silica formed in cells are often preserved in archaeological and paleoecological settings where pollen does not survive (Pearsall 1989, 2000; Piperno 1991;

---

[4]For example, the earliest pottery recorded thus far in the New World comes from Colombia, yet these technological developments did not spread to surrounding cultures until a millennia later (Bonzani and Oyuela-Caycedo 2006; Staller 2006c).

[5]Infrared spectroscopy offers the possibility for measuring different types of interatomic bond vibrations at different wavelength frequencies. Such approaches are particularly concerned with the analysis and identification of substances through the spectrum emitted from or absorbed by objects and IR absorption spectra shows what types of bonds are present in the sample. Spectroscopy approaches are used to identify and delineate archaeological remains through remote sensing (e.g., Pope and Dahlin 1989).

Pearsall and Piperno 1990). Such methodological approaches have been used extensively by paleoethnobotanists in the past 40 years to document ancient maize in the Americas, but only recently has such research been initiated in regions of Mesoamerica where teosinte is native (Piperno et al. 2007, 2009; Holst et al. 2007; Pohl et al. 2007). As revealed in the previous chapter, the reason that teosinte was largely dismissed as a possible progenitor relates in part to the hard fruit cases and ear morphology (Fig. 4.2). The teosinte spike holds the seeds or fruit cases. Teosinte seed dispersal, unlike maize is not dependent upon human agents. Since teosinte and the maize are both grass species, their stalks are very similar morphologically. Wind spreads pollen from the tassels onto the silk. When teosinte is grown in *milpas* they spread their pollen to the maize stalks making them more drought and insect resistant. However, maize and the annual teosintes also resemble each other in a number of anatomical features, most notably the tassel or male inflorescence morphology[6] (Fig. 4.3; compare to Fig. 2.13a).

Recent ground breaking research on the identification of plant microfossils from carbon residues have been particularly valuable in the study of early maize because they can be directly dated and statistically differentiate to specific maize landraces (Staller and Thompson 2002; Thompson 2006, 2007; Hart et al. 2007a; Matson and Hart 2009). How did these studies influence current debates? Particularly, the general consensus among many archaeologists and historians on the antiquity and spread of plant domestication, and the role of maize as a primary economic component and catalyst to social complexity? The most recent debates surrounding the antiquity and location of the original domestication event(s) have also been greatly influenced by various methodological approaches and recent technological innovations (Fritz 1994). These innovations and breakthroughs include direct AMS radiocarbon dating of macrobotanical remains, which was introduced in the previous chapter, and now plant microfossils from food residues in ancient pottery (Long et al. 1989; Staller and Thompson 2000, 2002; Hart et al. 2007a,b; Hart and Matson 2009).

In the past decade, direct dates on ancient cobs have indicated a more recent spread of maize through much of the Neotropics (Blake 2006). AMS dates taken directly from macrobotanical maize cob samples, recovered from the earliest levels of the Tehuacán caves in highland central Mexico, and the Guilá Naquitz rockshelter in Oaxaca have, in some specimens, produced younger dates by over two millennia than the initial radiocarbon assays from associated archaeological strata (Fritz 1994; Long and Fritz 2001; Blake 2006). Research (Michael Blake 2006; Chisholm and Blake; 2006; Blake et al. 1992; Bruce Benz 1999, 2006; Benz and Iltis 1990; Benz and Long 2000; Benz et al. 2006; Long et al. 2001) has indicated that dates by association have been largely unreliable, particularly in those areas and sites in the Neotropics where maize was believed to have originated. The most

---

[6]Teosinte is much more environmentally and morphologically flexible than maize. Annual teosinte has an architectural genetic locus called Teosinte Branched Locus (*tb1*), which makes the branch resemble a branched teosinte plant (see Fig. 2.2b). When teosinte grows in full sunlight without competition *tb1* lateral branching is suppressed resulting in long singular branch tipped by male inflorescences (see Iltis 2006, p. 28–32, Figs. 4.3–4.7; see also Benz 2006, Table 2.1, Fig. 2.1).

Fig. 4.2 Teosinte spike and a cob of Hopi blue maize showing the differences in morphology and ear size between contemporary teosinte spikelet and maize cob. Such morphological differences confounded many researchers and provide some insight as to why it took scholars so long to identify the progenitor of *Zea mays* L

recent radiocarbon evidence (conventional and AMS) indicates that direct dates on macro remains and food residues provide far greater precision and remarkably different chronologies than dates by association.

Guilá Naquitz rockshelter is located 400 km northeast of the Balsas River (Doebley et al. 2006). The three oldest known archaeological maize cobs Guilá Naquitz have in one case, two of the four morphological traits defining

Fig. 4.3 Teosinte male inflorescence or tassel. This feature is one of a number of anatomical similarities that the teosintes share with *Zea mays* L

domesticated maize, while the third sample, has three of the four defining characteristics by 5,450 years ago (see Staller 2003, p. 377; Benz 2001). Benz (2001, p. 2104) refers to these ancient cobs as domesticated *Zea* inflorescences, despite the morphological changes and the *tga1* mutation to soft glume architecture, they are in this time period examples of teosinte still evolving into fully domesticated maize. It was Dorweiler and Doebley (1997; Dorweiler 1996; Dorweiler et al. 1993) who demonstrated the molecular relationship between maize and the cupule and glume morphology of the cobs. Their genetic research revealed a single Mendelian locus, *teosinte glume architecture* (*tga*1), that controls several aspects of glume morphology, both macroscopic and microscopic, including the deposition of silica in the form of phytoliths and silicified epidermal cells (Dorweiler and Doebley 1997, p. 1314). Teosinte is not native to this valley nor was it identified near the rockshelter[7] (Piperno and Flannery 2001, p. 2102). Nevertheless, the morphological results from the ancient cobs indicated that maize was not yet fully domesticated by 5450 B.P. in this region and that teosinte was brought into the valley (Piperno and Flannery 2001; Benz 2001). Recent direct dating of seeds and other plant parts that exhibit well-documented morphological characteristics, indicating they are domesticated or dependent upon humans for their reproduction, is now being applied throughout the Americas (Tables 3.1, 4.1–4.3). These AMS dates produce more conservative timetables of domestication and agricultural origins and have considerably revised earlier chronological frameworks based on dates by association (Blake 2006; Benz and Staller 2006).

Having precise chronological information regarding plant domestication was critical to archaeological model building, since these developmental processes were visualized along a continuum (Smith, 1998, 2001). Under the new standard of directly dated remains, we may infer that the early domestication of maize in highland Mexico, and domestication of camelids in the Andes, appears (in the archaeological record) between about 5,500 and 4,000 years ago, more recent, in some cases by 2,000 years, than was previously assumed only a decade ago. These data, as was discussed in some detail in the previous chapter, have had profound implications as they indicate that all of the genetic mutations associated with fully domesticated maize were undeveloped until ca., 4500 B.P. (Benz et al. 2006; Jaenicke et al. 2003; Jaenicke-Després and Smith 2006). These data also have implications for interpretations regarding the early economic significance of maize to New World prehistory. Some archaeologists have already begun to explore other explanations for the seemingly rapid spread of this important New World domesticate.

4.2 Interdisciplinary Approaches to Domestication, Agriculture, and Adaptation

The methodological approaches and theoretical models presented in archaeology regarding the origins of agriculture have generally emphasized analyses which view this transformation in human adaptation in terms of a continuum, beginning with

---

[7]This may indicate that teosinte was brought from its native habitat into this region by archaic foragers, and that some of these morphological changes may have been related to adaptation to such environs. Ongoing analyses continue to document changes in the morphology of maize and the development of regionally distinct land races (Benz 2001; Benz et al. 2007; Latoumerie et al. 2006; Tuxill et al. 2009).

Table 4.1 Associated ^{14}C Dates with Zea Pollen Samples

Country/region	Site	Dated material	^{14}C method	Associated Zea	^{14}C years B.P.	Sample ID number	Reference
Mexico/Oaxaca	Guilá Naquitz	Organic materials	Conventional	Pollen (teosinte)	ca. 9500–6980 (Zone C-B1)	?	(Schoenwetter and Smith 1986, p. 229; Piperno and Flannery 2001, p. 2102)
Guerrero/highlands	Iguala valley	Pollen residue	AMS	Pollen	6290 ± 40	Beta-196250	(Piperno et al. 2007)
Tabasco/coast	San Andrés	Organic materials	AMS	Pollen	c. 6200	n/a	(Pohl et al. 2007, p. 6870)
Veracruz/coast		Pollen fraction	AMS	Pollen	4150 ± 50	Beta-130582	(Sluyter and Domínguez 2006, Table 1)
Veracruz/coast	Laguna Pompal	Pollen residue	AMS	Pollen	4250 ± 70	CAMS-1770	(Goman and Byrne 1998, p. 84–86)
México	Laguna Zoapilco	?	Conventional	Pollen	5090 ± 115	I-4405	(Niederberger 1979, p. 132–137)
Guatemala/Pacific coast	Sipacate	Wood	AMS	Pollen	4600	Not given	(Neff et al. 2002)
Belize/Caribbean Coast	Cob swamp	Wood	Conventional	Pollen	4610 ± 60	Beta-56775	(Pohl et al. 1996, p. 360–361)
Honduras	Lake Yojoa	Wood	Conventional	Pollen	<4770 ± 385	UGa-5380	(Rue 1989, p. 178)
El Salvador	Río Paz Valley	Lake sediment	AMS	Pollen	3470 ± 50	CAMS-44169	(Dull 2007, p. 131)
Costa Rica	Laguna Martínez	Associated charcoal	AMS	Pollen	4760 ± 40	not given	(Arford and Horn 2004, p. 112)
	Lago Cote	Associated charcoal	AMS	Pollen	3630 ± 70	not given	(Arford and Horn 2004, p. 112]
	Laguna Zoncho	Wood	Conventional	Pollen	2940 ± 50	Beta-115186	(Clement and Horn 2001, p. 422)

Region	Site	Sample	Method	Type	Date	Lab no.	Reference
Panama	Cueva de los Ladrones	Pollen residue	Convencional	Pollen	6860 ± 90	?	(Piperno et al. 1985, p. 873) (date given, 4919 ± 90 B.C.)
	La Yeguada			Pollen	4200	?	(Piperno et al. 1990)
	Gatun Lake	Charcoal?		Pollen	ca. 4000	UCLA-11354	(Piperno 1985, p. 17)
Colombia/inland	Hacienda El Dorado			Pollen	6680		(Bray et al. 1987)
	Hacienda Lusitania			Pollen	5150 ± 180		(Monsalve 1985)
Ecuador/Amazonia	Lake Ayauchi	Associated charcoal	AMS	Pollen	4570 ± 70	Beta-20956	(Bush et al. 1989, p. 304)
Amazonia	Lake San Pablo			Pollen	4000	Phase	(Pearsall 1999, p. 421)
highlands/Valley of Quito	Cotocollao	Associated charcoal	Conventional	Pollen	3545 ± 210	GX-4768	(Villalba 1988, p. 339)

Table 4.2 Dates on materials associated with Zea phytolith samples

Country/Region	Site Name	Dated Material	^{14}C Method	Associated Zea	^{14}C Years B.P.	Sample ID Number	Reference
Mexico/Tabasco Coast	San Andrés	organic materials	AMS	phytoliths	6208 ± 47	n/a	(Pohl et al., 2007:6872)
Highlands, Iguala Valley, Guerrero	Xihuatoxtla	charcoal	AMS	phytoliths	7920 ± 40	n/a	(Piperno et al., 2009:5021)
Guatemala/Pacific coast	Sipacate	wood	AMS	phytoliths	4600	?	(Neff et al., 2002)
Panama/inland	Monte Oscuro	bulk sediment	AMS	phytoliths	7500 ± 70	Beta-74292	Piperno and Jones 2003:81)
Inland	Cueva de los Ladrones	unknown	Conventional	phytoliths	6860 ± 90	?	(Piperno and Flannery 2001: 873) (date given: 4919 ± 90 B.C.)
Inland	Gatun Lake	charcoal?	Conventional	phytoliths,	4750 ± 100	UCLA-1354	(Piperno 1985:15)
Inland	Aguadulce	unknown	Conventional	phytoliths	4500	?	(Piperno 1985, 1988)
Inland	La Yequada	unknown	Conventional	phytoliths	4200	?	(Piperno et al., 1990)
Inland	Lake Wodehouse	unknown	Conventional	phytoliths	3900	?	(Piperno 1994)
Ecuador/Coast	Vegas site	shell	Conventional	phytoliths	7150 ± 70	Tx-3314	(Stothert 1985:618, 621)
coast	Loma Alta	unknown		phytoliths	5000	?	(Pearsall 2003:224)
coast	Loma Alta	direct date on starch grains	AMS	Phytoliths in residue	4570	Beta-198623	(Zarillo et al. 2008)
upper Amazon	Lake Ayauch	charcoal	AMS	Phytoliths	4470 ± 70	Beta-20956	(Bush et al., 1989:304)
coast	Real Alto	unknown		phytoliths	4450	?	(Pearsall 2003:225; Pearsall and Piperno 1990:330)
coast	La Emerenciana	direct date on carbon residue	AMS	phytoliths in residue	3860 ± 50	Beta-125107	(Staller and Thompson 2002:44)
coast	La Emerenciana	direct date on carbon residue	AMS	phytoliths in residue	3700 ± 50	Beta-125106	(Staller and Thompson 2002:44)
coast	La Emerenciana	charcoal	Conventional	phytoliths	3775 ± 165	SMU-2563	(Staller 1994; Staller and Thompson 2002:44; Tykot and Staller 2002:670)
Bolivia/highland	Copacabana peninsula	Direct date on food residue	AMS	phytoliths in residue	2750 ± 40	Beta-162147	(Chávez and Thompson 2006: Table 30.4)
Argentina/ highland	Quebrada Seca	wood	Conventional	phytoliths	4500 ± 100	?	(Babot 2004)

Table 4.3 AMS dates from La Emerenciana (OOSrSr-42)

Sample data	Measured ^{14}C Age	^{13}C/^{12}C Ratio	Conventional ^{14}C Age
(a) Beta-125106 Sample A Cat. No. 5480 Analysis: Standard AMS Material/pretreatment: (organic material/acid washes)	3720 ± 40 B.P.	−25.8 o/oo	3700 ± 40 B.P.
(b) Beta-125107 Sample B Cat. No. 5623 Analysis: Standard AMS Material/pretreatment: (organic material/acid washes)	3810 ± 50 B.P.	−21.9 o/oo	3860 ± 50 B.P.

hunting, gathering, and foraging at one end of the spectrum and resulting in fully developed agricultural societies on the other (Flannery 1969, 1972b; Gebauer and Price 1992; Smith 2006). Much of the archaeological research along these lines has been methodologically driven to focus on the cultural contexts into which fully domesticated plants were introduced to different regions of the Neotropics and the conditions under which indigenous plants were morphologically and genetically altered by prehistoric cultures (Ford 1985a). Maize, perhaps more than any other cultigens, has received the most attention by archaeologists and ethnobotanists. What archaeologists documented early on is where and when along the developmental continuum was maize integrated into a fully agricultural economy in different regions of the Americas. They used various lines of evidence to determine if food staples like maize were a necessary prerequisite for the development of an agricultural economy and social inequality (Smith 1998, 2005b). What the archaeological evidence suggested from the early studies at various sites throughout Mesoamerica was that human societies existed for thousands of years on a hunting and gathering life style before they became dependent upon certain food crops for their survival (MacNeish 1967b, 1985). Approximately 6,000 years ago, some societies across Mesoamerica, Central and Andean South America began to carry out subsistence strategies that involved an increased dependence upon domesticated plants such as maize, beans (*Phaseolus vulgaris*), squash (*Cucurbita pepo*), and the bottle gourd (*Lagenana siceraria*)[8] (Piperno and Flannery 2001; Benz 2001; Freitas et al. 2003; Smith 1997a, 2001, 2006; Piperno et al. 2009; Ranere et al. 2009). Botanists and archaeologists puzzled over the ancestry of maize for over 100 years. Part of the reason for this is morphological, in that domesticated varieties of cereals, grasses, and grains such as wheat, barley, and rice are nearly identical structurally to living

---

[8]The most important food plants in the New World include arrowroot, achira, chilli peppers (*Capsicum* spp.), *Xanthosoma* tubers such as cassava (*Manihot esculenta*), and sweet potatoes (Piperno and Pearsall 1998). The common bean (*Phaseolus vulgaris* L.) was first domesticated at about 500 BC (Smith 2005a, Table 1). Bottle gourds and squash were domesticated early in the Holocene by Native populations in the Neotropics (Smith 1997a, 2000).

wild species, but this is not the case with maize (Fig. 4.4). The principal difference is that, in the domesticated varieties of cereal grains and grasses such as maize, the edible seeds tend to remain fastened on the plant, while their wild progenitors have shattering inflorescences – shattering is a mechanism by which the seeds of the plant are dispersed naturally (Iltis 2000, 2006). For grasses and large seeds, a nonbrittle rachis, and loss of mechanical means of protection are but several of the most conspicuous of the many genetic changes that brought them into a dependent relationship with humans (Mangelsdorf 1974). How they were selected remains a critical question. Improved methodological and genetic approaches have in the analysis of the cultural ecology as well as technological innovations since the early 1970s, provided a basis for increasingly detailed paleoclimatic reconstructions. Recent studies have shown that plants selected by ancient foraging populations were changed morphologically and genetically long before such patterns became apparent in the archaeological record (e.g., Smith 1997a; Smith and Yarnell 2009). Such adaptive patterns were not only changing the plants associated with human adaptive patterns, but in various ways transformed the ancient landscape (Terrell et al. 2003).

Fig. 4.4 Maize cobs presented as the evolutionary sequence, resulting from management and conscious selection of certain traits by Archaic and ceramic period occupants in the Tehuacan Valley. The three cobs from *left to the right* are from San Marcos Cave, the final two right Coxcatlan Cave. (**a**) Early cultivated Abejas Phase maize; (**b**) Chapalote variety from the Palo Blanco phase; (**c**) Chalote variety from the Venta Salada Phase; and (**d**) Cinico maize from the Venta Salada Phase (from Mangelsdorf et al. 1967, Fig. 103)

Archaeologists focused on the early food production generated various lines of evidence to determine whether a dependence upon food crops led to population increases which induced environmental stress that ultimately *favored* food production and a dependence upon crops such as wheat or maize (Binford 1968; Flannery 1969). The management and cultivation of increasingly productive varieties of maize created conditions where a greater amount of food could be produced per land unit with the use of domesticated plants in general (at least in the short term), and there is good archaeological evidence to suggest that increased population pressure in some areas may have created the conditions which favored farming as a preferred adaptation. Agriculture and increased dependence upon certain crops is in this sense an adaptive response to either a reduction in the carrying capacity of the local environment and/or an increase in the human population (Binford 1968).

More recently, scholars have presented evidence that genetic changes in plants such as maize were critical to the shift towards an agricultural economy (Smith 2006; Jaenicke-Després and Smith 2006; Doebley et al. 2006). Other archaeologists and botanists believe that humans gradually came to be more and more dependent on the harvest of a limited number of species as a form of specialization, and therefore, domestication was seen as the foundation of civilization (Watson and Watson 1969, p. 94). These researchers have uncovered evidence of highly specialized technologies, such as storage and ways to process foods that was also of foremost importance (Rouse and Cruxent 1963). In other words, some believe that it was related to specific technological innovations that made agriculture necessary for human survival in some regions.

Studies of ancient DNA associated with archaeological maize from northeast Mexico and the American southwest indicate a possibility that scientists have the technology to track human selection for specific attributes that are not observable morphologically in the archaeological cobs (Jaenicke-Després and Smith 2006; Jaenicke-Després et al. 2003; Vigouroux et al. 2002, 2003). The seeming absence of any morphological indicators for human selection emphasizes the importance of combining genetic and archaeological research to reconstruct conscious and unconscious selection of crop plants since the beginning of the Holocene. Recent genetic research which traces the temporal and geographical radiation of maize from southern Mexico to the limits of cultivation in the Americas compare very closely to large scale efforts by archaeologists to track the gradual expansion of maize cultivation in different regions of the Neotropics to areas of the north (Matsuoka et al. 2002; Freitas et al. 2003; Blake 2006; Doebley et al. 2006). All these forms of plant manipulation were in use for thousands of years until societies created new food plants through deliberate or conscious selection. When humans began to consciously domesticate the landscape, they essentially created plant communities, which were to become the dominant components in their ancient diets (Smith 2006).

In the past 30 years, anthropologists and archaeologists have increasingly emphasized that the evidence indicated that domestication in general and the early domestication of maize in particular should not be seen as a kind of sudden innovation or idea, but that it was a gradual process which took place over hundreds if not thousands of years in each of the separate regions of the Americas (Smith 1998; Bellwood 2005). This idea was in line with the thinking of many plant geneticists such as Mangelsdorf and archaeologists such as MacNeish. archaeologists studying early plant and animal domestication in the Old and New World indicate that the transition from hunting and gathering to food production was very gradual in some regions and more rapid in others. Preagricultural cultures began to depend upon domesticated food staples because of a whole host of interrelated causes (Braidwood and Reed 1957, p. 19). What the research in the Tehuacán valley showed was that there are many variables that needed to be understood before ethnobotanists, archaeologists, and anthropologists were able to reconstruct the conditions under which maize agriculture was first regarded as profitable and under which maize became a primary food crop (MacNeish 1967b, 1985, 1992). archaeologists began to search for sites in which various factors such as population pressure, distribution of plants and animals, the rate at which the environment (or generic changes) climatic changes, and technologies then available all played an important part in making agriculture more important to human survival. Food production resulted, in a number of effects, which led away from the mobile lifestyles of hunting and gathering, to sedentary villages (Flannery 1972a, b, 1973; Kelly 1995). Higher than normal population concentrations means more people than are generally possible to be sustained by the environments. Thus, certain groups within the society become specialized upon exploiting specific species found in those regions (Flannery 1969; see also Binford, 1962, 1965).

Early agricultural sites are generally distributed in a linear fashion along major streams. Such settlement patterns show differences in the size and function of the habitations, and usually have burial grounds in association. Such cemeteries often functioned as territorial boundaries (Flannery 1972b). Several lines of evidence also suggest that the Amerindian societies who first brought these species under domestication may not have been seasonally mobile hunter-gatherers of higher elevation environments exclusively.[9] These explanations support the fact that most early research on the origins of agriculture and food crops like maize was carried out at rockshelters and caves. Early maize and other domesticates recovered from upland caves may reflect a transition to a farming way of life accomplished by societies occupying more sedentary settlements in river valleys. In some cases, upland caves containing domesticates may represent one component in the seasonal round of early food production by societies occupying nearby river valleys; in others, they may mark the subsequent expansion of food production economies out of rich highland river valley resource zones into adjacent upland environments (Flannery 1986c).

The increased dependence upon plants, mainly food crops, resulted in other changes in the archaeological record that involved technologies specifically geared

---

[9]According to MacNeish (1978), the Pre-ceramic societies of the Tehuacán Valley lived in small microbands that dispersed periodically. Some camps accommodated only a single nuclear family, while others sheltered much larger groups for part of the annual cycle (see also MacNeish 1999).

to their processing and consumption (Binford 1980, 1989). In other words, tool kits specialized at exploiting certain species of plants and/or animals (Fig. 4.5). The *metate* in the *milpa* depicted in this figure is from an old village site called Mopila, near Yaxcaba, Yucatan, Mexico. The village was abandoned in the early twentieth century, ca. 1930. The Maya families now live in Yaxcaba, but the descendants continue to cultivate maize there due to the highly fertile black soil enriched by centuries of previous habitation. The *metate* may date to an even older period of use. At nearby Mopila, there are the ruins of a church that dates back to the 1600s, and the presence of an accessible cenote suggests the area was almost certainly inhabited prior to Spanish arrival. Stone grinding and processing tools such as *manos* and *metates* were particularly important indicators of a dependence upon domesticated plants, particularly maize, and were often perceived as synonymous with its presence (Lathrap et al. 1975). Ultimately, plant production and an agricultural economy involves the deliberate manipulation or conscious selection of plants, intervening in their life cycle to make sure that some useful parts are available for consumption or for some other possible economic use (Rindos 1984; Heiser 1988; Zohary 2004). These preceramic societies often select certain characteristics, such as multiear stalks or larger kernels, more rows, and put those cobs aside to plant for the growing season, so that they reproduce more productive maize harvests (Mangelsdorf et al. 1967). The growing bodies of evidence in the last 40 years indicate that ancient foragers were domesticating not only plants, but also the landscapes to which they were adapting, that is, they were consciously harvesting certain food crops and in the process changing their surroundings (Terrell et al. 2003). The domestication and increased dependence upon maize and other plants was part of a long continuum of human interaction with the natural vegetation that started with the first settlements (Ford 1985a; Smith 1986; Smith 1997a, 2000, 2001, 2006). Biologists working on existing plant and animal populations and interdisciplinary archaeological investigations at ancient sedentary settlements have steadily intensified the search for evidence of plant and animal domestication. In the past decade, the initiation of a variety of new research approaches taken by biologists and botanists as well as archaeologists, have applied new standards of evidence that has radically changed our understanding of the adaptive shift from foraging to farming in such a brief period of time, that many researchers are only now coming to terms with the implications of these more recent results (see, e.g., Staller et al. 2006).

4.2.1 Plant Domestication and Cultivation

The recent published data on plant biology, genetics, and ethnobotany has changed and continue to influence the way in which archaeologists consider human/plant interaction with *Zea mays* L. and food crops in general in their interpretations of the archaeological record. The current consensus from the archaeological and biological sciences is that cultivation of wild plants and potential domesticates

Fig. 4.5 (a) Specialized tool kits involving grinding implements such as *manos* and *metates* at archaeological sites were seen as indirect and sometimes direct evidence for the presence of maize. (b) Metate in the Lol-Tun Cave in northern Yucatan, Mexico. A person holds the *mano* to show how maize and other plants would have been processed using such grinding implements. Lol-Tun cave is an important archaeological site with evidence of occupation extending back to the pre-Classic period. (Courtesy of John Tuxill) (Photograph by John Tuxill)

that are undergoing genetic change were not always necessarily designed or geared to produce stable crops (Wright et al. 2005). Most early cultivation appears to have been aimed at producing seasonal supplements to broad-based vegetable diets as a guarantee to a bountiful harvest (Smith 2006). Such human selection may have eliminated the need for wide-ranging searches for additional foods. The very act of gathering vegetable foods can lead to unintentional or unconscious tending of plants – accidental seed dispersal and trampling can benefit wild resources as well (Rindos 1984, p. 90–91; see also Terrell et al. 2003). Intensive gathering or selecting of larger seeds at the expense of smaller ones can also have unexpected genetic consequences, selecting against less desirable traits (Rindos 1984, p. 90–91; see also Terrell et al. 2003). These changes were in some cases maintained over time, even when intensive human exploitation ceases permanently or for parts of the annual cycle. For instance, such accidental genetic changes may have caused the condensing of the lateral branches and tiny cobs of teosinte (Z. *mays* ssp. *parviglumis*), the indigenous wild grass that is the ancestor of domesticated maize (Galinat 1985, p. 257–259; Brown 1999; Doebley et al. 1990; Camus-Kulandaivelu et al. 2008). In many regions of the Americas, archaic and Preceramic hunters and gatherers used fire to encourage the regeneration of grasses and edible plants (Simmons 1996; Redman 1999; Dull 2007). For example, California Indians used fire to eliminate plant competitors under edible acorn oak trees, also to encourage growth of *Corylus* sprouts, which they prized for making their baskets (Ford 1985a, p. 3). The setting of controlled fires for certain species is not generally considered to be an example of deliberate plant production, but it does show how such activities can have dramatic effects on the plant and animal species in the local ecologies (Lewis 1972; Redman 1999). Slash and burn cultivation is the primary technique used by early agricultural societies to cultivate domesticates such as maize, beans, and squash (Fig. 4.6a). In tropical environments, the cutting and burning of old growth forest resulted in dramatic ecological changes over relatively short periods of time (Fig. 4.6b). Such environmental alterations and changes can be identified and recorded using a variety of techniques, but such data does little to provide information on how plants were originally manipulated by archaic societies or how the various floral and faunal components of a given ecology result in an agricultural economy.

Richard Ford (1985a, p. 3–4, Figure 1.1) observed that food production begins with "deliberate care afforded the propagation of a species," what is referred to as "cultivation." Cultivation does not imply full domestication, but does infer that the life cycle of a plant has in some way been disrupted by human selection (Ford 1985a, p. 4; Rindos 1989, p. 111). Humans generally collect larger quantities of food because it enables them to obtain its products with greater ease in the course of their annual cycle (Binford, 1965). Cultivation proceeds in several ways; weeding, pruning, and otherwise tending plants are commonplace in many parts of the world (Weatherwax 1954; Ford 1985a). Tending is a casual rather than conscious activity, usually removing competing vegetation around root plants, weeding the soil near important medicinal species, as is done in the American SW and the Peruvian Amazon and usually results in a higher yield and therefore increased food

Fig. 4.6 (**a**) Slash and Burn cultivation is widespread throughout Mexico and Central America. A mature milpa is being burned in Northern Yucatan, Mexico. Such activities usually take place in April (**b**) A *milpa* or cornfield after burning. The ash and other sediments provide nutrients to the soil. When such *milpas* are located near bodies of water, the rain deposits such sediments and microfossils of the plants that were cultivated can be identified in pollen cores (Courtesy of John Tuxill) (Photograph by John Tuxill)

production (Ford 1985a, p. 4). Tilling the soil with digging sticks or simple hoes tends to encourage the germination of naturally dispersed seeds (Fig. 4.7). Grubbing the earth with simple digging sticks or hoes can increase the moisture retention or aerate the ground (Weatherwax 1954, pp. 60–62; Ford 1985a, pp. 4–5). Many foragers used digging sticks to obtain tubers, removing lateral roots or bulbs at an early stage in growth. One way that archaeologists and biological scientists working with the Tehuacán collections determined when and if such activities occurred, is by looking at plant remains in the various archaeological deposits, and then determining whether there was some selection occurring over time with certain species such as maize (see, e.g., Mangelsdorf et al. 1967; Kaplan 1967; Smith 1967; Benz 2001; Benz et al. 2006).

Domesticated plants are ultimately cultural artifacts in that they could not exist in nature without human assistance (Ford 1985a, p. 6). All these forms of plant manipulation and management were being carried out for thousands of years until people created new food plants through deliberate or conscious selection (Heiser 1988; Smith 2000, 2001, 2006). When humans began to consciously domesticate the landscape, they created plant communities that were essentially the dominant component in their ancient diets (Smith 1986, 2006). Some of these plants became totally dependent upon humans for their reproduction while others did not and either became extinct or returned sometimes in modified form to a wild state. This is particularly the case with members of the grass family and chenopods in the Eastern North America (see, e.g., Smith and Yarnell 2009).

Research on the evolution of plants in the past century has been generally focused on the appearance of new species arising from environmental and cultural

Fig. 4.7 Modern day Peruvian farmers using digging sticks to till the soil. Prehistoric farmers throughout the Americas used such implements to till the soil before cultivation. Tilling the soil with digging sticks or simple hoes tends to encourage the germination of naturally dispersed seeds (Courtesy of University of Illinois-Chicago)

selection and at the molecular level the genetic changes which resulted in domesticated plants (Ford 1985a, p. 11). The general genetic mechanisms underlying mutation and genotypic variability in domesticated plants have been of greatest scientific concern to researchers (Pickersgill and Heiser 1976; Doebley et al. 2006). Genetic approaches to crop domestication traditionally begin with detailed analysis of plant morphology or phenotype and extend back to how these traits are regulated by genes (Doebley et al. 2006). Researchers working with landraces of various kinds of food crops, on the other hand, employ a population genetic approach beginning with genes and determine whether these genes were targets of selection (e.g., Dorweiler and Doebley 1997; Dorweiler et al. 1993; Jaenicke-Després et al. 2003; Gallavotti et al. 2004). Such research involves comparing genetically modified plants with their wild counterparts. The genes controlling morphological and structural changes during domestication are referred to as transcriptional regulators and it is this class of genes that play a central role in the domestication and regulate the morphological development in plants (Doebley 1994; Doebley et al. 1983, 1984, 1997; 2006; Iltis 2006; Dorweiler and Doebley 1997; Dorweiler et al. 1993; Doebley and Lukens 1998). For example, research in molecular biology documented that the genetic locus, called *tga1*, controls induration, orientation, length, and shape of the glume architecture (Dorweiler 1996, p. 20). Perhaps the most significant breakthrough has been through analytical techniques at the molecular level of plant DNA (Matsuoka et al. 2002; Gallavotti et al. 2004). A series of recently published reports from molecular biology and plant genetics have also seriously challenged previous interpretations regarding the domestication and spread of maize in the Neotropics. Analysis of microsatellites on ancient maize DNA has demonstrated unequivocally that this event occurred in a single location (Matsuoka et al. 2002). The locus of initial domestication is now considered to have occurred in Balsas River drainage of Central Mexico. These data further indicate that *Z. mays* L. had a single evolutionary progenitor, the wild grass teosinte *Z. mays* spp. *parviglumis* and that its early spread out of Central America was along a highland corridor (Matsuoka et al. 2002, p. 6083; see also Freitas et al. 2003). Thus, the current consensus among geneticists and botanists is that the maize from highland Mexico along a highland corridor:

> Among archaeologists, there have been two models for the early diversification of maize. According to one, because the oldest directly dated fossil maize comes from the Mexican highlands, then the early diversification of maize occurred in the highlands with maize spreading to the lowlands at a later date. The second model interprets maize phytoliths from the lowlands as the oldest maize, and accordingly places the early diversification of maize in the lowlands. Our data suggest that maize diversified in the highlands before it spread to the lowlands (Matsuoka et al. 2002, p. 6083).

As noted in the previous chapter on the history of science surrounding the origins of maize, a central issue was the virtual absence of teosinte in Preceramic highland caves and rockshelters. It was at least in part the fact that teosinte was not exploited by humans for food that called into question how maize was originally domesticated, and how and when its evolutionary progenitor was first modified through human intervention and natural selection (Iltis 2000, 2006; see also Smalley and Blake 2003).

The innovative, and in some cases, iconoclastic published research from the biological sciences, particularly the work of Hugh Iltis and John Doebley and their associates, have redefined the way in which archaeologists interpret the domestication of maize. This particularly emphasizes its mutational properties and the rate at which it was domesticated and then modified by human selection. It is becoming increasingly apparent that these recently published results and those of various plant geneticists from different parts of the world have revolutionized the way in which the academic community has begun to rethink the biological evolution of maize and its biogeography in the New World. These various lines of evidence have also influenced the initial assumptions regarding the economic role of maize to ancient hunters and gatherers and later agricultural societies. It is increasingly apparent that these data have a multidimensional effect on the scientific community and are constantly challenging archaeologists, biologists, and ethnobotanists to reconsider previous assumptions about the evolution and role of maize in the prehistoric past. The basis for much of the archaeological, botanical, and molecular research on maize was derived at least in part by the pioneering research of archaeologists and botanists in highland Mexico in the Tamaulipas caves, and the Valley of Tehuacán and Oaxaca (Fig. 4.8).

Fig. 4.8 Map of Mesoamerica showing the location of the Tamaulipas cave, and the Tehuacán and Oaxaca Valleys

4.2.2   Approaches to Domestication and Cultivation in the Tehuacán Valley

The historical beginning of interdisciplinary research on early agriculture and on plant domestication have been greatly influenced by scholarship in the biological sciences and botany (e.g., Kaplan 1967; Smith 1967; Cutler and Whitaker 1967; Mangelsdorf et al. 1967). As apparent in the previous chapter, the research on teosinte, the progenitor of maize, had a profound influence on the archaeological interpretations of early agriculture. This is in part because maize was the primary economic staple in the nuclear areas of sociocultural development during the contact period, and secondarily because its progenitor was seen as morphologically different, and did not appear to have been consumed as a food crop. The interdisciplinary research on early agriculture in the New World not only influenced the kinds of plants that would be the focus of attention, but what kinds of environmental settings and prehistoric sites would be considered for such research.

It was a scholarly interest in the origins of maize that made the Tehuacán Valley research possible. MacNeish (1961, 1962) initiated fieldwork at a series of caves and rockshelters in the highland Tehuacán Valley of Puebla, Mexico, on the advice of Paul Mangelsdorf. Little was known about the Preceramic cultures of Mesoamerica or the beginnings of agriculture in the New World. The archaeological research in the semi-arid highlands of northeastern Mexico focused on many dry caves, because it was in these localities that plant parts along with the ancient tools for gathering and processing were best preserved (Smith 1967, 1986; Mangelsdorf et al. 1967). This is a region of the Americas that is known for its striking ecological diversity. Neotropical diversity is created in part by the mountainous terrain and valleys where, the altitude, rainfall, and soil differences create innumerable regional variations in the local ecology and geology. The rich volcanic soils and tropical ecology fostered an adaptation that was focused on the plant life from the very early periods. The interdisciplinary research of MacNeish has produced archaeological evidence of a sequence of adaptations, from hunting and gathering to fully agricultural societies (MacNeish 1978, 1992). These studies documented in considerable detail the gradual processes and adaptations that ultimately led to sedentary villages and an agricultural way of life.

MacNeish (1961) chose to search for the origins of maize (*Z. mays* L.) in the relatively small Tehuacán Valley because of the arid climate and its positive effects on the preservation of faunal and macrobotanical remains. Preliminary excavations unearthed fragments of basketry and plant materials in limestone cave deposits. MacNeish (1947, 1958) already had recovered small 5,000-year-old cobs in cave deposits in both the northeastern Mexican state of Tamaulipas and the southern state of Chiapas (see Fig. 4.8). Indirect association through conventional ^{14}C dating of organic materials from the same stratigraphic layers generally determines the antiquity of these archaeological cobs. The initial results suggested that these earliest domesticates dated to between 7000 and 10000 B.P., which was roughly contemporaneous with such developments in the Old World (MacNeish 1992, p. 77–78). The 7,000-year-old date became associated with the origins of maize by many archaeologists and was the generally accepted chronology until the advent of AMS dating. It should be noted that maize appeared relatively late in the archaeological deposits in the Tehuacán Valley (Johnson and MacNeish 1972, p. 17). Richard MacNeish (1967a, p. 3) maintained that research surrounding early agriculture and the origins of plant domestication was best explored by cooperative research between botanists and archaeologists. He hypothesized, on the basis of the botanical and genetic evidence, that the earliest maize was of even greater antiquity and would be found in the highland valley of Tehuacán, situated between Tamaulipas and Chiapas (MacNeish 1967b, p. 14–15). The maize cob Coxcatlan Cave radiocarbon dated by association to 3610 BC [M-1089] indicating that this highland valley had great potential for answering questions regarding the origins of maize and early agriculture (see also Johnson and MacNeish 1972, p. 17). In order to test this hypothesis, MacNeish (1967b, Fig. 2, 1978, Fig. 2.2) designed an interdisciplinary methodological approach in the Tehuacán valley. This research was focused on examining what variables ultimately led to the domestication of food crops such as maize, and if these changes in adaptation provided the foundation for later Mesoamerican civilization (MacNeish 1992).

The research in the Tehuacán Valley was regional in scope, involving a settlement survey that located the remains of more than 450 pre-Hispanic sites over the 1,500 km^2 (575 sq. miles) of the valley (MacNeish 1967b, p. 22–24). Given the primary goals of the project, excavations were focused on a series of 12 caves and open-air rockshelters. On the basis of this archaeological research and stratigraphic excavations, a large number of ^{14}C dates were recorded and produced evidence of a continuous occupation of the valley spanning 10,000-years (MacNeish 1967b, p. 18–19). Various archaeological projects over the past 50 years have documented the gradual transitions in different regions as people moved from nomadic but intensive exploitation of wild foods to settled cultivation, and ultimately to a dependence upon a few domesticated plants. The Tehuacán cultural sequence was the longest recorded in the New World at that time. On the basis of the artifacts (tool kits), macrobotanical and faunal remains, the researchers were able to reconstruct the seasonality of resource availability and the scheduling of resource extraction (MacNeish 1978, pp. 146–148, 152–153). Seasonality and scheduling were found to be critical data sets from which to reconstruct the annual preceramic subsistence cycle. The Tehuacán research suggested that the problem confronting New World hunters and gatherers about 8,000 years ago was how to cultivate and collect a set of food plants that provided sufficient nutrition and a well-balanced diet through the annual round (MacNeish 1978, 1992).

The results indicated that the early inhabitants of Tehuacán scheduled their seasonal movements to coincide with the periodic availability of local plant and animal species adapted to the riverbanks at the foothills to the mountains (MacNeish 1978, 1992). These Preceramic foragers hunted terrestrial mammals such as rabbits and deer and the supplemented plants in the diet (Flannery 1967, pp. 134–135). During the May–October rainy season, edible plants were more abundant, and a diversity of seeds, cactus fruits, and berries were exploited, in addition to the bountiful seedpods of the mesquite tree. Small game, birds, and lizards, were hunted and consumed at this time, and the band sizes of human groups was generally larger. Although some fruits were still available during the early part of the dry season (January to April), cactus leaves and deer apparently were the staples during the dry season (Fig. 4.9a, b). The primary sources of meat protein among Mesoamerican populations were deer and turkeys, while in the Central highlands and along the Gulf and south Pacific coast they were aquatic resources (Chisholm and Blake 2006; Parsons 2006). Although this way of life persisted for almost 6,500 years, from 8000 to 1500 BC, several important dietary changes did take place (MacNeish 1967c). A wild ancestor of the domesticated squash was used 8,000 years ago, probably as a container or for its protein-rich seeds[10] (Whittaker et al. 1957; Culter and Whittaker 1967; Smith 2005a) (see Fig. 4.1a,b). The radiocarbon and archaeological evidence indicates that squash was domesticated about 4,000 years before maize. Squash was one of the so-called "Three Sisters" cultivated by Native Amerindians, that is, the three main indigenous plants associated with early agriculture: maize, beans, and squash, which were often cultivated together in agricultural fields (Mt. Pleasant 2006).

It was only after this 6,500-year period that domesticated varieties of squash, avacado, zapotes, chili peppers, and the earliest known maize appeared in the Tehuacán Valley (Smith 1967, Table 26; Mangelsdorf et al. 1967, Fig. 96). The archaeological evidence suggests that these early domesticates represented a minor portion of the diet, which largely consisted of wild plants and animals (Flannery 1967, pp. 156–162). It was also during the Coxcatlan Phase that storage

---

[10]Archaeological evidence suggests that squash (*Curcurbita pepo* L.) may have been first cultivated in Mesoamerica ca. 8,000–10,000 years ago (Roush 1997; Smith 1997a).

Fig. 4.9 (**a**) Present day Mexican populations still depend upon hunting to varying degrees for their meat protein. Yucatecan hunters take home their prize. Deer was and still is a major source of meat protein (Courtesy of John Tuxill) (Photograph by John Tuxill). (**b**) Turkeys were essential to the ancient Mesoamerica diet. They were often penned in and kept in villages as well as sold in the Pre-Columbian markets. They along with migratory fowl were important sources of protein in the ancient Mesoamerican diet. These turkeys were from Sayil, Yucatan (Courtesy of Michael D. Carrasco) (Photograph by Michael D. Carrasco)

technologies begin to play a larger role in the subsistence round. Thus, these initial experiments in cultivation that in some cases led toward plant domestication occurred among a population that was largely mobile and remained so for thousands of years (MacNeish 1978, pp. 146–151). The Tehuacán research created a set of ethnobotanical and archaeological data from which to analysis human/ environmental interaction over very long span of time, in a region where maize and other crops appear very early in archaeological sediments (Smith 1967; Flannery 1967). In his report on field research, MacNeish (1967b, Fig. 2) maintained that the archaic diet was primarily made up of plant resources (see also Callen 1967). Their results indicate that the diet composed almost entirely of vegetables, fruits, nuts, and berries, with very little meat protein, other than turkey and a native breed of dog. Tropical lowland and highland zones traded products peculiar to each – cacao from the tropical lowlands and avocados from the highlands (MacNeish 1967, 1978). Later breakthroughs in stable carbon isotope and strontium isotope analysis would provide more precise data regarding paleodiet and the role of certain plants, such as maize in the prehistoric diet (e.g., Tykot 2006; Tykot et al. 2006; Burger and van der Merwe 1990; Tykot and Staller 2002).

The research results on domestication with respect to maize (Z. mays L.) suggests that this was achieved relatively late, about 5,400 years ago and even then it was an unimproved variety, good only for chewing for the juices (Mangelsdorf et al. 1967). Their excavation and survey results indicate that more productive varieties had been developed and adapted to nearly all Mesoamerican climates by 1600 BC (Mangelsdorf et al. 1967, Fig. 103–106; MacNeish 1978, p. 178). The macrobotanical evidence from the various Tehuacán Cave sites indicate a gradual increase in the overall proportion of both wild and domesticated plant foods being harvested (MacNeish 1978, p. 179). The wild ancestors of the major Mesoamerican cultigens – maize, beans, and squash – are all highland plants. Thus, it is not surprising that the earliest archaeological evidence for Mesoamerican agriculture has been found in highland valleys like Tehuacán and Oaxaca (Whitaker et al. 1957; MacNeish 1961, 1962, 1967b; Flannery 1986). The dry caves in these upland valleys are recognized for their superb archaeological preservation. Significantly, some of the earliest sedentary villages in Mesoamerica are established in the coastal lowlands, where the highland cultigens eventually were incorporated into a subsistence economy that featured marine resources and lowland plants (Chisholm and Blake 2006). A single circular pithouse, the earliest in Mesoamerica when reported, was identified in a 5,000-year-old level at an open air site in the region (MacNeish 1978, p. 154). Data from the Tehuacán Valley uncovered very early evidence of cultivation and the adaptations surrounding early plant domestication among societies that remained residentially mobile for thousands of years (MacNeish 1985).

The Tehuacán sequence also reveals an increase in population and a decrease in the residential mobility. Based on the size and number of sites, the total population density for the Tehuacán Valley may have increased several fold during this period, but sedentary villages only appeared 4,000–3,000 years ago (MacNeish 1978, pp. 154–156). The occurrence of such sites coincides with the spread of more productive maize varieties to different sites in the valley. The conventional ^{14}C dates for maize at Tehuacán appear in cave deposits dating to the end of the sixth millennium BC (MacNeish 1967b, 1978). These early ears were no more than 3 or 4 cm in length, with no more than four to eight rows of kernels (see Fig. 4.4 see also Fig. 2.7) (Mangelsdorf 1967, Fig. 103). The highlanders in the Tehuacán Valley adopted maize in small-organized seasonally mobile societies rather than habitations in sedentary villages (MacNeish 1978, pp. 154–155). These groups added maize to their diet without radically changing their social or economic behavior (MacNeish 1978, 1985).

In the last 30 years, the antiquity of many of the proposed food crops in the Americas, as well as their contexts of domestication, have been reexamined and consistently produced more recent dates than had been initially published on the basis of associated dates (Blake 2006, p. 68). The recent direct AMS dates on the early maize cobs from the Tehuacán cave cluster around the mid-third millennium BC is approximately two millennia younger than previously reported (Long et al. 1989; Benz and Long 2000; Blake 2006), and are generally consistent with the reported increase in the population density and changes in settlement patterns (MacNeish 1978, 1999). The AMS dates provide a revised timetable for the arrival of maize and more productive varieties in the highland Tehuacán Valley (Table 3.1).

The ground breaking research in the Tehuacán Valley fostered cooperation among a diverse group of scientists from a variety of disciplines to the question of the origin of maize and agriculture. The project brought together archaeologists, zoologists, botanists, and geneticists to solve the mystery of the origin of maize and in the process, gathered data that generated a cultural sequence of considerable time depth. These data included Archaic occupations in those Preceramic periods when the processes of domestication were changing this arid highland landscape. The multidisciplinary research in the Tehuacán valley generated a large body of data about prehistoric adaptations and interactions with plants and animals, and about the genetic and morphological processes underlying conscious and unconscious human selection (Mangelsdorf et al. 1967).

Research on early agriculture by MacNeish has spanned the tropical lowlands of Belize to semi-arid highlands of northeastern Mexico, as well as cave sites in the Andes mountains (MacNeish 1992; MacNeish et al. 1981). Since most of his research was involved with dry cave sites, the excavations he directed have produced a large body of data, primarily plant parts and ancient tools used for gathering and processing. This difficult research has documented the primary features associated with the adaptive shift to a greater dependence upon domesticated food crops. His results indicate that such developmental processes were for the most part gradual, that is, spanned long periods of time and generally resulted in a shift in settlement toward village agricultural life. There is little in the way of documented evidence to suggest a great leap forward or a rapid transition as envisioned by some archaeologists and theorists in the beginning of the last century. His results also indicate that each region had its own inventory of native plants that varied slightly or greatly from that of other zones (MacNeish 1978, 1992). Therefore, the first steps toward domestication of plants were accompanied by regional trading of plants. By this means, selection of desirable traits and hybridization were accelerated. Eventually, each region of Mesoamerica emerged with a large set of native and imported plants that were suitable to its altitude, rainfall, and soils. Viable combinations of food plants were achieved in some precocious zones by 2000 BC and in most regions by about 1500 BC. Botanists and geneticists working with archaeological collections provided an indispensable foundation for further research on the potential progenitors of domesticated plants as well as the early adaptive changes generally associated with an agricultural economy (MacNeish 1978, 1992).

The introduction of methods and techniques initially developed in the biological sciences and botany greatly influenced the archaeological research on early agriculture after the Tehuacán Valley project and particularly after publication of the fieldwork by Flannery and his associates in Oaxaca in the mid 1980s. The use of Scanning Electron Microscope (SEM) and the identification of plant microfossils had a major impact on the research surrounding early plant domestication, particularly research involving the origins and spread of maize. The Tehuacán Valley research was carried out before such technological breakthroughs, but the systematic and careful excavation and use of flotation, fine screens, and botanical identification were considered major methodological breakthroughs at that time in archaeology (Willey and Sabloff 1980). As evident from the previous chapter, botanists and plant morphologists laid the foundation for later research by archaeologists. They reported on the potential ancestors of an array of domesticated plants and described their behavior and biogeography. Such research also provided a basis for analyzing the mutational steps that led to domestication and specified the associated phenotypic changes. They have also played an important role in reconstructing the prehistoric environments and identified habitats where potential domesticates survived and provided technical identifications of plant remains recovered in ever increasing detail by archaeologists.

4.2.3   Approaches to Domestication and Cultivation in Oaxaca

Later interdisciplinary research in Oaxaca by the University of Michigan, Ann Arbor on the early domestication and cultivation also provided important evidence of early maize and the domestication of various food crops. Kent V. Flannery directed the field research and had previously spent several field seasons as a faunal analyst on the Tehuacán interdisciplinary team. Thus, the Oaxaca research was modeled to varying degrees after the Tehuacán Valley research. The interdisciplinary researchers in Oaxaca focused their efforts on a preceramic cave at Cueva Blanca, and thus, with support from the Smithsonian Research Foundation and National Science Foundation, The Prehistory and Human Ecology in Valley of Oaxaca Project was initiated. Like the Tehuacán Project, the archaeological research in Oaxaca was also first and foremost, focused on the origins of maize.[11] The interdisciplinary research in Oaxaca was also regional in scope involving survey and excavations at a number of sites. Caves and rockshelters were again a focus of excavation because the low soil moisture provided ample macrobotanical evidence of early domesticates and better preservation of the ancient plant remains (Flannery 1986a, 1986d).

---

[11]The research in Oaxaca produced an incredible body of macrobotanical data, including ancient samples of major food crops such as maize.

The small Guilá Naquitz Rockshelter[12] was the focus of extensive excavations since the ground surface remains included chipped stone flakes and half of a projectile point, indicating the presence of preceramic occupation. Flannery and his field crew carefully peeled away the layers of occupation floors, with the earliest dating back to 8750–6670 BC (Flannery 1986d). Based on careful retrieval and analysis of the floral, faunal, and artifactual remains found in the cave strata, they concluded that the rockshelter was occupied during the dry season between August and December.

The archaeological contents of the Oaxaca cave strata indicated that a diversity of plant foods, such as acorns and the roasted beans of maguey plants (the source of tequila and mescal as well as cloth) were exploited and collected from the surrounding thorn forest. In the course of their annual round plant foods, such as mesquite pods and hackberries were brought back to the rockshelter (Smith 1986). A small part of the Guilá Naquitz diet, came from squash (*Cucurbita pepo*) and bean (*P. vulgaris*) plants, which may have been tended or cultivated in the disturbed terrain around the site (Flannery 1986a, p. 6–7). Consumption and cultivation of the wild squash may have been a first step toward eventual domestication (Flannery 1986a, p. 8–9). A variety of nuts, seeds, fruits, and cactus eaten during late summer and early autumn were supplemented by a small amount of venison and rabbit meat (Flannery 1986c, pp. 313–315). Although beer and rabbit bones appeared in small numbers in their excavations, they nevertheless provided much of the protein consumed by archaic and preceramic societies at the Guilá Naquitz rockshelter (Flannery 1986c, pp. 314). It appears that seasonally abundant plant foods may have been collected from the immediate vicinity of the rockshelter (Flannery 1986c). Neither maize cobs nor kernels were identified in these ancient levels (Flannery 1986a,Table 1.1). However, years later a direct AMS date on one of the early cobs at Guilá Naquitz produced the earliest assay recorded thus far in Mesoamerica at 5420 B.P. or dendrocalibrated at 2σ age ranges to 4340–4228 CAL. BC (Piperno and Flannery 2001, Table 1).

The Preceramic collecting strategy was interpreted as a broad spectrum adaptation, associated with an increase of storage facilities to extend the seasonal availability of food crops such as maize and beans [13](Flannery 1986a, p. 13–14; Binford 1980, p. 18, 1989). The greater dependence upon plant resources to the Preceramic diet in these regions is related to biodiversity, the smaller body size of terrestrial mammals in the Neotropics, and the semi-arid climate, which restricts the growing season (Flannery 1986a; Binford 2001). Storage does not appear to become a major

---

[12]The Guilá Naquitz Rockshelter has a small overhang and artifactual surface remains diagnostic of the pre-ceramic and archaic periods (Flannery 1986a). In the Native Zapotec language, Guilá Naquitz means, "white cliff".

[13]Binford (1980) and Flannery (1969, 1986a) emphasize a broad spectrum adaptation in association with procurement strategies in the shift to agricultural production, in both the Old and New World. However, the Neotropics are characterized by greater biodiversity and terrestrial mammals of the smaller body size reflected by a greater dependence upon plant resources than is evident in archaeological remains from the Old World (see Binford 1989, 2001).

factor in extending seasonal availability of plant resources until late in the sequence, in association with the appearance of maize (Flannery 1986a, p. 13).

Archaeologists reported that the archaic occupants of Guilá Naquitz were organized into small groups, or microbands, composed of a series of mobile nuclear families living in several different camps during the course of their yearly activities (Flannery 1986c). The gradual process of interdependence between humans and certain plants related to domestication was recorded in these data. The research also provided valuable evidence on early maize morphology and taxonomy, as well as specified the necessary phenotypic changes that resulted in the domestication process (e.g., Benz 2001). They reconstructed prehistoric environments associated with the various Preceramic layers, as well as, suggested habitats where potential domesticates would have been exploited (Flannery 1973, 1986a–c; Kirkby et al. 1986).

Flotation recovery has become widely employed throughout the Americas, and dramatically increased the recovery of the fragmentary carbonized remains of both wild and domesticated plants from caves and rockshelters, as well as open air archaeological sites in river and stream valley alluvial (floodplain) settings (Smith 1986; Pearsall 1989, 2000; Piperno and Pearsall 1998). Botanists and archaeologists have benefited from the recovery of archaeological plant parts for hypotheses testing issues surrounding crop evolution and biogeography. Interdisciplinary research at Guilá Naquitz provided data on the collection and processing of plant foods and the butchering and consumption of animals, stone tool manufacture, the digging of pits to store acorns, the use of fire pits to prepare food, and even the collection of leaves for bedding in the cave. The subsistence adaptation remained stable and changed very little over the millennia of intermittent occupations (Flannery 1986c, pp. 315–316). These research projects illustrate how archaeologists can reconstruct the events of the past into a detailed picture of ancient life just before the advent of agricultural economies in the highlands of ancient Mexico.

The methodological approaches and macrobotanical collections generated in these two important archaeological projects provided a basis for later genetic allozyme and DNA analysis, which resulted in the identification of the wild progenitors of a number of important domesticated food crops (Smith 1997a; Doebley et al. 2006). Allozyme and DNA analysis comparing modern domesticates and wild populations from sites such as the Guilá Naquitz rockshelter and Tehuacán valley caves and rockshelters have provided evidence of the wild progenitors of squash (*Cucurbita pepo* L.), common beans (*P. vulgaris*), Lima beans (*P. unatis*), and maize (*Z. mays* L.) (Smith 2006). Because they were carried out near the region that maize was first domesticated, their role in understanding the adaptations and processes that underlay the domestication of *Z. mays* have made them critical to archaeologists as well as scholars in the biological sciences. The present day geographical range of these progenitor populations has, in turn, suggested possible centers for the domestication of these crop plants different from those initially identified on the basis of archaeological evidence.

The general focus of research in the highland Valleys of Tehuacán and Oaxaca and the large body of macrobotanical remains have to some extent biased the archaeological record with regard to early agriculture and the origins of maize

(Piperno and Pearsall 1998, p. 31). In the absence of any parallel evidence of early crop food plants from lowland river valley sites, due largely to lack of preservation in the tropical and subtropical soils, it was generally assumed that such plants were first domesticated in upland environments in proximity to caves and rockshelters rather than in the more fertile, better watered soils of riverine settings (Mangelsdorf 1974; Smith 1977, 1986; MacNeish 1978). The more recent use of charred food residues to the identification of early cultigens has been particularly critical to the identification of maize in lowland and early coastal settings (Thompson 2006; Staller and Thompson 2000, 2002; Thompson and Staller 2001).

Excavation in riverine settings has uncovered settlements with deep cultural deposits that appear to have been occupied throughout, much if not all of the year, over a long period, that is, sedentary villages (Flannery 1972; MacNeish 1992, p. 286; Blake et al. 1992). These open-air settlements have produced evidence of early domesticates as ancient and often of greater antiquity than those recovered from upland caves and rockshelters, one notable exception in this regard is maize (Piperno and Flannery 2001). What research in such arid upland valleys as Tehuacán and Oaxaca demonstrated is that these caves and rockshelters primarily represented seasonal camp sites of small family groups (MacNeish 1978, 1992; Flannery 1986b, c). One of the more surprising findings of the Tehuacán and Oaxacan regional surveys was that many early sedentary agricultural societies were still dependent upon wild plants and animals at least to some extent for their subsistence (Flannery 1986c, 2002). This suggested that the transition to food production and a dependence upon food crops was related to some extent on the environmental setting and seasonal availability of resources (Binford 1964, 1965, 1968; Flannery 1986a).

Advances in the analysis of plant remains from archaeological sites reopened the consideration of the temporal, environmental, and cultural context of agricultural origins in the Americas (Ford 1985a; Flannery 1986c; Smith 1986; MacNeish 1992). The results produced a large body of evidence on early foraging adaptations, plant cultivation, and early agriculture, in a more humid upland valley setting (Kirkby et al. 1986, p. 48). The research in the Tehuacán valley and Oaxaca generated a considerable body of the primary data on early plant domestication and agriculture in Mesoamerica. These data essentially formed the basis of much of the archaeological theory and model building on the origins of agriculture as well as the origins of a number of important food crops such as maize up to the present (e.g., Benz 2001; Benz and Long 2000; Smith 1997b, 2000, 2005a; Piperno and Flannery 2001; Blake 2006; Smith and Yarnell 2009). These important interdisciplinary projects established the need for detailed information on plant morphology, particularly early cultigens, as well as large well-preserved macrobotanical remains as critical data for understanding the early agriculture and the process of domestication (Ford 1985a). Consequently, as more archaeological projects incorporated research from ethnobotanists and plant morphologists, and more type collections were generated with flotation recovery, there was an incremental increase in knowledge regarding the morphology of cultivated and domesticated plants and their wild progenitors. The selection process that transformed certain species such as maize into economically productive landraces is as yet not well documented by social and biological scientists (Wilkes 1989, p. 441). The general assumption, until the maize varieties studies initiated by Mangelsdorf (1974) over 40 years ago was that there was a conscious selection to ever more productive landraces (e.g., Wellhausen et al. 1952). However, the conscious and unconscious maintenance of landraces by indigenous farmers over long periods of time challenged their overall assumption of increased grain yield as apparent in the ear of the plant, and how this was reflected in the phylogenetic relationships of the various landraces (Fig. 4.10). The previous chapter made quite clear the problems inherent to using

Fig. 4.10 The general tendency for indigenous farmers to select for ever more productive landraces is complicated by recent linguistic and ethnographic evidence, which indicates the human selection for maintaining, kernel color, and shape, as evident by this nal t'eel variety from the northern Yucatan. The cob has orange-red colored kernels and measures about 10 cm in length (Courtesy of John Tuxill) (Photograph by John Tuxill)

morphological traits of the ear, kernel shape, etc., to reconstruct phylogenies (see, e.g., Doebley 1994; Iltis and Doebley 1980).

These data underscore the need for more exacting standards of evidence of plant domestication throughout the Americas (Smith 2001, 2006). In a landmark study, Harlan (1975) and others have identified both the range of likely morphological markers of domestication in seed plants and the specific human actions that caused these morphological and genetic changes (Dorweiler 1996 Dorweiler et al. 1993; Jaenicke-Després et al. 1993; Doebley et al. 1997; 2006; Gallavotti et al. 2004). These changes from the morphology of wild forms are primarily greater seed size, thinner seed coats, and loss of natural seed dispersal mechanisms such as nonbrittle rachis in maize (Iltis 2006; Doebley et al. 2006).

These changes, when documented by SEM or other microscopic analysis in specimens constitute the primary and essential class of evidence for the domestication of seed plants in the Americas. The Tehuacán Valley and later Oaxaca research projects established that these processes occurred from the beginning of the Holocene and continued after the appearance of sedentary villages, consequently much subsequent research on the origins of food production was focused on sites dating to before 1500 BC where the Amerindian societies brought plants under domestication, and advances in the analysis of plant remains from archaeological sites reopened consideration of the temporal, environmental, and cultural context of agricultural origins in the Americas. The major focus of the literature surrounding early plant domestication and agriculture in the Americas has been centered on maize. As evident from previous chapters of this book, such research was largely inspired by the perception among scholars and scientists that it provided the economic basis for the rise of civilization in the New World.

4.3 Ethnobotanical Approaches to Early Agriculture and Biogeography

Technological innovations such as scanning electron microscopy (SEM), digital enhancement through optical stereology, and accelerator mass spectrometer ^{14}C dating (AMS) of plant microfossils in residues are dramatically changing our understanding of the chronology and biogeography of early agriculture in the Americas. These techniques and technologies have also expanded those regions of the Americas where archaeologists and ethnobotanists have been able to look for evidence of such domesticates and have been particularly influential in our previous understanding of the spread and antiquity of maize in the Neotropics. In the past 30 years, archaeologists and paleobotanists have looked for evidence of such plants, particularly maize, in the coastal regions and tropical forest lowland settings. Intense research for evidence of early agriculture in the coastal and tropical lowland environmental zones of Mexico and the south–central Andes is relatively recent, when compared to the early studies focused on the macrobotanical remains in the Mexican highlands, Neotropics, and American SW (e.g., Sauer 1952; MacNeish 1948, 1958, 1962; Cutler 1952). A number of archaeological sites with associated dates and purported evidence of early plant domestication and spread of agriculture in the Neotropics have been sampled for microfossil remains of plants. These data include pollen extracted from sediments and carbon residues in pottery, desiccated coprolites (coprolites can also contain phytoliths, but their preservation in tropical regions are rare), phytoliths from archaeological sediments, and starch grains, usually identified on the surface of ancient processing tools and is not a new technique in archaeology (Ugent et al. 1984, 1986; see also Thompson 2005, 2006, 2007). However, the identification of food crops such as maize and manioc from the edges of grinding stones (primarily, *manos* and the surfaces of ancient *metates*) is more recent (Rossen et al. 1996; Chandler-Ezell et al. 2006). The vast majority of site microfossil data reported thus far consist of pollen cores and phytoliths from archaeological soils at sites in Mexico, Guatemala, Belize, Costa Rica, Panama, Colombia, and particularly Ecuador.

Despite the relatively poor preservation in such tropical environmental settings, paleoethnobotanists have documented the presence of food plants, particularly economic staples, through the identification of microfossils such as pollen, phytoliths, starch residues and remote sensing (e.g., Pohl et al. 1996; Piperno and Pearsall 1998; Rahman et al. 1998; Hastorf 1999; Hastorf and DeNiro 1985; Pope and Dahlin 1989; Pope et al. 2001; Perry et al. 2006, 2007). This renewed interest in these environmental settings and their associated cultural contexts for domestication and early plant cultivation has also spread to other regions of the Middle America (Horn 2006; Horn and Kennedy 2001; Dull 2006). The questions surrounding ethnobotanical approaches in lowland Neotropical settings, however, goes beyond the problem of preservation to most recently discussing and addressing the strengths and limitations of such approaches[14] (Reber and Evershed 2004; Haslam 2004; Rovner 2004; Holst et al. 2007). Most microfossil data refer to the presence and absence in archaeological sites, thus equating higher numbers of pollen, phytoliths, and particularly starch grains with artifact use, or importance to ancient diet can be problematic at best, since such approaches are generally used to document presence/absence (see Haslam 2004).

When such new technological innovations and approaches were first being developed, they were believed to have the ability to address some of the major scientific questions surrounding plant and animal domestication, and the role of an agricultural economy in the development of social inequality (Rovner 1971, 1983; Pearsall 1979, 2000; Piperno and Pearsall 1998; Hastorf 1999; Perry et al. 2006, 2007; Zarillo et al. 2008). In recent years, questions have arose surrounding ethnobotanical approaches using microfossils, because of context, where they are

---

[14]Haslam (2004:1717) states that the range of starch residue preservation on such grinding stones is between 75% and 80% for buried artifacts and 35% for surface finds according to his experiments.

found, and how they are analyzed and differentiated from other domesticates and their wild progenitors (see, e.g., Staller and Thompson 2001; Staller 2003; Rovner 2004; Rovner and Gyuli 2007). Others have provided evidence which indicates that organic residues, such as starch grains and lipids breaks down starches usually in a short period of time (>100 years) depending upon the soil environment where they are found (Haslam 2004, pp. 1720–1722) and Reber and Evershed (2004, p. 401) state that organic residues even common lipids degrade rapidly and differentially depending upon the buried environment (see also Evershed et al. 1992). They caution that identifying starchy grains, maize lipids from other lipids, and lipids from other starchy grains is problematic, and may be related to the abundance of nonspecific species compounds. Haslam (2004) goes into considerable detail to show how they survive best in soils with high clay contents, and in sheltered environments like caves and rockshelters – precisely those environments where macrobotanical remains are best preserved. Maize produces a great deal of starch residues. Root crops like manioc do not, and when they do, they are very small so they are not easily identified and are often confused with transitory starch grains in the process of decomposition. There are therefore contextual issues in rainy environments they move around, nor are looking for root crops on stone implements very good indicators of what is being processed (Haslam (2004), p. 1727). Reber and Evershed (2004, p. 400) mention that there are no species-specific biomarkers for starchy grains though there are plant biomarkers (see also Evershed et al. 1992). Babot (2003) points out that some starch residues may be damaged, presumably due to processing, making their identification problematic. Researchers are still studying the extent to which this may also be related to the decomposition and transitory starches, thus, decomposition and damage to starch grains in archaeological soils may also be related to soil chemistry or decomposition rather than on how the plant was processed (see Haslam 2004). These methodological constraints with regard to identification and protocol have encouraged ethnobotanists to apply multiproxy methodological approaches in the identification and documentation of food plants in ancient archaeological sites.

Ethnobotanists have long maintained the necessity for interdisciplinary approaches (Ford 1985a, b) using a diversity of methodologies for evidence gathering, citing that the long term emphasis by some scholars of early agriculture upon macrobotanical remains are placing disproportionate reliance on such data (e.g., Fritz 1994; Smith 1997b, 2000, 2001, 2005a; Parsons 2006). Moreover, ethnobotanists have thus far generated important ecological and environmental data obtained from where ancient societies often cultivated crops using such approaches. Sediments from these "off-site" contexts, that is lake and sediment cores, were found to contain identifiable microscopic indicators of human modification of the surrounding vegetation that reflect evidence of former land clearance and agricultural plots (e.g., Colinvaux and Bush 1991). These interdisciplinary approaches to early plant domestication and the early development of food production provided increasing lines of robust evidence, and data in tropical areas of the hemisphere where there is incredibly high species diversity and subsistence alternatives may favor a diverse array of adaptations where organic remains may be modified by paleoclimatic or environmental conditions. Such modifications provide important clues regarding human adaptation and plant domestication (e.g., Colinvaux 1993; Colinvaux et al. 1996a, b, 1997). However, recent genetic research and direct AMS dates of carbon residues have challenged previous phylogenies and early associated dates obtained from pollen and phytolith research (Tables 4.1, 4.2). This is particularly true with regard to the identification of maize and the ability to differentiate *Z. mays* L. microfossils from other wild grasses (Staller 2003; Rovner 2004; Staller and Thompson 2000, 2002; Matsuoka et al. 2001; Freitas et al. 2003; Vigouroux et al. 2002, 2003; Holst et al. 2007).

4.3.1 Classes of Ethnobotanical Evidence

Ethnobotanical evidence involving plant microfossils associated with tropical food production are primarily of four types: (1) Botanical remains from archaeological deposits, both macrofossil (seeds, tubers, wood, plant remains, and plant fragments) and microfossil (pollen, phytoliths, and starch grains); (2) Vegetational records obtained from perennially humid regions in the Neotropics (primarily pollen and phytolith evidence), from lake cores, also from bogs and swamps, usually with associated evidence of archaeological remains in the immediate vicinity; (3) Evidence of plant microfossils in charred organic residues or starch grains from ancient ceramic pots or processing tools; and (4) Molecular markers that provide evidence that extant crop plant species are genetically derived from a particular wild ancestor.

The genetic evidence has been found to be very detailed in terms of the phylogenetic relationships between domesticated and wild plants, and in recent years, particularly with respect to maize origins, have set the limits of where a particular crop plant was originally cultivated or modified by human and natural selection, and if these modifications occurred more than once in prehistory (see, e.g., Matsuoka et al. 2002; Jaenicke-Després et al. 2003; Jaenicke-Després and Smith 2006). Some ethnobotanists asserted that the records of changes in vegetation from cores and sediments in the Neotropics at the close of the Pleistocene can serve as proxies for the resource density and distribution over time and provide a basis for estimating the degree of dependence on certain food crops during the late Pleistocene and early Holocene periods (Pearsall and Piperno 1998, p. 31). Different classes of botanical remains provide different kinds of evidence. Differences in the character of the different classes of plant remains relate mainly to the deposition and preservation of such microfossils and how analysts evaluate such data as indicators of early agriculture. Contradictions among various lines of ethnobotanical evidence as indicators for the antiquity of agriculture in the Neotropics are, in large part related to the recovery bias and chronological ambiguities regarding the associated dates (Smith 1998; Staller and Thompson 2002; Rovner 2004). Some ethnobotanists have suggested that recovery bias was reflected in the excellent preservation of organic materials in dry caves, as is evident from the previous discussion of the research in the Tehuacán Valley and Oaxaca. However, the paleoethnobotanical recovery bias can also be related to the focus on subsistence plants to exclude other plants in the archaeological remains. Although more recent evidence has indicated that such data can serve as an inferential evidence of the presence of certain food crops and agricultural activity, recent questions regarding their identification and differentiation from other related species and/or with wild progenitors, and contextual integrity have arose, particularly regarding associated dates versus directly dated microfossils derived from sediment and lake cores (Smith 1998, 2001; Staller 2003; Staller and Thompson 2001; Rovner 2004; Piperno et al. 2007; Holst et al. 2007). Contextual issues related to the recovery of macrobotanical remains in dry caves is related to the fact that undigested food and organic remains in general are potential foods for burrowing animals and therefore can be repositioned in archaeological deposits.

4.3.2   Pollen Analysis and the Spread of Early Cultigens

Ethnobotanical approaches involving plant microfossils are generally based on paleoclimatic and paleoecological reconstruction, initially dealing with geological time scales, and in the case of food crops, from the early Holocene (Horn 2005, Fig. 27-4; see also Schoenwetter and Smith 1986; Wright et al. 1984; Sluyter and Dominquez 2006). Fossil pollen analysis involves the study of morphologically distinct microscopic pollen grains, which refer to ancient plant assemblages (Fig. 4.11). Occurrence is directly measured by the presence and/or ubiquity of particular microfossils at archaeological sites over time. However, the shift to interdependence on certain food crops is generally associated with the landscape modification related to the increase in cultivation yields (Pearsall 1994, p. 269). Ethnobotanic evidence from pollen analysis on the spread of maize and other food crops from southeastward Mexico have exponentially increased in recent years with the identification of fossil pollen from lake and sediment cores as well as archaeological soils. The spread of early domesticates in cases where the progenitor was known has made such reconstructions relatively straight forward compared to maize (see Sluyter 1997; Sluyter and Dominguez 2006).

The quantitative standards for differentiating maize from other wild grasses were initially established by Whitehead and Langham[15] (1965). Subsequently, Sluyter (1997) conducted experiments to neutralize the effects of mounting medium

---

[15]Whitehead and Langham (1965) established the various size ranges for maize, teosinte, as well as gamma grass or *Tripsacum* using modern specimens mounted in silicone oil. Whitehead and Sheehan (1971) also developed protocols for identification of maize pollen using measurement. Later, Sluyter (1997) conducted experiments in an attempt to normalize the effects of microscopic slide mounting media in order to facilitate comparative analysis of maize pollen grains mounted in silicone oil, glycerine jelly, and a new type of acrylic resin mounting medium.

Fig. 4.11 SEM image of miscellaneous pollen from common plants such as sunflower (*Helianthus annus*), praire hollyhock (*Sidalcea malivflora*), and morning glory (*Ipomea purpurea*). Cross-pollination in maize occurs when wind-borne pollen is carried from the tassels of one plant to the silks of another and affect the phenotypic characteristics of the maize plant

on the size of pollen analyzed from slide samples. A somewhat limited understanding of the distributions of pre-Columbian populations of teosinte complicates the archaeological and paleobotanical reconstructions of early agriculture and the spread of maize from sedimentary pollen. In the modern period, subspecies of teosinte range as far south as the Gulf of Fonseca in northern Nicaragua (Iltis and Benz 2000; Matsuoka et al. 2002). If the Pre-Columbian teosinte populations had a geographic range that closely reflects what exists in the present, then pollen analysis in the regions of Central and South America should be less problematic, however, the lack of preservation of pollen in some of these regions requires other approaches such as starch grain and phytolith analysis be employed (see, e.g., Pearsall 1979; Holst et al. 2007; Piperno et al. 2009).

A primary concern for pollen analysts working in areas where teosinte is known to have ranged is whether ancient maize pollen may in fact represent the teosinte pollen (Dull 2006, pp. 358–359). Ancient maize microfossils can hypothetically produce pollen grains that are in the size range of modern teosinte (Mangelsdorf 1974; Mangelsdorf et al. 1978; Beadle 1981). Pollen grains extending back to the middle Holocene have been identified as maize (Rust and Leyden 1994; Pope et al. 2001). The relatively large size of maize pollen grains compared with most other pollen makes documenting their presence in sediment samples easier, but some researchers have suggested that the size of maize pollen may be related to cob size (Galinat 1961; Mangelsdorf et al. 1978; Beadle 1981). Given that the early maize cobs found in caves in the Tehuacán Valley and Guilá Naquitz rockshelter, were relatively small compared to later domesticated samples, the probability that pollen grain sizes overlap with teosinte subspecies would hypothetically be increased, thus making such data at best problematic for scholars studying the origins of the food crop (Dull 2006, p. 359; Sluyter and Dominquez 2006). Moreover, some maize varieties are selected to maintain small cob size and particular kernel color (see Fig. 4.10). More recently, researchers have found that pollen grains from teosinte overlap in size with those of maize to a much greater degree than had been previously reported, making the differentiation of wild (teosintes) and domesticated maize in palynological studies difficult. Holst and her associates (2007, pp. 17608–17609, Table 1) recently examined a large number of modern pollen grains and starch granules of teosinte (wild *Zea* spp.), maize (*Z. mays* L.), and closely related grasses in the genus *Tripsacum* to assess whether existing protocols were useful for studying the origins and early dispersals of maize. Their research later indicated that there is no valid method for separating maize and teosinte pollen on a morphological basis. Thus, analysis of fossil pollen data pertaining to the origins and early spread of maize is no longer tenable using existing protocols, since pollen grains of fully domesticated maize (*Z. mays* L.) can overlap in size with those of the teosintes (*Z. mays* ssp. *parviglumis* Iltis and Doebley), (*Zea perennis* Hitchc.),[16] and other *Zea* subspecies (Mangelsdorf 1974, pp. 182–183; Horn 2006, p. 368; Holst et al. 2007). This is also the case with *Tripsacum* and teosinte and maize, as these are the three New World grasses that have the largest long axis pollen diameters (Dull 2006, p. 358). Pollen researchers distinguish *Zea* from other grass genera on the basis of the maximum length diameter of the grain using light microscopy and more recently SEM analysis.

Although large maize pollen size is advantageous when counting and identifying grains under a microscope, this size difference directly affects the dispersal of the grain and may affect its representation in pollen records. Since maize is wind pollinated, its comparatively heavy pollen grains do not generally travel far under normal atmospheric conditions; and some researchers have reported that over 90% of maize pollen grains would disperse within 60 m of the parent plant, or only common in lake core sediments if maize were cultivated in the immediate vicinity of the lake shore (Raynor et al. 1972, p. 425; Islebe et al. 1996). Robert Dull (2006, p. 359) has observed that there is no single standard for maize pollen identification, that is accepted and practiced by all pollen analysts in the published literature on fossil *Zea* pollen. This has resulted in multiple *Zea* pollen identification procedures and has also made some claims on maize pollen, particularly those in the very early archaeological deposits are questionable or problematic at best. Many archaeologists have rather unwittingly accepted the past and recent claims of the antiquity of maize, and consequently an agricultural

---

[16]The teosinte subspecies *Zea mays* ssp. *parviglumis* was initially identified by Iltis and Doebley (Iltis 2000). The perennial teosinte subspecies (*Zea perennis* Hitchc.) was identified by Mangelsdorf and Reeves (Mangelsdorf 1974).

economy on the basis of pollen data that were based on questionable *Zea* pollen identification criteria (Robert Dull (2006, p. 359). Horn (2006, p. 368–369) has suggested that the reliability of pollen-based reconstructions on the spread of food crops such as maize would be greatly enhanced by controlled studies in which fossil pollen cores would be prepared and measured in the same laboratory using the identical chemical procedures, mounting medium, microscope, and measuring system. Although such standardization in methodological procedure would provide much greater reliability in the pollen identification particularly maize pollen, they would not resolve the problem of differentiating maize pollen from those of other related grasses.

Robert Dull has posed the important question of whether scholars can assume that the existing biogeography of the teosintes clearly reflect their prehistoric distribution – that it may be inappropriate to use the modern distributions of the teosintes, *Tripsacum,* and other related grasses (cf. Horn 2006, p. 369). This is particularly problematic in Tabasco and the Veracruz lowlands where Colonial and modern European landscape alteration, particularly the introduction of cattle, have dramatically transformed the landscape.

As more reference collections of various wild grasses are Cataloged and ethnobotanists learn more about the modern biogeography of maize and its relatives, it may be necessary to reconsider earlier identifications of maize pollen and to reassess its relative antiquity in different regions of Mexico and Central America. All such ethnobotanical evidence with regard to maize assumes that large grass pollen grains, which have been shown to have considerable overlap in size range with maize pollen, represent maize. Thus, recently published associated radiocarbon and directly AMS dated microfossils asserted to represent that maize are somewhat problematic (e.g., Pope et al. 2001; Arford and Horn 2004; Sluyter and Dominguez 2006). The pollen evidence of maize in highland Oaxaca at Guilá Naquitz is dated to about 6980 B.P. (Schoenwetter and Smith, 1986, p. 229). Piperno and her associates (2007, p. 11874) obtained similar dates for pollen cores taken in the central Balsas River drainage (Table 4.1).

The maize cobs from Oaxaca do not have all of the morphological characteristics of domesticated maize (Benz 2001; see also Staller 2003). Previous and more recent pollen core studies indicate the presence of *Z. mays* L. pollen on the Gulf Coast of Veracruz and Tabasco, Mexico between 5000 and 7000 CAL B.P. (Goman and Byrne 1998; Pope et al. 2001; Pohl et al. 2007), in the Peten lowlands of Guatemala and Belize at ca., 5500 and 4600 CAL B.P. (Pohl et al. 1996), and in central Honduras by 4500 CAL B.P. (Rue 1987). Dated early pollen records from the Pacific coast region of Costa Rica record the maize pollen at ca. 5500 CAL. B.P. (Arford and Horn 2004; cf. Dull 2006, Fig. 26-2). This is not to imply that the application of sedimentary pollen core analysis or phytoliths from archaeological sediments cannot provide compelling evidence to our understanding of early maize biogeography when maize pollen occurs with other agricultural indicators such as charcoal (evidence of slash and burn) and plant species adapted to disturbed environmental settings, these data taken together provide evidence of early agriculture as well as maize cultivation (Fig. 4.11).

4.3.3 Phytolith Analysis and Maize Biogeography

Various specialists in the biological sciences and paleobotany have, until the recent evidence from molecular biology, promoted the hypothesis of multiple domestication events for maize in different regions of the Neotropics. These studies have promoted the belief that modern maize lineages evolved from a variety of wild grass progenitors at different times in prehistory. However, this is untenable with the advent of the recent results from molecular biology. Recent genetic research has indicated that paleobotanists are now able to identify distinct maize lineages or clades statistically (Thompson 2006, 2007; Hart and Matson 2009). These techniques initially involved testing both macrobotanical samples and plant micro-fossils, and comparing them to modern reference samples. These data will, in the near future, permit specialists to retrace the evolutionary relationships of distinct races of maize through time and space, and ultimately answer the important questions on how and where the various clades or landraces diversified. Phytolith analysis on the origin of maize and its dispersal to different regions of the Neotropics now comprises a significant and widely cited literature. Opal phytoliths are composed of amorphous silica exuded by plants (Staller and Thompson 2002, p. 34).

Plants in the natural world take up monosilicic acid from the soil in the process of obtaining nutrients through their roots. While most nutrients are absorbed as organic compounds used by the plants, silica is not. Plants deposit silica within and between cells in a variety of forms (Thompson 2006). Ethnobotanical studies involving plant microfossils were initially concerned with the classification of phytolith taxonomies as indicators of past environments (e.g., Rovner 1971, 1983; Piperno 1985, 1988, 1991). One of the critical aspects of any analysis of opal phytoliths is the development of taxonomy, by which to classify the various microfossils. Comparing phytolith assemblages recovered from modern lineages of maize, archaeological cobs, as well as food residues requires a phytolith taxonomy flexible enough to allow description of the types of phytoliths recovered from the maize chaff from a number of genetic and environmental backgrounds (see Thompson 2006, 2007). Several phytolith taxonomies have been generated on the basis of morphological features, plant type, and tissue of origin. In an effort to develop a taxonomic scheme useful in classifying the disaggregated assemblages of phytoliths, Mulholland and Rapp (1992) classified phytoliths recovered from Graminae based solely on their morphological characteristics. This classification scheme proved effective in describing phytoliths recovered from sediments and plants, allowing the statistical comparison of assemblages. This microfossil research was later modified into a functional classification scheme, incorporating more three-dimensional variation observed through analysis of literally thousands of maize cob phytoliths (Thompson 2006). Mulholland and Rapp (1992) initially generated the taxonomy for the identification of the silica bodies associated with grasses based on a three-dimensional morphology observable through microscopy. Phytolith categories were generally broken down into subcategories through more detailed analysis of morphological traits. Thompson redefined the subcategory of rondels published by Mulholland and Rapp (1992) somewhat differently and in greater detail (Thompson 2007; Thompson and Staller 2001).

Lipids have also been used to analyze food residues, as have carbon isotopes, and both methodological approaches have been used to trace the presence of maize (Hastorf and DeNiro 1984; Heron et al. 1991; Letts et al. 1994; Reber and Evershed 2004; Reber et al. 2004). Food residue analysis has been conducted in various ways to derive information about the uses of pottery and the existence of plants in archaeological contexts. Maize flour has been shown to contain abundant silica bodies from maize chaff. Maize cob chaff can be found in food residues of ancient pottery, resulting in more reliable, if not unquestioned, cultural context (Staller 2003; Thompson and Mullholland 1994; Thompson and Staller 2001). Rovner (1983) reported early on that a method of phytolith recovery called "dry ashing" involved the incineration of the portion of the plant from which the phytoliths were to be obtained, that is the glumes and cupules of the maize plant produce abundant silica. Thus, opal phytoliths can withstand the heat of cooking, and may therefore be derived from carbon residues sometimes present in ancient cooking pots (Thompson 2006, p. 83).

Methodological approaches developed by Thompson and Mulholland (1994) and associated with the identification for rondel phytoliths produced by maize inflorescences, that is the cob chaff assemblage have generated "profiles" identified in food residues in ancient pottery (Thompson 1993, 2005, 2007; Thompson et al. 1995; Hart and Matson 2009). The methodology involved in the identification of cob phytoliths (rondels) from carbon residues in pottery has been used in archaeological contexts in both North and South America with some success (see, e.g., Reber 2006; Thompson 2005, 2006, 2007; Thompson et al. 1995; Staller and Thompson 2000, 2002; Hart et al. 2003, 2007a). Dorwieler and Doebley (1997) have demonstrated that silica deposition in the chaff (cupules and glumes) of maize cob is under genetic control – including subspecific variation in deposition (see also Dorweiler 1996; Dorweiler et al. 1993; Thompson 2007; Staller and Thompson 2002, Table 2). Since silica deposition from maize cob cupules and chaff phytoliths differs on a subspecific basis, these characteristics have been used to identify maize at the subspecies level, in other words, differentiate different landraces as well as maize from its progenitor teosinte as has recently been demonstrated by various ethnobotanists and archaeologists (Thompson 2007; see also Chávez and Thompson 2006; Hart et al. 2003, 2007a; Laden 2006; Lusteck 2006; Hart and Matson 2009). Since the *tga1* gene has major effects on phytoliths present in the flowering parts of the plant, and, unlike maize, would show a wide range of different variants among teosinte, the rondel "profiles" should be distinctive between teosinte varieties. Thus, the "configuration" of the rondel profiles of teosinte samples should reflect their biological classification (Hart and Matson 2009, p. 75). The protocol for phytoliths assemblage profiles from maize and non-maize grass inflorescences developed by Thompson (2007) are based on the comparison of archaeological phytolith assemblage "profiles" with modern phytoliths from maize cob cupules and chaff, in which silica deposition differs on a subspecific basis (Thompson 2007; see also Dorweiler and Doebley 1997; Dorweiler et al. 1993). The methodology employed by Thompson use over 200 variables that are categorized and classified using multivariate statistical analysis and can differentiate between maize and non-maize grasses as well as maize and teosinte (Thompson 2007; Hart and Matson 2009).

Pearsall and Piperno (Pearsall 1978, 1979; Piperno 1984, 1988) initially reported a three dimensional morphology classification technique based on the identification of phytolith forms reported to be only produced by maize, and focused on extra large crosses from the leaves of grasses. The classification and protocol developed by these researchers was initially based on the leaf phytoliths identified in archaeological sediments rather than on the cob phytoliths or rondels from carbon residues in ancient pottery.[17] Their pioneering research with plant microfossils was first and foremost focused on maize biogeography. Using cross-shaped phytoliths, described as unique to maize, they identified maize microfossil in early archaeological contexts in lowland Central America and Northwestern South America (Pearsall 1992, 1999). Fundamental to the interpretations regarding early spread of maize into South America is the rapid radiation of this cultigen into lowland Central America. Piperno and Pearsall (1998) report a very rapid spread from its origin in Southwestern Mexico to the lowlands of Central America, and is well established in coastal Ecuador by the early formative pottery culture of Valdivia by 5400 B.P. (Pearsall 1999, 2002, Pearsall and Piperno 1990, 1993; Piperno et al. 1985). Their research indicated that maize spread early on into South America and is based on relatively few phytolith forms, recovered from archaeological soils at a variety of early pottery sites in this region, particularly at Real Alto, Loma Alta, and the site of San Isidro. In phytolith assemblages, obtained from archaeological soils, any silica-producing, existing or ancient plant in the vicinity of the site must be considered as a potential source for a portion of the recovered assemblage. The identification of phytolith assemblages in food residues, however, provides greater contextual reliability as one could reasonably expect to identify only plants that were actually cooked or processed in the pots, greatly reducing the number of plants which need to be examined in a given study (Thompson 2007; Staller and Thompson 2002). Furthermore, methodological and contextual weaknesses inherent to their approach have been a subject of contention by various researchers specialized in the origins of maize (see, e.g., Fritz 1994; Smith 1998; Rovner 2004; Russ and Rovner 1992; Staller 2003). For example, some ethnobotanists have stated that phytolith typologies utilizing the three dimensional morphology technique for cross-bodies could not be replicated but can be misinterpreted statistically with other grasses[18]

---

[17]Pearsall and Piperno previously focused on bilobates, cross-bodies, and critical cross-body variants or subtypes, thought to be exclusive to maize but have more recently been found to be present in other grasses (see e.g., Pearsall and Piperno 1993, pp.14–15, Tables 1–4; Piperno and Pearsall 1993, Table 7).

[18]Wild grass reference data published by Piperno (1988, Table 3.3) also appears to indicate a lack of replicative precision. The mean values for Variant 1 crosses the four Panama populations of *Cenchrus echinatus* range 13.3–14.0 microns while those from the Belize population has a *mean* value of 15.1, significantly outside and above the *range maximum* of the four Panama replicates (see Piperno 1988, p. 76).

(Dolittle and Fredrick 1991; Rovner 1995, 2004; Rovner and Russ 1992; Russ and Rovner 1989). Dolittle and Fredrick (1991, pp. 182–183) report that they could not find cross-bodies in their reference samples, stating that the definition provided by Pearsall obfuscates rather than clarifies the identification of bilobate and cross phytoliths (see also Staller 2003, p. 374). Pearsall and Piperno (1990, pp. 330–331) noted the statistical overlap in their discriminant-function values from the Validivia site of Real Alto and attributed this to the presence of a specific wild grass, *Cenchrus echinatus*. This wild grass is found in coastal Ecuador that is still used in some areas as roof thatch. They state that the decay of roof thatch from *C. echinatus* "easily could mask light maize occurrence resulting from decay of husks or cob residue" (Pearsall and Piperno 1990, p. 131). Pearsall (2000) subsequently modified the three dimensional morphological algorithm when it was found that some cross variants could statistically overlap with wild grass species (see also Piperno et al. 2001). Piperno et al. (2004) later reported that only one of their subtypes, variant one cross phytoliths, provided a clear distinction between maize and other wild grasses. Variant 1 cross-bodies are reported to have the largest sized mean width of the eight variant types identified by these researchers in various publications. The implication is that as in maize pollen, the size is the primary characteristic defining maize leaf phytoliths.[19] In response to these problems in the classification of phyotliths from archaeological soils, Pearsall et al. (2003) identified maize microfossils using rondels or cob phytoliths from archaeological soils (rather than from carbon residues in pottery) and most subsequent research by these ethnobotanists and their associates has involved cob phytoliths from either archaeological soils or starch grains in residues or on grinding stones (see also Pearsall 2003; Piperno 2006; Chandler-Ezell et al. 2006; Zarillo et al. 2008).

Another perhaps more serious problem regarding the use of bilobates and cross-body variants and more recently "wavy top" rondels, cob phytoliths from archaeological soils has to do with contextual reliability and relative antiquity as measured by associated dates rather than direct dates (Staller 2003, p. 374; Thompson and Staller 2001, pp. 8–9; Staller and Thompson 2002, p. 34; see also Fritz 1994; Smith 1998; Blake, 2006). Since such leaf and cob phytoliths are derived from archaeological sediments and dated by association, the ^{14}C dates reported thus far from many sites, have now been shown to be in some cases several thousand years earlier than the directly dated macrobotanical remains reported from the sites such as Guilá Naquitz (see Tables 3.1, 4.2). The mixing of sediments and movement of phytoliths

---

[19]Piperno (1988) made similar claims for mean values of cross-body and bilobate phytoliths of South American maize. With respect to Variant 1 cross-bodies – which are reported to have the largest sized mean width of the eight variant types, Piperno states: "*It is clear that from these data and analysis of single specimens of many races...that the production of numerous Variant 1 crossbody shapes with mean sizes between 12.7 and 15 um is a fundamental characteristic of Central and South American maize leaves.*" (Piperno 1988, p. 78). This would imply that mean size values for primitive maize are as large or larger than the mean value for modern maize. Piperno et al. (2009, p. 5023, Table 1) make similar claims for their phytolith assemblages from Central Balsas at Xihuatoxtla rockshelter.

in archaeological soils appears to also pertain to the macrobotanical remains from such contexts. The disparity between directly dated macrobotanicals and such remains by association has been a matter of some concern to researchers attempting to understand the biogeography of maize (Blake 2006, Tables 4.1, 4.4, Figures 4.1, 4.3 Benz and Staller 2009). Zarillo et al. (2008, p. 5007) report finding rare carbonized maize kernels in lower levels of the Early Formative site of Loma Alta in coastal Ecuador and obtained a 2730 CAL B.P. [Beta-103315], which they interpreted as too young[20] (see Table 3.1). They state that, "there was some mixture of small remains between the occupation layers" (Zarillo et al. 2008, p. 5007). In order to alleviate these contextual and chronological concerns some researchers are now dating starch grains from grinding stones (Zarillo et al. 2008, Table 1; Perry et al. 2006, 2007; Chandler-Ezell et al. 2006). The possibility of multiple domestication events is no longer tenable since the publication of the maize DNA research by Matsuoka and his associates (2002). Although these authors have placed the origins of maize at about 9,000 years ago, the earliest direct dates on maize macrobotanical remains are dated to c. 5400 B.P. (c. 6200 CAL B.P.) (Matsuoka et al. 2002, p. 6084; see also Piperno and Flannery 2001, Table 1). Michael Blake (2006) has demonstrated in considerable detail using the existing radiocarbon evidence the inconsistencies regarding the earliest presence of maize, based on dates by association (see also Bruhns 1994; Fritz 1994; Smith 1998). Since rondel phytoliths in carbon residues can be directly dated, they provide more precise chronological information on the spread of maize than can be obtained from associated dates of phytoliths found in archaeological sediments (Tables 4.3, 4.4) (see, e.g., Staller and Thompson 2002; Chávez and Thompson 2006; Hart et al. 2003). Moreover, rondel phytoliths can also be taken from dental calculus of ancient skeletons, providing another independent line of evidence and basis for contextual

Table 4.4 Calibrated conventional and AMS dates from La Emerenciana

Sample data	Corrected ^{14}C Age B.P.	Calibrated ^{14}C 1-δ Age Range B.C.
Beta-125106	3720 ± 40 B.P./3700 ± 40 B.P.	2137–1979 cal B.C.
Beta-125107	3810 ± 50 B.P./3860 ± 50 B.P.	2240–2201 cal B.C.
^{14}C No. SMU-2225 (charcoal)	3707 ± 148 B.P.	2288–2245 cal B.C.
^{14}C No. SMU-2226 (charcoal)	3400 ± 220 B.P.	1941–1428 cal B.C.
^{14}C No. SMU-2241 (charcoal)	3361 ± 246 B.P.	1935–1323 cal B.C.
^{14}C No. SMU-2563 (charcoal)	3775 ± 165 B.P.	2459–1922 cal B.C.

Note: SMU conventional and Beta AMS dates are corrected for ^{13}C/^{12}C fractionation and calibrated using Calib 4.1.2 (Struiver et al. 1998), with a minus 24-year Southern Hemisphere atmospheric sample adjustment and are at the one sigma range. All ^{14}C assays are taken from Stratum 5 (Floor 2) except the 3361 ± 246 B.P. (SMU-2241) date, which is from the uppermost layer Stratum 6

---

[20]The site of Loma Alta pertains to the Valdivia culture and has Early Formative Period occupations spanning to between 5350 and 4240 years ago with some of the earliest ceramics along the eastern Pacific. However, Loma Alta is multicomponent, with a large Guangala Phase (c. 2350–1500 B.P.) site on its eastern periphery (see Staller 2001a, Table II).

integrity (Ugent 1994, pp. 217–218; Thompson and Staller 2001, pp. 8–9; Staller and Thompson 2002, pp. 35–36, 38–40; see also Pearsall et al. 2003).

4.3.4 Interdisciplinary Approaches to the Analysis of Phytolith Assemblages

Direct AMS dating of food residue samples routinely includes carbon isotope analysis, which has proven useful in recognizing the remains of different types of C_4 grasses in such residue samples (see, e.g., Hastorf and DeNiro 1984; Kelly et al. 1991; Staller and Thompson 2002; Hart et al. 2007b). Since grasses "breathe" in different ways, they accumulate δ ^{13}C from the atmosphere differentially (Kelly et al. 1991). The amount of δ ^{13}C deteriorates at a known rate after death, resulting in what are referred to as C_3 and C_4 plants. C_3 plants accumulate relatively less δ ^{13}C during their lives than do C_4 plants (Schwarcz 2006; Morton and Schwarcz 2004). The amount of δ ^{13}C present can be measured after carbonization of plant remains, and is not affected by charring (DeNiro 1987). Morton and Schwarcz (2004) developed an algorithm for estimating the percentage of C_4 plants in a cooking residue sample that assumed a direct linear relationship. Using large number of residue samples on pot sherds from southern Ontario, they concluded that C_4 plants, specifically maize, were not commonly found in the cooking pots. However, in contrast to the $^{13}C/^{12}C$ ratios from the cooking pots, human bone collagen suggested maize was significant to the diet. They hypothesized that maize was eaten in other ways besides being cooked in pots. John Hart and his collaborators (2007b) have, however, challenged the assumption of a linear relationship between the δ $^{13}C/^{12}C$ values and the percentage of C_4 plants cooked in the pot because in several previous analyses, maize cob phytoliths (rondels) were identified in the residues even though the $^{13}C/^{12}C$ values were very low (see, e.g., Staller and Thompson 2002, p. 38, Table 9a; Hart et al. 2003, Table 1). Hart et al. (2007, pp. 809–811, Fig. 6a,b, 7) have shown that with 60% maize in the cooking pot, the $^{13}C/^{12}C$ values ranged from about $-28‰$ to $-14‰$ in their samples depending on whether the maize is dry or fresh, or if it is cooked with deer meat, wild rice, or *Chenopodium*. When the maize is dry (flour), its value of about $-28‰$ would lead to the conclusion that little if any maize was cooked in the pot, while in the case of fresh (fresh kernels) maize its value of about $-14‰$ would indicate that maize constituted between 60% and 90% of the food being processed (Hart et al. 2007, pp. 809–811, Fig. 6a,b, 7, Table 1). Ironically, in either of these scenarios, one might note the same numbers and types of maize phytoliths in the carbon residues (Hart et al. 2007b). These researchers conclude that phytoliths are important in documenting the presence of maize in the cooking pot, but that the issue of how much maize was being cooked, let alone how significant it was in the diet based on stable isotope measurements of cooking residues alone (Hart et al. 2007b, p. 811). While phytoliths, pollen, and starch grains can document the *presence* of maize, only coprolites and microfossils from dental calculus or carbon residues of cooking pots or *ollas* are direct indicators of what was actually being processed and/or consumed (Thompson 2005; Staller and Thompson 2002, p. 37). Food residue phytolith assemblages are culturally created artifacts, unique in their characteristics and analysis of the charred encrustations on the interiors of ancient pottery, coprolites, or calculus deposits removed from teeth provides a set of approaches and challenges that are different from those faced when using assemblages of phytoliths from sediments (Thompson 2006, 2007).

Recent research by John Hart and R. G. Matson (2009) using cob phytoliths and testing the Thompson protocol have demonstrated that they can statistically discriminate between maize and non-maize types. These authors also stated that they found the Thompson (2007) protocol based on 209 variables rather cumbersome. Moreover, Hart and Matson (2009, p. 75) found that it is not primarily focused on looking at maize and non-maize grasses and that the Thompson references are primarily geared to comparison with modern maize profiles. This is a critical point because the utility of this protocol would be infinitely more valuable if it is able to make such distinctions rather than solely identify maize microfossils from other grass phytoliths. Applying statistics using Euclidean distance scaling and discriminant function analysis, they reduced the variables to seven, including three sets of morphological size variables (Hart and Matson 2009, pp. 77–79, Tables 2–4). They conclude that the seven variables identified by their stepwise discriminant analysis produced results largely similar to those in the original cluster analysis using 209 variables (Matson and Hart 2009, p. 81). The implications of these results are that the Thompson (2007) protocol for rondel phytoliths can be replicated, which was not the case for earlier methodologies with cross-types (see Doolittle and Fredrick 1991; Staller 2003; Rovner 2004). Hart and Matson (2009, p. 82) caution that their statistical protocol should not necessarily be used as a replacement for the initial protocol using cluster analysis with 209 variables.[21] These authors are now in the process of carrying out blind tests to determine if other ethnobotanists can replicate their statistical analysis. If proven successful, such data will have broader implications for the analysis of cob phytoliths from food residues in ancient pottery for our understanding of the origins and biogeography of early maize.

4.3.5 *Ethnobotanic Approaches to the Origins of Maize: Central Balsas*

Recent cob phytolith research has been suggested to be effective in discriminating the female reproductive structures of maize, teosinte, and *Tripsacum* (Pearsall et al. 2003; Piperno 2006). These researchers have now focused on the ruffle and wavy-

---

[21]Hart and Matson (2009:83) also state that these statistics further support direct dates of 2270 ± 35 B.P. at the Vinette site in New York State – the earliest recorded date on maize microfossils in NE North America. They assert that they also support a continual presence of *Zea mays* L. in this region thereafter (see Hart et al. 2007a).

top cob phytoliths found in maize husks and tassels and to a lesser extent on other structures, as well as or in combination with starch grain analysis in their most recent fieldwork (Pearsall et al. 2003; Piperno et al. 2001; see also Pearsall 2003; Chandler-Ezell et al. 2006). Piperno (2006, p. 56) independently examined a broader array of morphological characteristics to identify large-sized cross-bodies and using discriminant function analysis, classified eleven assemblages of archaeological phytoliths as either pertaining to maize or wild grass, maize and non-maize grasses[22] (Piperno 2006, pp. 145–148). The most recent ethnobotanical research has found that cob phytoliths and starch grains protocols are more productive than pollen in discriminating the teosintes from maize and therefore have application to the study of its origins and early maize biogeography (Piperno et al. 2007, 2009; Holst et al. 2007). The utility of such methodological approaches has been said to be related to human selection involving the improvement of plant productivity, food quality, and to facilitate food preparation (Holst et al. 2007, p. 17612). It is already known that *tgal* gene for soft glume architecture, plays an important role in phytolith formation and morphology in both wild and domesticated *Zea*, and is a factor for the morphological differences in such microfossils (Dorweiler 1996; Dorweiler and Doebley 1997; Dorweiler et al. 1993; Thompson 2006; Piperno 2006; Hart and Matson 2009). The most recent research has, in part by necessity, applied multiproxy microfossil protocols to investigate the earliest stages of maize domestication and dispersals (e.g., Piperno et al. 2007, 2009). The necessity for applying multiproxy approaches is related in part to previously mentioned issues regarding the classification, identification, and replication of leaf phytoliths assemblages pertaining to maize from archaeological sediments (see, e.g., Doolittle and Fredrickson 1991; Staller 2003; Rovner 2004; Rovner and Gyuli 2007), as well as the more recent innovations regarding the identification of rondel phytoliths from carbon residues in ancient pottery, dental calculus and coprolites (Mulholland 1989; Thompson 2005, 2007; Thompson and Mulholland 1994; Staller and Thompson 2002; Hart and Matson 2009; Hart et al. 2007a).

Recent research from the Iguala River Valley in Central Balsas at the Xihuatoxtla rockshelter has reported evidence of maize (*Z. mays* L.) and squash (*Curcurbita argyrosperma* Huber) dated by association to c.a. 8700 CAL B.P. (Piperno et al. 2009, p. 519; Ranere et al. 2009, Figs. 1, 2). The associated date is the earliest yet recorded for maize in the Neotropics. The identification of these cultigens involved the application multiproxy methodological approaches; analysis of starch grain from ancient grinding stones and cob phytoliths from archaeological sediments (Piperno et al. 2009; Ranere et al. 2009; see also Piperno 2006; Piperno et al. 2007; Pearsall

---

[22]Piperno (2006) reduced the morphological variables to three– length, width, and aspect ratio, correlating width and aspect ratio at 87%, thus the independent variable with regard to the morphological characteristics are not entirely independent since the aspect ratio is length divided by width. Moreover, botanical size measurements even at the microfossil level are highly susceptible to systematic error due to ecological factors. It has been demonstrated that size measurements often do not replicate even within a single taxon, since Darwinian natural selection favors variation rather than bell-shaped curves (Rovner and Gyulai 2007, pp. 155–157).

et al. 2003). They report no evidence of pollen in their archaeological excavations, although pollen cores were taken from nearby lakes (Piperno et al. 2007, 2009). Starch grains were reportedly recovered from 19 grinding stones and 3 chipped stone tools, and cob phytoliths were the dominant starch type in every tool, accounting for 90% of all grains recovered (Piperno et al. 2009, Table 1; Figure 1). Eight of the grinding stones from which maize starch was recovered were securely stratified deep in preceramic levels, "well below" an associated date of 4730 B.P. (5590 ± 5320 CAL B.P.) on charcoal (Piperno et al. 2009, p. 5021, Table 1). The early radiocarbon assay was also indirect and yielded an age of 7920 ± 40 B.P. (8700 CAL B.P.), – the earliest associated date from highland Mexico reported thus far in the literature (Table 4.2). Detailed study of their results indicates that the 7920 B.P. (8700 CAL B.P.) date in Layer D is only 16 cm below the 4730 B.P. sample from Layer C, i.e., 49 cm versus 65 cm below surface (Ranere et al. 2009, Fig. 3). Moreover, in stark contrast to the deep cave deposits in Tehuacán and Oaxaca, the layers representing the archaic period occupations at Xihuatoxtla rockshelter from which these dates were derived, were only 8–12 cm (Layer C) and 8–10 cm thick, respectively (Ranere et al. 2009, p. 5016, Fig. 3). Above both these occupation layers (Layer B), a silty clay of angular roof-fall was reported (Ranere et al. 2009, p. 5016). Angular roof-fall was also encountered in Layer C, where the associated date of 4730 B.P. was derived. Layer A contained bottle glass and pottery sherds as well as obsidian blades. Maize starch grains, contemporary with or below the [14]C 7920 B.P., were recovered from both sediments and stone tools throughout the sequence, including those below the associated charcoal sample. Piperno et al. (2009) appear to emphasize the presence and absence of typologically distinct forms in their strata rather than ratios of phytoliths forms. Thus, the presence of a few "ruffle or wavy top" rondel phytoliths is interpreted as an evidence of maize. Although this is not stated in their published reports, this appears to be related in part to the assumption that sediments dating to the early Holocene should not contain large quantities of maize microfossils, and teosinte phytoliths would not be expected because as has been demonstrated in Tehuacán and Oaxaca, the wild grass is rarely if ever found in highland rockshelters or caves. The mean averages for rondel phytoliths and starch grains pertaining to maize are statistically close if not identical to modern maize and squash microfossils.

Recent research by Hart and Matson (2009) has demonstrated and various ethnobotanists have reported that there should be relatively little variation in maize rondels and teosinte phytoliths when maize was first domesticated. This is because the rondel phytolith size is believed to have reference to the cob size since phytoliths are indirectly formed in the cells of plants, and thus conform to varying degrees of the growing cell. Geneticists have indicated that the pleiotropic effects suggest that *tga1* may represent a regulatory locus. Dorweiler (1996, p. 20) states:

We have investigated several features of glume development to understand how tga1 controls glume induration (hardening). We compared the effects of the maize and teosinte alleles in the maize inbred W22. In this background, increased induration of the glumes in teosinte homozygotes (tga1 + teosinte/tga1 + teosinte) is attributable to a thicker abaxial mesoderm of lignified cells. Silica deposition in the abaxial epidermal cells of the glumes is also affected. The standard W22 line (Tga1 + maize/Tga1 + maize) has high concentrations of silica in the short cells of the epidermis of the glume, but the long cells have virtually no silica. In contrast, teosinte allele homozygotes deposit silica in both the short and long cells of the glume epidermis. Silica deposition also appears to be affected by genetic background. The teosinte background modifies the phenotype of tga1 plants towards a more uniform distribution of silica, whereas the maize background modifies the phenotype of tga1 plants toward concentration of silica in the short cells.

This was consistent with results obtained through analysis of the deposition of silica in the glumes of the three types. Dorweiler and Doebley (1997, pp. 1320–1321) state:

> The effects of *tga1* on silica deposition have some archaeological relevance. Long after most plant material has decomposed, the insoluble silica crystals from within the cells, called phytoliths, remain. Phytolith size and three-dimensional structure can be analyzed to determine the species, and relative proportions of plants that were growing in a particular area at a given time... There is even some evidence... that maize and teosinte may be distinguishable by the relative proportions of each phytolith type (Piperno 1984)... Our results showing the effects of tga1 and genetic background on the deposition of silica in the glume indicate that it will be important to analyze glume phytoliths in an archaeological context.

As the macrobotanical evidence from Tehuacán and Oaxaca has shown, the earliest cobs were very small in some cases no more than 5 cm in length (see Benz 2001; Iltis 2006). Surprisingly, recent results from central Balsas indicate that the microfossils approximate and are in some cases indistinguishable from modern maize[23] (see Piperno et al. 2009, pp. 5022–5024). The research in Central Balsas distinguished maize and teosinte grain starch granules upon infraspecific distinctions made on size of the grains. They state that maize grains are slightly larger than teosinte –irregular in shape and facet – maize granules are irregularly shaped and present compression facets (Piperno et al. 2009, pp. 5022–5024). Evidence documenting the use of teosinte grains is derived from an unspecified number of cob-type, and wavy and ruffle-top rondel phytoliths from Zones D and E as well as approximately 200 stratigraphically associated starch grains from ground and chipped stone tools (Piperno et al. 2009, pp. 5022–5024). Surprisingly, no macrobotanical evidence, i.e., charred cobs, stems, fruit cases, or other recognizable plant parts were recovered for either teosinte or maize, this is in stark contrast to the caves and rockshelters in Tehuacán and Oaxaca (see, e.g., MacNeish 1967c; Mangelsdorf et al. 1967; Smith 1986; Smith 2000). These data suggest otherwise, as it is evident that there was no long-term occupation at this locality. The associated dates and the context of the findings described by these authors pose problems similar to those discussed in some detail in previous chapters, but nevertheless, require some explanation, as do the preservation of starch grains in these shallow deposits in

---

[23]Maize and teosinte inflorescences are also distinguished on the basis of morphology, certain of short-cell phytoliths (rondels and surface sculpturing) and the relative proportion of larger cross-body phytoliths. Maize and teosinte stalks are distinguished based on the occurrence of deeply notched bilobate phytoliths and morphological characteristics seen in maize but not teosinte (Piperno et al. 2009, p. Table 1).

the time range discussed by these authors, particularly since the average annual rainfall presented by these researchers is relatively high compared to other regions where such research has been carried out[24] (see Haslam 2004).

It is evident from the relatively shallow deposits and absence of macrobotanical remains that hunters and gatherers who periodically occupied this rockshelter during their annual round were few in number and the duration of occupancy short. The stone tools and grinding implements, as well as the charcoal found in these sediments could hypothetically have been charcoal residues from the angular blocks and cobbles of roof-fall reported from the dated layers analysis. This begs numerous questions regarding the site formation processes since the possibility of migration of the materials in such contexts cannot be dismissed. It is possible that the direct dates recorded may reflect the soot and charcoal collected on the roof over years of brief occupation at this locality and provides little in the way of compelling evidence for the presence of maize two millennia earlier than the earliest directly dated cobs from other rockshelter and caves in highland Mexico (Benz and Staller 2009; see also Smith 2001, 2005a). These results speak of the importance of applying interdisciplinary evidence that provides greater chronological precision and the need for independent lines of evidence that speak more directly to the paleodietary importance of these cultigens.

4.3.6 Isotope Analysis, Paleodiet, and Geochemical Approaches

Recent innovations in stable carbon isotope analysis of ancient skeletons have provided direct evidence of diet and consequently have had important implications for our understanding of the roles of ancient cultigens to pre-Hispanic economies. Stable carbon isotope analysis has its basis in radiocarbon dating (Lippy 1955), but the significance of such data to archaeological reconstruction and the role of domesticates to ancient economies is more recent (e.g., van der Merwe and Vogel 1978; White and Schwarcz 1989). Isotope research involves careful analysis of the chemical pathway and differential fractionation of atmospheric carbon as a product of photosynthesis (Calvin and Benson 1948; Tykot 2006). Subsequently, research-ers identified multiple photosynthetic pathways, commonly referred to as C_3, C_4, and CAM (crassulacean acid metabolism)[25] (Ransom and Thomas 1960; Hatch and Slack 1966). The discovery that carbon isotope values provided paleodietary information was developed in association with stable carbon isotope analysis of marine plants and animals (see Parker 1964). Nik van der Merwe and J.C. Vogel (1971) tested an Iron Age Khoi skeleton from the Transvaal of South Africa and

---

[24]Rainfall averages between 1,000 mm and 1,400 mm annually and is highly seasonal, with 90% falling between June and October; thus, the area has a marked 7–8-month dry season (Ranere et al. 2009, p. 5014; see also Piperno et al. 2007).

[25]The earliest ^{14}C dating of bone involved demineralization of bone and extraction of humic and fulvic acids to produce much more accurate dating results on bone collagen, and this was the specific sample material tested by Vogel and van der Merwe (1977).

reported that paleodietary study on human skeletons could provide dietary information. Their pioneering research with bone collagen (protein made of multiple amino acids), carbon isotope values indicated a dependence upon sorghum (or other C_4 plants) in the Transvaal Lowveld (van der Merwe 1982). Most of the previous published syntheses on isotopic analysis regarding paleodiet in the New World have focused on maize (see, e.g., Tykot and Staller 2002; Chisholm and Blake 2006; White et al. 2006; Gil et al. 2006). The general emphasis on maize is related in part to the discovery that grasses from hot or arid environments follow the Hatch-Slack, or C_4 photosynthetic pathway, whereas the majority of plants from temperate regions, be they wild or domestic, show the Calvin-Benson, or C_3 pathway (Schwarcz 2006, p. 315; see also Sage et al. 1999). Average ratios for $\delta^{13}C$ of C_3 plants is around $-26‰$, whereas C_4 plants have $\delta^{13}C$ values averaging around $-12‰$, with a pure maize diet at $-7‰$ (see Tykot and Staller 2002, p. 669). The distinction in the photosynthetic pathway generates strikingly different signatures in $\delta^{13}C/^{12}C$ ratios with C_4 versus C_3 plants as well as provides researchers with information on other biochemical properties, including proteins that were consumed by ancient peoples (Tykot 2006, Fig. 10.1–10.2; Tykot and Staller 2002, p. 669). Carbon and nitrogen isotopic ratios in human bone may be used to reconstruct prehistoric diet because of differential fractionation of atmospheric carbon dioxide during photosynthesis and nitrogen during fixation or absorption (Sage et al. 1999; Katzenberg 2000; Tykot 2006; Tykot and Staller 2002). Isotope analysis provides another line of quantitative evidence to complement ethnobotanic, ethnohistoric, and archaeological data about paleodiet and therefore has direct reference to the economic importance of food plants like maize to ancient New World economies (Tykot 2006; Schwarcz 2006; Tykot and Staller 2002; see also Ubelaker et al. 1995; Ubelaker and Bubniak Jones 2002).

Recent isotopic research has enabled researchers to identify the presence of maize in various contexts. As mentioned above, isotope signatures have been used with residues in pottery to detect what kinds of plants were cooked or stored, with directly AMS dated residues, tooth calculus as well as ancient skeletons found in archaeological sites (Thompson 2006, 2007; Staller and Thompson 2000, 2002; Hart and Matson 2009; Hart et al. 2007a, b; Howie et al. 2009). Most isotopic analysis has involved ancient skeletons from archaeological sites. Ancient bone preserves at least two important molecules whose isotopic composition can be measured: collagen, the most abundant protein in living bone, and the carbonate (CO_3) molecule, which forms as bone mineral, hydroxyapatite[26] (Schwarcz 2006, p. 316). In addition, ancient bone contains lesser amounts of cholesterol and lipids, non-collagenous proteins whose $\delta^{13}C$ values are also inherited from the diet (Schwarcz 2006, p. 316). Stable isotope analysis of nitrogen developed in the early 1980s provided quantitative evidence of clear differences in the effects of

---

[26]The $\delta^{13}C$ value of CO_3 in hydroxyapatite (HA) of bone $(\delta^{13}C_{ap})$ is believed to represent the total $\delta^{13}C$ of the diet, that is, all the C atoms that are consumed and contribute to caloric value Collagen, which is a protein found in bone, whose amino acids are present in the diet and represent the most widely used methodology in paleodiet studies (Vogel and van der Merwe 1977).

trophic levels, particularly marine ecosystems (DeNiro and Epstein 1981). Such recent biochemical research has also increased our understanding and interpretation of nitrogen isotope ratios, and the effects of climate and environment on both plant and animal values, and trophic level increases in both terrestrial and marine ecosystems (Tykot 2006, pp. 133–134, Fig. 1, 2; see also DeNiro and Schoeninger 1983; Jakes 2002; Staller et al. 2006). These and other recent innovations in biochemical research on stable carbon isotopes have provided direct evidence on ancient diets and consequently have provided an independent line of evidence of the role of domesticates like maize to the rise of New World civilizations (see, e.g., Tykot 2002; Tykot and Staller 2002; Tykot et al. 2006; White et al. 2000; Chisholm and Blake 2006; Vierra and Ford 2006; Finucane 2009). The analyses of stable carbon isotopes, as in the case of analysis of carbon residues, are methodological approaches that actually provide direct evidence of what was consumed. Analysis of carbon and nitrogen ratios found in bone collagen and direct AMS dating of isotopes and ancient bone provide the most compelling evidence available at present to our understanding of variability in diet among and between ancient cultures, and are now documenting with increasing precision such dietary variation to further our understanding of the role of maize in the development of complexity and biogeography (see, e.g., Tykot 2006; Schwarcz 2006; White et al. 2006).

Another important isotope approach, developed more recently, involves oxygen and strontium isotopes found in soils (Barba and Ortiz 1992; Ortiz and Barba 1993; Price et al. 2000, 2002; Dahlin et al. 2007, 2009). Strontium in soils have different isotopic ratios (^{87}Sr/ ^{86}Sr) making it possible for researchers to identify if plants such as maize or animals were brought to different regions or represented "exotics" in a given area (Benson et al. 2006; Freiwald 2009). Oxygen isotopic values relate directly to the local climate, temperature, and humidity (Kolodny et al. 1983; Luz et al. 1984) and are thus data that refer to the seasonality of various species, and their consumers, as well as provide insights into climate and mobility with appropriate changes in dietary patterns (Schoeninger et al. 2000; White et al. 2000). Isotopic ratios of strontium, which does not isotopically fractionate like biological C, N, and O, directly represent the geographic area of food production/acquisition, and thus the mobility of dietary resources and/or their consumers (Ericson 1985, 1989; Price et al. 2002). Strontium isotope analysis has been applied to identify nonlocal species, be they human or animal, and with reference to humans as a basis for tracing ancient migrations and pilgrimages or verifying the presence of foreign or nonlocal artisans and craft specialists at major Mesoamerican centers such as Teotihuacan and Tikal (Price et al. 2002; Hodell et al. 2004), as well as to the identification of activity areas and ancient markets (Barba et al. 1987; Barba and Manzanilla 1987; Ortiz and Barba 1993; Dahlin and Ardren 2002; Dahlin et al. 2007, 2009; Freiwald 2009).

Recent pioneering geochemical research on archaeological soils have demonstrated that trapped chemical compounds are directly associated with specific kinds of activities, often activities that were performed repeatedly in a given locale, and that such traces can be identified even beneath earthen and stucco floors even when such activities are not present or visible archaeologically or architecturally (Barba et al. 1987; Barba and Manzanilla 1987; Manzanilla 1987, 1996; Manzanilla and

Barba 1990; Dahlin et al. 2007, 2009). Soil phosphorus (P concentrations) present in archaeological sites are chemically identified concentrations of organic matter, and have thus far, been found to provide answers for a whole host of questions surrounding ancient economies, and the presence of food crops like maize, and even provide direct evidence of ancient markets (Dahlin et al. 2007, 2009). As organic materials are processed, consumed, and disposed, phosphorous constituents released from the organic matter become fixed and adsorbed in soil particles on the surface, where they can remain for centuries (cf. Dahlin et al. 2009; see also Barba and Ortiz 1992; Parnell et al. 2001, 2002a, 2002b). Modern-day activities associated with high levels of such organic soil concentrations include gardening, waste disposal, and sweeping, which tend to push organic material to the peripheries of concentrated activity areas (Parnell et al. 2001; Dahlin et al. 2007, 2009). Thus, high concentrations of phosphorus in soils and on floors may be associated with prehistoric food preparation, consumption, storage, and disposal (Barba and Ortiz 1992; Fernández et al. 2002). Interdisciplinary evidence combined with oxygen and strontium isotope analysis and the identification of mineral and phosphorus concentrations from soils have the potential to also identify ritual and funerary activities areas and spaces (Barba et al. 1995; Manzilla 1997). Many metallic ions can also remain stable in soils for long periods in the form of adsorbed and precipitated ions on clay surfaces, or as insoluble oxides, sulfates, or carbonates (Lindsay 1979; Wells et al. 2000). Trace metal extraction and ICP/MS or AES (inductively coupled plasma mass spectrometry or atomic emission spectroscopy) analyses of soil and floor samples at various Mesoamerican sites such as Piedras Negras, Cancuén, and Aguateca, Guatemala, have provided evidence of a whole host of activities, that are often not visible archaeologically (Cook et al. 2006; Parnell et al. 2002b; Terry et al. 2004; Wells et al. 2000).

The methodological approaches to isotopic and biochemical analysis of residues in bone, tooth calculus, residues in pottery as well as in soils and stucco floors suggest that such data can provide precise information on what was consumed, paleodiet, the movements of consumables, and human and animal populations as well as distinguish patterns of heavy use areas involved in food preparation, consumption, and disposal. The geochemical signatures of organic residues and minerals in soils and ancient architecture appear to have the capacity to identify such activity even from relatively low use areas and not just high traffic areas that were deliberately kept clean, as is common in ceremonial centers and plazas (see, e.g., Dahlin et al. 2009). Such data are also having a profound effect on our perceptions of the organization and structure of ancient New World economies, and the role of maize within and on such sociocultural developments and forms of social organization.

4.4 Multidisciplinary Approaches to Maize Biogeography

Multidisciplinary research was conducted at the late Valdivia earthen mound at La Emerenciana in El Oro Province, Ecuador has uncovered evidence of early agriculture and the presence of maize (Fig. 4.12). The research involved both regional

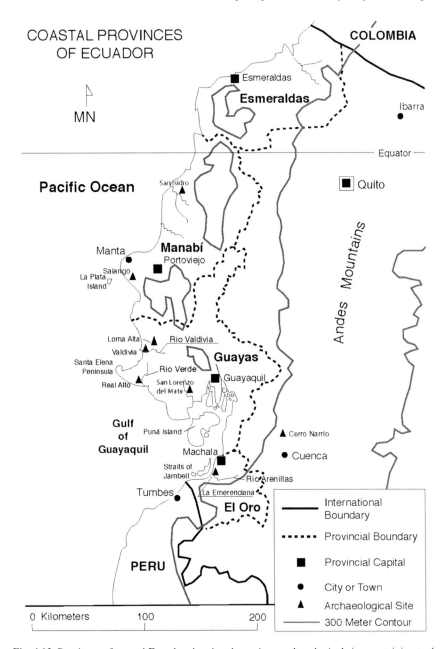

Fig. 4.12 Provinces of coastal Ecuador showing the various archaeological sites pertaining to the Valdivia culture, which have been the subject of microfossil research on early maize and also late Valdivia sites with ceramic affinities to what has been identified in the Arenillas River Valley at La Emerenciana, in El Oro Province. Chronology of coastal Ecuador based upon uncalibrated radiocarbon dates

settlement survey and large-scale excavations and multidisciplinary lines of evidence have been documented and indicate the presence of maize between 4200 and 3800 B.P. (Staller 1994, 1996, 2003, 2007a; Staller and Thompson 2000, 2002; Tykot and Staller 2002; Ubelaker and Bubniak Jones 2002). Valdivia occupations in this region correspond to the final portion of the Early Formative Period (Fig. 4.13). The southern coast of Ecuador represents a barrier island estuarine environment of slow moving undulating rivers and a progradational geomorphology (Staller 1994, 2000, 2001a,b). The Guayas Estuary is a conduit for a variety of smaller coastal streams that empty into the main estuary channel, the Canal de Jambelí. The Río Arenillas and Buenavista are located in the southern portion of the Guayas Estuary in the Gulf of Guayaquil (Staller 1994, 2000). Coastal El Oro represents an ecotone or transitional environmental zone. Some ecologists have maintained that ecotones are potentially suitable for agricultural innovation by early agriculturalists (Harris 1972).

Coastal El Oro represents the southernmost extent of the moist tropical environments in the Guayas Basin, and the northernmost extent of the dry desert coasts of Peru. The Pampas de Cayanca and adjacent coastal savanna near Huaquillas are the driest areas of coastal El Oro (Fig. 4.14), with an average annual rainfall of only 129 mm at Zorritos, just across the Peruvian border. The areas to the southeast extending to the political border with Peru are drier and therefore experience an increase in evapotranspiration behind the mangrove forest, creating slightly greater salt accumulation on the intertidal salt flats, than in the area between the Río Arenillas and Buenavista. The lowlands along the Río Buenavista are periodically flooded, forming a complex network of freshwater swamps. Explorations along the immediate margins of the Río Buenavista indicate the presence of late Valdivia sites located upstream about 5 km inland and sherds collected by local villagers suggest such sites extend to the foothills of the Andes, which are 15 km from the coast (Staller 1994). La Emerenciana is situated on the landward edge of the *salitral* or intertidal salt flats along the western banks of the Río Buenavista directly adjacent to the existing stream channel (Fig. 4.15). The prehistoric midden is on a fossil beach ridge about 2 km south of the active shoreline (Staller 1994, p. 202). The fossil beach ridge under La Emerenciana is one of a series of such topographic features and represents the earliest and highest of the ridges identified in survey in this area of southern Ecuador (Staller 1994, p. 202). Large-scale aerial photos indicated that the site was within immediate access of the mud flats, lagoons, ponds, and mangrove forest as well as fresh water swamps and salt marshes on both sides of the Río Buenavista between Puerto Jelí and the town of Santa Rosa (see Fig. 4.14). The close proximity of the Río Buenavista also makes this location favorable for seasonal cultivation, although the overall setting implies an economic focus based on resources from the estuary and mangrove forest (Staller 1994, p. 202). The climate of coastal El Oro is classified as "semi-arid," and distinguished by a marked annual variation of wet and dry seasons (Ferdon 1950; Parker and Carr 1992). The region has a 9-month long dry season restricting plant cultivation, and the societies in this region have depended to a great extent on hunting terrestrial mammals and exploitation of maritime and aquatic resources from the coastal lagoons and streams (Fig. 4.16). The mangrove forest along the Río Arenillas

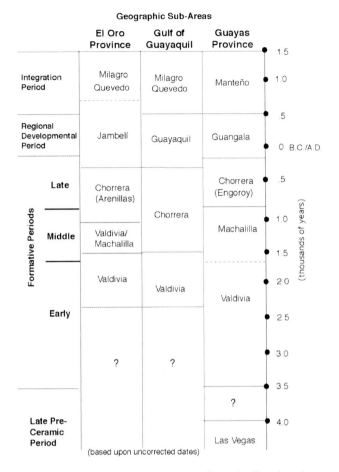

Fig. 4.13 Chronology of coastal Ecuador based on uncalibrated radiocarbon dates

provides a shelter for breeding and is an important spawning ground for a number of species of fish, shrimp, and crustaceans (Parker and Carr 1992, p. 18).

Late Valdivia occupations in coastal El Oro Province are distinguished by the presence of the earliest stirrup-spout and single spout bottles identified thus far in the Andes (Staller 1994, 1996, 2001b). Certain vessel forms and their associated stylistic attributes, such as open bowls and red on white banded motifs, are emblematic of the earliest pottery complexes of southern highland Ecuador and northern Peru as well as the Initial Period of pottery along the coast (Fig. 4.17a, b). Stirrup spout and single spout bottles are important to archaeological reconstruction because they are diagnostic of later Andean effigy and funerary vessels, and their distinctive shape and presence have been used to trace the spread of pottery technology in the Andes and other regions of the Americas (Ford 1969; Estrada et al. 1964; Meggers et al. 1965; Lathrap 1970, 1974; Lathrap et al. 1975; Holm 1980; Coe 1994b; Bruhns 1994; DeBoer 2003; Lunniss 2008). The research in

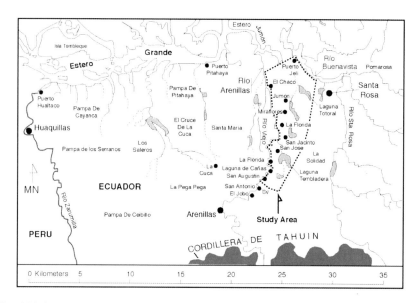

Fig. 4.14 Southern coastal Ecuador showing the study area and the various toponyms of the region. This region of the Ecuadorian coast represents a dry tropical estuary protected by barrier-islands largely made up of mangrove trees and xerophytic vegetation. Massive shell mounds, some reaching 30–40 m in height, were identified near the Peruvian border before recent shrimp farming. Such shell middens and faunal remains from excavations even at more inland sites from various time periods suggest a long-term dependence upon aquatic and maritime resources.

southern El Oro Province represents the earliest presence of such bottle forms, as well as pedestal bowls, and these formal and stylistic patterns appear to have influenced later ceramic traditions (Staller 1994, 1996, 2007b; see also Holm 1980; DeBoer 2003; Lunniss 2008). There is archaeological evidence to indicate that such ceramics and stylistic patterns spread with maize and *Spondylus* and *Strombus* shell objects in the context of a religious cult (Staller 2007b; see also Collier 1946; Collier and Murra 1943; Hocquenghem 1993). Thorny Oyster (*Spondylus* spp.) is currently perceived as synonymous with the Andean word "*mullu*," Blower (2001) presents ethnohistoric and ethnographic evidence indicating it was a "form" of *mullu*, possibly the most important element in sacrificial offerings left at *huacas* (Blower 2001, pp. 209; Hocquenghem 1993; Hocquenghem et al. 1993). *Spondylus* has a multifaceted role in Andean cosmology. It has female symbolic associations, it is a symbol of sexuality, fertility (both agricultural and human), and rain, and it was often offered at *huacas* especially springs and rivers as a sacrificial offering (Murra 1975 (1972); Paulsen 1974; Davidson 1981; Hocquenghem 1991, 1993; Burger 1992; Pillsbury 1996; Reinhard 1998). There also appears to be an interrelationship between *Spondylus* and concepts surrounding the *vagina dentada*, female sex, water, fertility, and *mullu* (Pillsbury 1996, pp. 323, 331–333; Hocquenghem 1993, pp. 702–703; Blower 2001, pp. 218). The archaeological research from this region of coastal Ecuador has generated evidence of early

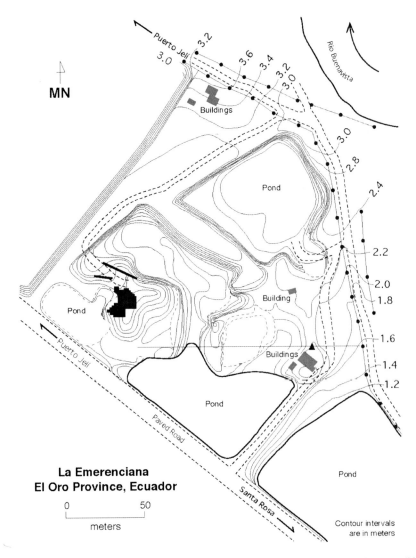

Fig. 4.15 Topographic map of La Emerenciana. Portions of the earthen mound had been modified by shrimp pond construction. The site is situated on a fossil beach ridge on the inter-tidal mud flats or salitral. The beach ridge is one of several identified in regional survey, and rises 2.5 m above sea level. Large-scale excavations involving vertical trenches and meter excavation units were focused over the course of several field seasons upon the NW platform mound. A smaller SE earthen mound is situated under the buildings, south of the site datum

agriculture and the presence of maize, and the ceramic diagnostics pertaining to the Valdivia occupations in this region have provided compelling evidence of being a major source region for the spread of ceramic technology to surrounding regions of the Andes and the origins of Andean civilization (Staller 2001a,b, 2007a,b).

Fig. 4.16 Most habitations in this region of coastal Ecuador were made of cane with thatch roofs. Traditional habitations located in the midst of the lagoons are generally situated on natural hills or rises, and built on logs made from red mangrove (*Rhizophora* spp.). One reason for this adaptive pattern is that the region is subject to the sometimes-catastrophic effects of El Nino Southern Oscillation (Photograph by John E. Staller)

Specifically, the spread of various ceramic vessels associated with the consumption and preparation of maize and Thorny Oyster (*Spondylus princeps*) and Strombus conch (*Strombus galeotus*), shell species that play a central role in ancient Andean religious belief from the beginnings of complex sociocultural development to the arrival of the Spaniards (Staller 1994, 2007b; Lunniss 2008; see also Hocquenghem 1991; Hocquenghem et al. 1993; DeBoer 2003).

4.4.1 Ethnobotanic and Isotopic Research at La Emerenciana

Archaeological excavations on the earthen mound at La Emerenciana involved the documentation of the architectural details and archaeological features associated with an earthen ceremonial mound on the NW sector of the site (Fig. 4.18) (Staller 1994, pp. 249–283). Excavation at the earthen mound at La Emerenciana was by natural stratigraphic layers in intervals corresponding to the vertical and horizontal extent of the layers and differentiated on the basis of the physical properties of the strata following the conformities and contours of the sediments (see also Staller 1994, 2001b). Two living floors were identified and all of the residue samples that produced evidence of maize, as well as the skeletons from the burials were taken from either floor one or two (Staller 1994, 2001b; Staller and Thompson 2002; Tykot and Staller 2002; Ubelaker and Bubniak Jones 2002). Carbon residue samples from the late Valdivia pottery at La Emerenciana have produced the earliest directly dated microfossil evidence for maize in the Andes (Staller and Thompson 2002, Table 9a; Thompson and Staller 2001). A total of ten sherd samples were analyzed and microfossil samples were also taken from the

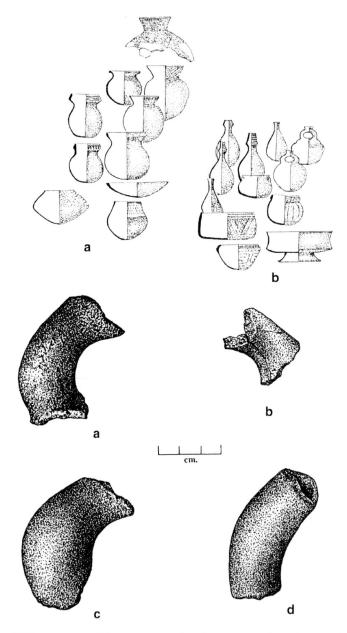

Fig. 4.17 (**A**) Diagnostic vessel shape categories based on reconstruction of excavated ceramic at La Emerenciana and materials from excavations and regional survey at other Valdivia sites. The stirrup and single spout bottles represent the earliest identified thus far in the Andes. (**a**) These ceramics represent various utilitarian wares that functioned either as serving vessels or were used for processing food and drink. (**b**) These ceramics are interpreted as ritual vessels associated with high status individuals that include composite forms, that is vessels made up of distinct component parts, such as the bottle forms and pedestal bowls. Most early Valdivia pottery is coiled.

dental calculus of two of the La Emerenciana skeletons (Fig. 4.19). The dental samples provide an independent line of evidence that indicate maize was actually consumed by Valdivia occupants of the site. In combining multiple lines of evidence to answering questions directed at the presence and consumption of maize, these data directly address some of the methodological and contextual concerns outlined in previous chapters regarding the recovery of plant microfossils from archaeological soils. Since only small amounts of carbon were required for processing, two of the organic residue samples were directly AMS dated (see Tables 4.3 and 4.4). The ^{13}C to ^{12}C ratios of -21.9 and -25.8 would normally suggest that maize played a minor to insignificant role in the diet if not for the fact that rondel phytoliths were identified on the dentition of two of the La Emerenciana burials and the food residues analyzed from archaeological features also pointed to C_4 plant consumption (Staller 2003; Staller and Thompson 2002). Given the recent research by Hart et al. (2007b) on stable isotope analysis of cooking residues, the isotopic signature could be explained on the basis of containing green or fresh kernels. Kernels would have been kept moist in such constricted containers until they spouted and it would have been the spouted kernels (*joras*) that would have been used to make the beer. These results support previously published by the author that the stable isotope and ethnobotanical results as reflecting the primary consumption of maize as a vegetable (rather than a flour) and in the form of a fermented intoxicant, that is maize beer or chicha (Staller 2003, 2006b, 2008b, 2007a, b; Staller and Thompson 2002; Tykot and Staller 2002).

Thompson also used paper chromatography to trace the chemical composition for the presence of phenolic compounds in the charred botanical samples (Staller and Thompson 2002, p. 40, Fig. 10). Amino acid composition of charred remains was compared using a modification of the technique developed by Ugent (1994, pp. 217–218). Thus, opal phytolith analysis obtained from carbon isotopes was compared to results on the chemical composition of the phytolith (Staller and Thompson 2002).

These different analytical laboratory techniques verified the presence of maize at La Emerenciana, and assessed its importance to the ancient subsistence economy. Archaeological features identified on the earthen mound in association with the primary occupation layer (Floor 2) were either architectural features associated with the construction of retaining walls funerary deposits or represented the material remains of ritual offerings (Table 4.5, Fig. 4.15). The ceremonial mound appears to have been kept meticulously clean, and there was no evidence whatsoever of domestic activities. Seven of the ten sherd samples are from the late Valdivia living floor 2 (Stratum 5) (see Tables 4.5 and 4.6) (Staller and Thompson 2002, Fig. 6). One of the residue samples was taken from a constricted jar with red

---

Fig. 4.17 (Continued) Composite vessels are characteristic of many later and contemporaneous early ceramic traditions in highland and coastal Ecuador and the earliest pottery in coastal Peru. (B) Sherds from stirrup spout vessels associated with late Valdivia occupations in coastal El Oro Province. Evidence of single and stirrup spout pottery was identified in nine of the eleven Valdivia sites identified in regional survey

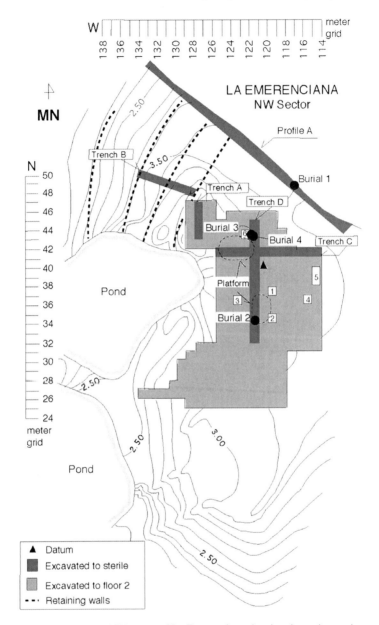

Fig. 4.18 Excavations on the NW sector of La Emerenciana showing the various units, trenches, and stratigraphic profiles. All three burials and samples from archaeological features involved in residue analysis were taken from trenches C and D

on white banded decoration on the body and broad line incisions on the rim exterior (Fig. 4.20). The constricted jar from feature 65 was in a clay-lined pit and appears to have been left as a ritual offering. It showed the strongest evidence for maize and

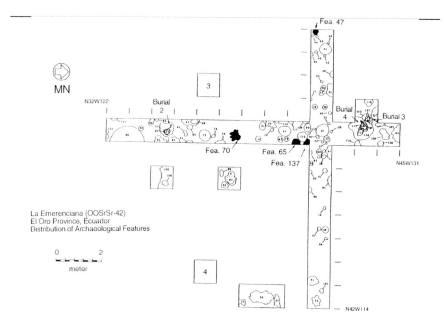

Fig. 4.19 Distribution of Archaeological features at La Emerenciana: Trench and unit excavations showing horizontal distribution of archaeological features and the locations of the various late Valdivia burials and those features from which carbon residues from pottery were taken are highlighted

was deposited upside down in the clay-lined pit and smashed in place and then the pit was sealed by a hard packed clay (Staller and Thompson 2002: 44, Table 9a). Three food residue samples from other archaeological features on the mound show clear evidence of maize opal phytoliths are all associated with the living floor 2 occupations (Fig. 4.21a, b). All four of the late Valdivia burials from La Emerenciana were upright bundle burials, fully articulated primary interments that were interred in burial pits lined with what is interpreted as junco grass (Staller 1994, pp. 304–312; Ubelaker and Bubniak Jones 2002). Such burial patterns are consistent with contemporaneous internments at preceramic sites in coastal Peru (see Staller 1994). The burials are all stratigraphically associated with Stratum 5 (floor 2), the gray ash layer (see Staller and Thompson 2002, Table 6). Burial 1 was an incomplete skeleton, with the upper torso and the cranium missing, while the dental calculus from Burials 2 and 4 produced maize phytoliths, the subadult Burial 3, did not (Staller and Thompson 2002). Each of the ten samples was composed of less than 0.1 g of food residue. Table 4.5 provides information on the provenience of the food residue samples and Table 4.6 a detailed description of the stratigraphic layers in which the Valdivia burials were placed. Figures 4.18 and 4.19 show their relative locations in the excavations (see also Staller and Thompson 2002).

Ethnobotanic evidence generated from the pottery residues at La Emerenciana documenting the presence of maize, is complemented and supported by stable isotope results from three of four burials uncovered in the earthen platform mound

Table 4.5 Proveniences of La Emerenciana food residue samples

A. Catalog No. 5480
    Provenience: N37W122 trench D
    Stratum: 5
    Feature: 1 (South platform) (pottery
    smashed on north edge of the platform)
	B. Catalog No. 5485
	    Provenience: N38W122 trench D
	    Stratum: 5
	    Feature: 70 (offering of an upright olla found on
	    the surface of the middle smashed olla filled with
	    shell, faunal, and organic remains)

C. Catalog No. 5623
    Provenience: N41W122 Trench D
    Stratum: 5
    Feature: 65 (pit offering with smashed
    and upturned constricted jar)
	D. Catalog No. 5430
	    Provenience: N41W126 Trench C
	    Stratum: 3
	    Feature: 47 (post impression with smashed and
	    burned pot at the base)

E. Catalog No. 5618
    Provenience: N41W122 trench D
    Stratum: 3
    Feature: 137 (storage pit)
	F. Catalog No. 4268
	    Provenience: N37W124 Cut 3
	    Stratum: 6
	    Sherd from surface of stratum 5

G. Catalog No. 5546
    Provenience: N44W122 trench D
    Stratum: 5
    Feature: 114 (clay lined offering pit)
	H.Catalog No. 5534
	    Provenience: N34W122 trench D
	    Stratum: 5
	    Feature: 90 (clay lined offering pit)

I. Catalog No. 4135
    Provenience: N40W116 Cut 5
    Stratum: 5
    Feature: 56 (mounded offering of shell,
    faunal remains, and smashed pottery)
	J. Catalog No. 5429
	    Provenience: N42W123 Cut 6
	    Stratum: 5
	    Feature: 117 (Burial pit, burial 4)

and these data indicate that maize was a minor component of the diet (Tykot and Staller 2002). Analysis of over fifty formative skeletons from other Valdivia sites such as Real Alto and Loma Alta, as well as the type site and later formative period sites indicated that maize does not become essential to the diet until the Late Formative Period (see Fig. 4.13) (see van der Merwe et al. 1993). Numerous microfossil researches have been carried out in this region of Ecuadorian coast, and many of these studies have provided important data on maize biogeography and the role of maize in early pottery cultures in this area of the Neotropics (Pearsall 1978, 1999, 2002; Pearsall et al. 2003; Pearsall and Piperno 1990; Chandler-Ezell et al. 2006). The stable isotope results represent quantitative data sets which speaks directly to diet and thus its early role in the Valdivia subsistence diet. The stable isotope residues of the pottery residues appear to support previous research with Valdivia skeletons, as well as burial remains from other formative sites from Andean South America (Tykot and Staller 2002: Table 3, Figure 5; see also van der Merwe et al. 1993; Tykot et al. 2006). Maize does not appear to have played a major dietary role in the Andean economy until much later than had been previously reported using approaches with associated rather than direct dates with evidence for maize.

The multidisciplinary evidence generated by the research at La Emerenciana and its ceramic and skeletal collection is compelling in that it provides independent lines of internally consistent data that together, speak directly of the importance of applying multidisciplinary methodological approaches to our understanding of

Table 4.6 Primary stratigraphic layers at La Emerenciana, El Oro province, Ecuador

Stratum	Depth	Horizon	Color	Description
6	0–55 cm.	A	10yr 5/3–10yr 5/4	Brown fine silty loam, loosely consolidated in the upper levels more dense in the lower levels, with evidence of bioturbation. Artifact and shell remains in the uppermost levels (Living Floor 3) (fluvial deposit)
5	15–93 cm.	B	10yr 6/1–10yr 5/1	Homogeneous gray ashy loam, loosely packed, very fine texture, the consistency of talc, fine quartz inclusions with artifact and shell remains in the uppermost levels of the stratum. (Living Floor 2). (ethnostratigraphic)
4	36–92 cm.	C	10yr 8/3	White dune sand, finely textured very loosely consolidated, with calcium carbonate inclusions in the upper levels. (eolian deposit)
3	78–145 cm.	Bwn	7.5yr 6/4–7.5yr 7/4	Pink quartz sand finely textured well consolidated, free of inclusions. (Living Floor 1) (ethnostratigraphic)
2	64–134 cm.	Bwk	2.5y 8/6–2.5y 8/8	Yellow sand finely textured, loosely consolidated, with calcium carbonate and small (3 mm 1 cm.) beach pebble inclusions. (eolian deposit)
1	97 cm	C	5y 8/2–5y 8/4	Olive white sand, finely textured, moderately packed with small (3 mm–2 cm) beach pebble and calcium carbonate inclusions. (fluvial deposit)

Note: All soil colors are classified using the Munsell Soil Color Chart 1975 Edition. Depths are given as below datum, and indicated as minimum and maximum levels which of course varied in different areas of the excavations. (after Staller 1994, Table 14)

maize biogeography and economic importance (Benz and Staller 2006; see also Staller et al. 2006). The archaeological, ethnobotanical, isotopic, and settlement pattern data are consistent in breath and scope to what has been presented in the early pioneering research on maize origins and spread, and at various levels directly challenges earlier and subsequent microfossil research on the antiquity and role of maize to Early Formative Period cultures of coastal Ecuador[27] (Meggers 1966;

---

[27]The archaeological soils of coastal Ecuador have not been conducive to the preservation of pollen or macrobotanical remains (see Pearsall 1978; Pearsall and Piperno 1990). Some maize macrobotanicals have been identified, but with few exceptions, most are from more recent archaeological contexts (see Zevallos et al. 1977; Staller 1994; Pearsall 2003).

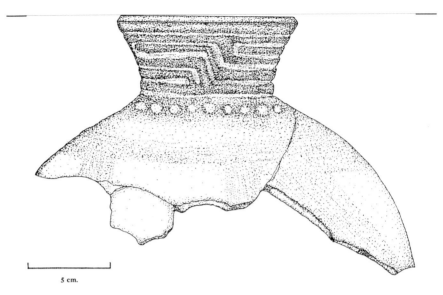

5 cm.

Fig. 4.20 One of the residue samples was taken from this constricted jar, which was placed upside down in a pit Feature 65 and smashed in place as part of a termination ritual offering. Red on white banded pottery styles are later integrated into a wide variety of the earliest ceramic traditions identified archaeologically in highland Ecuador and northern highland and coastal Peru.

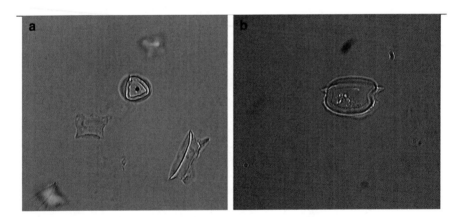

Fig. 4.21 (**a**) Rondel cob phytoliths identified in food residues from late Valdivia pottery at La Emerenciana, Ecuador. The image shows a decorated rondel in planar view and at 1,000X magnification. (**b**) Tilted rondel shows the constriction in the middle of the phytolith, and the projections from one face and is at 1,000X magnification. (Courtesy of Robert G. Thompson)

Meggers et al. 1965; Lathrap et al. 1975; see also Pearsall 1978, 1999, 2002; Pearsall and Piperno 1990, 1993; Piperno and Pearsall 1998). Ethnobotanical remains from carbon residues in excavated late Valdivia pottery, and stable isotope analysis of human collagen and apatite from pre-Hispanic burials provided quantitative data sets for the role and importance of maize and its early introduction to this region of the Neotropics (Staller 2003, 2007a,b). Settlement pattern data provide additional independent lines of evidence supporting the interpretations regarding the dietary significance of maize and its role in the ancient economy (Staller 2000). Settlement patterns of all prehistoric sites in the study area indicated the existence of a two-tiered hierarchy of site size and function including sites with artificial earthen mounds and occupations situated further inland from the coast along the coastal streams beginning in the formative periods (Fig. 4.22a, b) (Staller 1994, 2000). Early formative settlement patterns in the region are primarily located along the fossil beach ridge and around the lagoons with only three of the nine identified situated along the Arenillas River. These patterns are consistent with a broad spectrum subsistence adaptation, as suggested by the stable carbon isotope evidence (Tykot and Staller 2002), and mentioned by various scholars for other regions of the Americas for early agricultural societies. Furthermore, the Late Formative Period settlement patterns show a clear shift to riverine locations and direct access to alluvial soils, typical of agricultural societies in other regions of the Neotropics. Moreover, there is a significant increase in the number of sites and site size, although La Emerenciana was the largest pre-Hispanic site identified for any time period in this region (Staller 1994). Changes in earthen mound architecture between Early and Late Formative sites indicates a shift from earthen mound sites with two oval mounds delimited by retaining walls for the Valdivia settlements, to groupings of four circular earthen mounds in the Late Formative Period patterns (Staller 2000; see also Lunniss 2008). An increase from nine early formative to twenty-four late formative period settlements, further supports the multidisciplinary evidence from La Emerenciana that the adaptive shift to agriculture did not occur until this time in this region of coastal Ecuador. The settlement data also indicate an increase in carrying capacity associated with an agricultural economy in the Late Formative Period and such evidence is consistent with settlement patterns of early agricultural sites throughout the Americas. These settlement results are also consistent with the stable isotope evidence on paleodiet from other regions of the Andes (Tykot and Staller 2002: Table 3, Figure 5; Tykot et al. 2006) which indicate that maize was a minor component in the diet until about 300 B.C. in Peru and 1000–800 B.C. in coastal Ecuador (see also van der Merwe et al. 1993).

4.4.2 Advantages to Multidisciplinary Approaches

The ever intensifying pace of field and laboratory research in the social and biological sciences in the last three decades has, as evident in this and previous chapters of this book, produced a wide variety of scientific evidence bearing particularly on the issue of the origins, spread, and economic role of maize in the New World prehistory and the rise of civilization. The previous chapters make clearly evident that despite the intensity of such research, there still remains significant disagreements over the chronology, contextual integrity, and the positive identification of maize pollen as well as opal phytoliths from archaeological

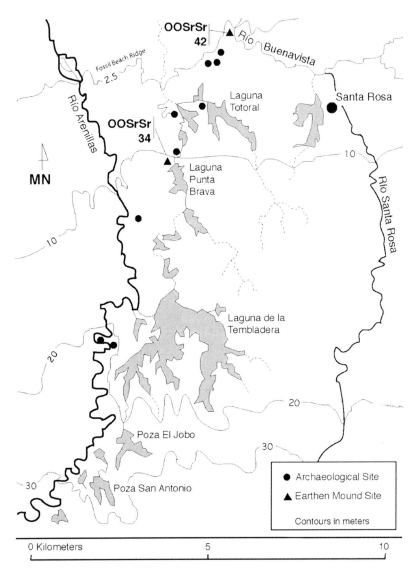

Fig. 4.22a (**a**) Early Formative Period settlement patterns in southern coastal El Oro Province, Ecuador. All Valdivia settlements in this region correspond to the final portion of the culture sequence and are dated between 2200 and 1450 B.C. on the basis of ^{14}C dates and stylistic attributes of the ceramics. (**b**) Late Formative Period settlement patterns in southern coastal El Oro Province, Ecuador. In contrast to other regions of coastal Ecuador, occupations pertaining to this time period correspond chronologically to between c. 1400 and 500 B.C. There was no evidence of Machalilla occupations and all Valdivia/Machalilla transitional pottery was dominated by red on white slipped banded wares, similar to what has been reported by archaeologists in the nearby southern highlands of Ecuador

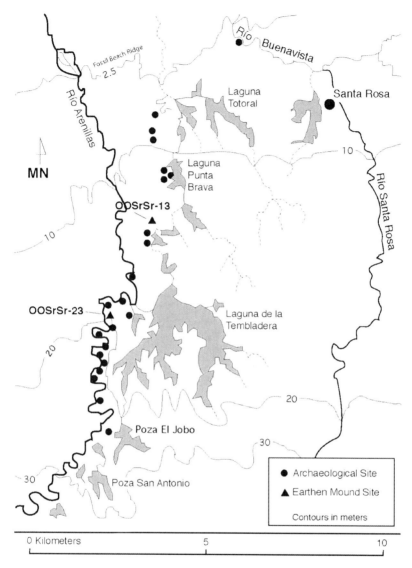

Fig. 4.22b (Continued)

sediments as well as macrobotanical remains. Food residue analysis appear to alleviate much of the chronological and contextual issues that have typically surrounded analysis of plant microfossils from lake cores and archaeological soils, and collagen and isotope analysis of ancient skeletons directly address the question of the role of maize in such ancient economies (Thompson 2006, 2007; Tykot 2006; Tykot et al. 2006; Tykot and Staller 2002; Holst et al. 2007). When such lines of evidence are combined with regional settlement patterns of the Formative Period sequence, they provide a robust body of multidisciplinary data from which to address these important scientific questions surrounding Z. *mays*. (Staller 2003; Staller and Thompson 2000, 2002; Tykot and Staller 2002; Tykot et al. 2006).

Despite the enduring controversy over the phylogeny, origins, chronology, macrobotanical identification, and routes of dispersal of maize, most scholars of Andean prehistory would agree that the pre-Hispanic subsistence patterns indicate considerable chronological and spatial variability, particularly when terrestrial, marine, or riverine fauna were regularly available for exploitation (Staller 2000; Staller and Thompson 2002). A general review of the radiocarbon evidence by Michael Blake (2006) has indicated that many of these discrepancies regarding the early presence of maize are related to whether dates were associated or direct. His research has indicated that when reviewing the earliest associated dates from both the highland and lowland regions of the Neotropics, they are generally more ancient than directly dated macrobotanicals or macrofossils. On the other hand, in areas of the Americas where maize appears later, direct dates tend to be older than dates by association (Blake 2006; see also Hastorf et al. 2006; Chávez and Thompson 2006; Hart et al. 2003, 2007a). These published reports emphasize the importance of direct dating of maize macrobotanicals and microfossils. The evidence from La Emerenciana and regional settlement data from coastal El Oro Province reaffirms the advantages of direct dates from residues, particularly in light of the fact that many of the associated dates published from this region are as much as 3,000 years older (e.g., Pearsall 1999, 2002, 2003, Pearsall and Piperno 1990; Pearsall et al. 2003; Piperno et al. 2009).

The ethnohistoric accounts clearly indicate that European perceptions of the New World and the various plants and animals which existed in this hemisphere at that time were largely conditioned by sixteenth century perceptions based on what was known at that time from pre-Linnaean herbals. Those documents convey a very different perception of the origin and to a lesser extent the role of maize in the New World economies. In fact, many Europeans thought that maize came from the Far East. It is apparent from the analysis of the history of science surrounding the origins of maize and from subsequent archaeological research and theory on its role in ancient New World economies that these earlier perceptions may have played a larger role in our current worldwide reliance on maize than had been previously supposed or even considered. Over the past 30 years, the methodological and technological breakthroughs in the study of maize origins and biogeography have greatly expanded our understanding of its role and early spread, this is particularly true for the research from molecular biology. The diverse and different data sets generated by these more recent approaches, have provided a more comprehensive and complex synthesis to our understanding of early plant and animal domestication, and particularly to the role of maize in such developmental processes. These recent data have considerably set the limits for how future scholars will pursue research and scientific questions surrounding the antiquity and location(s) of where early domestication event(s) occurred and how these developmental processes were related to the spread of maize to different regions of the Americas. It is evident from the previous analysis that direct dating of macrobotanical remains and plant microfossils should be a prerequisite for future studies, particularly in light of the discrepancies that have appeared in the published literature in the past three decades regarding the earliest presence of maize.

As evidence in the previous chapter and again above, these recent archeobotanical advances in the study of pollen, opal phytoliths from archaeological soils, and more recently from carbon residues in ancient pottery has dramatically modified our understanding of the biogeography of and antiquity of maize in the Americas (Thompson 2006, 2007; Thompson and Staller 2001a,b, 2003; Staller and Thompson 2000, 2002; Chavéz and Thompson 2006; Sluyter and Dominguez 2006; Lusteck 2006; Hastorf et al. 2006). It is now readily apparent that when maize cobs or microfossils are dated by association, as in the case of pollen and phytoliths from archaeological soils, the chronologies and presence of maize is quite different than when samples are directly AMS dated as in the case of macrobotanicals or food residues from ancient pottery (see, e.g., Long et al. 1989; Benz and Long 1999; Blake 2006; Thompson 2006, 2007; Thompson and Staller 2001a, b, 2003; Staller and Thompson 2000, 2002; Chavéz and Thompson 2006).

Data from the biological and molecular sciences has clearly had a profound influence on archaeologists and ethnobotanists working on domestication and origins of maize. The morphological difference between maize cobs and teosinte spikes appears to have had a major role in explaining why it took scholars so long to establish these evolutionary relationships. One of the most significant breakthroughs has been molecular research on maize DNA as these studies have set the limits on what can be said scientifically with regard to early maize morphology and biogeography. The maize genome project and the breakdown of microsatellites at the level of DNA have not only produced compelling evidence for the origins of maize, but the research on su1 and tga1 genes have had on maize morphology has been startling and iconoclastic on a number of different levels (Dorweiler 1996; Dorweiler and Doebley 1997; Doebley et al. 1987, 1990, 1997, 2006; Matsuoka et al. 2002; Whitt et al. 2002; Freitas et al. 2003; Jaenicke-Després et al. 2003; Jaenicke-Després and Smith 2006). The earlier genetic studies involved in understanding the molecular basis for maize morphology have now made it necessary for scholars to rethink their previous perceptions of the bottleneck phenomenon and the role of human selection to the creations of the incredible diversity of maize landraces existing in the present day (Doebley and Wang 1997, Doreweiler 1996; Dorwieler and Doebley 1997; Dorweiler et al. 1993; Eyre-Walker et al. 1998; Benz 2001; Staller 2003; Thompson 2006, 2007; Hart and Matson 2009; Hart et al. 2007a, b; Hart and Matson 2009). The breakthroughs in recent research on maize are truly remarkable and most scholars are only recently beginning to absorb the implications of these data. It is my hope that this book will provide readers and scholars who are interested in this fascinating and remarkable plant, a basis for beginning to appreciate the broader implications of what this research tells us about early agriculture and plant domestication.

References

Acosta J (1961 [1590]) In: O'Gorman E (ed) Historia Natural y Moral de las Indias, 2nd edn. Fondo de Cultura Económica, Mexico, DF

Adams KR (1994) A regional synthesis of *Zea mays* in the Prehistoric American Southwest. In: Johannessen S, Hastorf CA (eds) Corn and culture in the prehistoric new World. Westview Press, Boulder, CO, pp 273–302

Alt KW (2005) Ethnic anthropology. Ethnohistory. Ethnic interpretations of early-historical archaeology. History, fundamentals and alternatives. Homo J Comp Hum Biol 56(1):23–36

Albornoz de C (1967) [1570-1584]. Instruccioón para descubririr todas las guacas del Pirú y sus camayos y haziendas. In: Duviols P (ed) Albornoz y espacio ritual andino prehispánico. Revista Andina2(1):169-222.Cuzco

Ammerman AJ, Cavalli-Sforza LL (1971) Measuring the rate of spread of early farming in Europe. Man 6:674–688

Ammerman AJ, Cavalli-Sforza LL (1973) A population model for the diffusion of farming into Europe. In: Colin Renfrew A (ed) The explanation of culture change: models in prehistory. Duckworth, London, pp 343–358

Ampuero G (1982) El Norte Chico y su connotación en el área Andina Meridional. Actas IX Congreso Arqueología Chilena, La Serena, pp 179–186

Anderson E (1943) A variety of maize from Río Loa. Mo Bot Gard 30:469–476

Anderson E (1944) Homologies of the ear and tassel in Zea mays. Ann Mo Bot Gard 31:325–340

Anderson E (1947a) Popcorn. Nat His 56(5):227–230

Anderson E (1947b) Corn before Columbus. Pioneer Hi-Bred Corn Company, Des Moines

Anderson E (1952) Plants, man and life. University of California, Berkeley, CA

Anderson E, Cutler H (1942) Races of *Zea mays* I. Their recognition and classification. Ann Mo Bot Gard 29:69–88

Anderson AJO, Berdan F, Lockhart JM (1976) Beyond the codices: the Nahua view of colonial Mexico. University of California Press, Berkeley, CA

Andrew D, Fernandez-Armesto F, Novi C, Williams G (eds) (2001) [1789-94]. The Malaspina expedition 1789-1794- The journal of the voyage by Alejandro Malaspina: Volume 1 Cadiz to Panama. Introduction by Donald C. Cutter. London, The Hakluyt Society & Madrid, The Museo Naval.

Arford MR, Horn SP (2004) Pollen evidence of the earliest maize agriculture in Costa Rica. J Lat Am Geogr 3:108–115

Arriaga Fr. PJ de (1968) [1621] The extirpation of idolatry in Peru. Translated by C. Keating (ed). University of Kentucky Press, Lexington

Ascherson P (1875) Ueber *Euchlaena mexicana*. Schradscreift Botanica Vereins Provence Brandenburg 17:76–80

Aveni AF (2000) Empires of time: calendars, clocks, and cultures. Tauris Parke Paperbacks, New York

de Azara F (1809) Voyages dans l'Amérique Méridionale. Paris 1:146–148

Babot MP (2004) Tecnología y utilización de artefactos en el noroeste prehispánico. Ph.D. Dissertation. Universidad Nacional de Tucumán, Tucumán, Argentina

Bailey LH (1892) A new maize and its behavior under cultivation. Cornell Agric Expedition Stat Bull 49:333–338

Baird ET (1993) The drawings of Sahagun's Primeros Memoriales. University of Oklahoma Press, Norman

Barba L, Ortiz A (1992) Análisis químico de pisos de ocupación: un caso etnográfico en Tlaxcala, Mexico. Lat Am Antiq 3:63–82

Barba L, Ludlow BM, Manzanilla L, Valadez R (1987) La vida doméstica de Teotihuacan: Un estudio interdisciplinario. Cien y Desarro 7:21–33

Barba L, Manzanilla L (1987) Estudio de areas de actividad. In: Manzanilla L (ed) Coba, Quintana Roo Analysis De Dos Unidades Habitacionales Mayas. Universidad Nacional Autonoma de Mexico, Mexico, DF

Barber RJ, Berdan F (1998) The emperor's mirror: understanding cultures through primary sources. University of Arizona Press, Tucson

Barker G (1985) Prehistoric farming in Europe. Cambridge University Press, Cambridge

Bar-Yosef O, Belfer-Cohen A (1992) From foraging to farming in the Mediterranean Levant. In: Gebauer AB, Price TD (eds) Transitions to agriculture in prehistory, Monographs in World Archaeology No. 4. Prehistory Press, Madison, WI, pp 21–48

Beadle GW (1939) Teosinte and the origin of maize. J Hered 30:245–247

Beadle GW (1972) The mystery of maize. Chicago. Field Mus Natl His Bull 43(10):2–11

Beadle GW (1978) Teosinte and the origin of maize. In: Walden DW (ed) Maize breeding and genetics. Wiley, New York, pp 113–128

Beadle GW (1980) The ancestry of corn. Sci Am 242(1):112–119

Beadle GW (1981) Origin of corn: pollen evidence. Science 213:890–892

Bellwood P (2005) First farmers: the origins of agricultural societies. Blackwell, London

Beaumont P (1932 [1700s]) Crónica de Michoacán, vol 3. Publicaciones del Archivo General del Nación, México D.F

Bennetzen J, Buckler E, Chandler V, Doebley J, Dorwheiler J, Gaut B, Freeling M, Hake S, Kellogg E, Peothig RS, Walbot V, Wessler S (2001) Genetic evidence and the origin of maize. Lat Am Antiq 12:84–86

Benson L, Stein J, Taylor H, Friedman R, Windes TC (2006) The agricultural productivity of chaco canyon and the source(s) of pre-Hispanic maize found in Pueblo Bonito. In: Staller JE, Tykot RH, Benz BF (eds) Histories of Maize: Multidisciplinary approaches to the prehistory, linguistics, biogeography, domestication, and evolution of maize. Elsevier, San Diego, CA, pp 290–314

Benz B (1999) On the origin, evolution and dispersal of maize. In: Blake M (ed) Pacific Latin America in prehistory. The evolution of archaic and formative cultures. Washington State University Press, Pullman, pp 25–38

Benz BF (2001) Archaeological evidence of teosinte domestication from Guilá Naquitz, Oaxaca. Proc Natl Acad Sci USA 98(4):2104–2106

Benz BF (2006) Maize in the Americas. In: Staller JE, Tykot RH, Benz BF (eds) Histories of maize: multidisciplinary approaches to the prehistory, linguistics, biogeography, domestication, and evolution of maize. Elsevier, San Diego, CA, pp 9–20

Benz BF, Iltis HH (1990) Studies in archaeological maize. I. The "wild" maize from San Marcos Cave reexamined. Am Antiq 55:500–511

Benz BF, Long A (2000) Early evolution of maize in the Tehuacán Valley, Mexico. Curr Anthropol 41(3):459–465

Benz BF, Staller JE (2006) The antiquity, biogeography, and culture history of maize in the Americas. In: Staller JE, Tykot RH, Benz BF (eds) Histories of Maize: Multidisciplinary approaches to the prehistory, linguistics, biogeography, domestication, and evolution of maize. Elsevier, San Diego, CA, pp 647–673

Benz BF, Staller JE (2009) The antiquity, biogeography, and culture history of maize in Mesoamerica. In: Staller JE, Tykot RH, Benz BF (eds) Histories of maize in mesoamerica: multidisciplinary approaches. LeftCoast Press, Walnut Creek, CA, pp 267–275

Benz BF, Cheng L, Leavitt SW, Eastoe C (2009) El Riego and early maize evolution. In: Staller JE, Tykot RH, Benz BF (eds) Histories of maize: multidisciplinary approaches to the prehistory, linguistics, biogeography, domestication, and evolution of maize. Elsevier, San Diego, CA, pp 85–94

Benz B, Perales H, Brush S (2007) Tzeltal and Tzotzil farmer knowledge and maize diversity in chiapas, Mexico. Curr Anthropol 48(2):289–300

Bercaw LO, Hannay AM, Larson NG (1940) Corn in the development of the civilization of the Americas; a selected and annotated bibliography Compiled under the direction of Mary G. Lacy. B. Franklin, New York

Berdan F (1982) The Aztecs of Central Mexico: an imperial society. Harcourt Brace Jovanovich, Fort Worth, TX

Berdan FF, Anawalt PR (eds) (1992 [1541–1542]) The codex Mendoza, vol 4. University of California Press, Berkeley, CA

Berlin B (1992) Ethnobiological classification: principles of categorization of plants and animals in traditional societies. Princeton University Press, Princeton

Berlin B, Breedlove DE, Raven PH (1974) Principles of Tzeltal plant classification. Academic, New York

Betanzos J de (1968 [1551]) Suma y narración de los Incas, vol 209, Biblioteca de Autores Españoles. Atlas, Madrid

Betanzos J de (1987) [1551] Suma y narración de los incas. In: C.M. Rubio (ed). Madrid.

Betanzos J de (1996 [1557]) Narrative of the Incas. Translated and edited by R. Hamilton and D. Buchanan from the Palma de Mallorca Manucript. University of Texas Press, Austin, TX

Binford LR (1962) Archaeology as anthropology. Am Antiq 28(2):217–225

Binford LR (1964) A consideration of archaeological research design. Am Antiq 29(4):425–441

Binford LR (1965) Archaeological systematics and the study of culture process. Am Antiq 31:203–210

Binford LR (1968) Post-pleistocene adaptations. In: Binford SR, Binford LR (eds) New perspectives in archaeology. Aldine, Chicago, IL, pp 5–33

Binford LR (1980) Willow smoke and dog's tails: hunters and gatherer settlement systems and archaeological site formation. Am Antiq 45(1):4–20

Binford LR (1989) Debating archaeology. Academic, San Diego, CA

Binford LR (2001) Constructing frames of reference: an analytical method for archaeological theory building using hunter-gatherer and environmental data sets. University of California Press, Berkeley, CA

Bird R. McK (1966) El Maíz y las divisiones étnicas en la sierra de Huanuco. Cuadernos de Investigación 10:134–144

Bird R McK, Bowman DL, Durbin ME (1988) Quechua and maize: mirrors of central Andean culture history. J Steward Anthropol Soc 15:187–240

Blacker IR, Rosen HM (eds) (1960) The golden conquistadores. Bobbs-Merrill Company, Indianapolis, IN

Blake M (2006) Dating the initial spread of maize. In: Staller JE, Tykot RH, Benz BF (eds) Histories of Maize: Multidisciplinary approaches to the prehistory, linguistics, biogeography, domestication, and evolution of Maize. Elsevier, San Diego, CA, pp 55–72

Blake M, Chisholm BS, Clark JE, Voorhies B, Love MW (1992) Prehistoric subsistence in the soconusco region. Curr Anthropol 33(1):83–94

Blake M, Chisholm BS, Clark JE, Voorhies B, Michaels G, Love MW, Pye ME, Demarest AA, Arroyo B (1995) Radiocarbon chronology for the late archaic and formative periods on the Pacific coast of southeastern Mesoamerica. Anc Mesoamerica 6:161–183

Bogucki P (1996) The spread of early farming in Europe. Am Sci 84:242–253

Bonafous M (1836) Histoire naturelle, agricole et economique du Mais. Madame Huzard, Paris

Bonzani R, Oyuela-Caycedo A (2006) The gift of the variation and dispersion of maize: Social and technological context in Amerindian societies. In: Staller JE, Tykot RH, Benz BF (eds) Histories of Maize: Multidisciplinary approaches to the prehistory, linguistics, biogeography, domestication, and evolution of maize. Elsevier, San Diego, CA, pp 344–356

Boturini L (1746) Idea de una nueva historia general de la América Septentrional. Madrid.

Braidwood R (1952) From cave to village. Sci Am 187:62–66

Braidwood R (1960) The agricultural revolution. Sci Am 203:130–148

Braidwood R, Reed CA (1957) The achievement and early consequences of food production: a consideration of the archaeological and natural historical evidence. Cold Spring Harb Symp Quant Biol 22:19–31

Bray W, Herrera L, Schrimpff MC, Botero P, Monsalve JG (1987) The ancient agricultural landscape of Calima, Colombia. In: Denevan WM, Gregory Knapp KM (eds) Pre-Hispanic agricultural fields in the Andean region. BAR International Series 359, Oxford, pp 443–481

Bretting PK, Goodman MM, Stuber CW (1987) Karyological and isozyme variation in West Indian and allied American mainland races of maize. Am J Bot 74(11):1601–1613

Brooks FJ (1995) Motecuzoma Xocoyotl, Hernan Cortez, and Bernal Díaz del Castillo, the construction of an arrest. Hisp Am Hist Rev 75(2):149–184

Brown CH (1999a) Lexical acculturation in Native American languages. Oxford University Press, New York

Brown CH (2006a) Glottochronology and the chronology of maize in the Americas. In: Staller JE, Tykot RH, Benz BF (eds) Histories of Maize: Multidisciplinary approaches to the prehistory, linguistics, biogeography, domestication, and evolution of maize. Elsevier, San Diego, CA, pp 647–663

Brown DF (2005) The chontal maya of tabasco. In: Sandstrom AR, HugoGarcíaValencía E (eds) Native peoples of the Gulf Coast of Mexico. University of Arizona Press, Tucson, pp 114–138

Brown TA (1999b) How ancient DNA may help in understanding the origin and spread of agriculture. Philos Trans R Soc Lond B Biol Sci 354:89–98

Brown TA (2006b) Differing approaches and perceptions in the study of new and old World crops. In: Staller JE, Tykot RH, Benz BF (eds) Histories of Maize: Multidisciplinary approaches to the prehistory, linguistics, biogeography, domestication, and evolution of maize. Elsevier, San Diego, CA, pp 3–8

Brown WL (1953) Maize of the West Indies. Trop Agric 30:141–170

Brown WL (1960) Races of maize in the West Indies, National Academy of Sciences, National Research Council Publication 792. Government Printing Office, Washington, DC

Bruhns KO (1994) Ancient South America. Cambridge World archaeology. Cambridge University Press, Cambridge

Bruman HJ (2000) Alcohol in ancient Mexico. Forward by Peter T. Furst. University of Utah Press, Salt Lake City, UT

Burger RL (1992) Chavín and the origins of Andean civilization. Thames and Hudson, London

Buckler E, Holdsford T (1996) Zea Systematics: ribosomal ITS evidence. Mol Biol Evol 13:612–622

Burger RL, van der Merwe NJ (1990) Maize and the origin of highland chavín civilization. Am Anthropol 92:85–95

Burland CA, Forman W (1975) Feathered serpent and smoking mirror. The gods and cultures of ancient Mexico. Orbis Publishing, London

Bush MB, Piperno DR, Colinvaux PA (1989) A 6000-year history of Amazonian maize cultivation. Nature 340(6231):303–305

Callen EO (1967) Analysis of the Tehuacan coprolites. In: DS Byers (ed) The prehistory of the Tehuacan Valley, vol. 1, Environment and subsistence. University of Texas Press, Austin, TX, pp. 261–89

Calvin M, Benson AA (1948) The path of carbon in photosynthesis. Science 107:476–480

Campbell L (1997) American Indian languages: the historical linguistics of Native America. Oxford University Press, New York

Camus-Kulandaivelu LC, Chevin LM, Tollon-Cordet C, Charcosset A, Manicacci D, Maud I (2008) Tenaillon patterns of molecular evolution associated with two selective sweeps in the *Tb1–Dwarf8* region in Maize. Genetics 180:1107–1121

Candiolle A de (1855) Géographie Botanique Raisonnée: ou, Exposition des faits principaux et des lois concernant la distribution géographique des plantes de l'epoque actuelle. V. Masson, Paris

Carlsen RS, Prechtel M (1991) The flowering of the dead: an interpretation of highland Maya culture. Man 26:23–42

Carmack RM (1973) Quichean civilization; the ethnohistoric, ethnographic, and archaeological sources. University of California Press, Berkeley, CA

Carmack RM, Gasco J, Gossen GH (1996) The legacy of Mesoamerica: history and culture of a Native American civilization. Prentice Hall, Upper Saddle River, NJ

Carneiro RL (2002) Herbert Spencer's principles of sociology: a centennial retrospective and appraisal. Ann Sci 59:221–261

Carrasco D (1999) City of sacrifice: the Aztec empire and the role of violence in civilization. Beacon Press, Boston, MA

Carrasco MD (2009) From field to hearth: an earthly interpretation of May and Mesoamerican creation myths. In: Staller JE, Carrasco MD (eds) Pre-Columbian Foodways: Interdisciplinary approaches to food, culture and markets in Mesoamerica. Springer, New York, pp 601–634

Carrasco MD, Hull KM (2002) The cosmogonic symbolism of the corbeled vault in maya architecture. Mexicon 24(2):26–32

Carraso P (1975) Social organization in Ancient Mexico, vol X, Handbook of Middle American Indians. University of Texas Press, Austin, TX, pp 349–375

Chandler-Ezell K, Pearsall DM, Zeidler JA (2006) Root and tuber phytoliths and starch grains document manioc (*Manihot esculenta*), arrowroot (*Maranta arundinacea*), and llerén (*Calathea* sp.) at Real Alto site, Ecuador. Econ Bot 60:103–120

Chávez SJ, Thompson RG (2006) Early maize on the Copacabana peninsula: implications for the archaeology of Lake Titicaca basin. In: Staller JE, Tykot RH, Benz BF (eds) Histories of Maize: Multidisciplinary approaches to the prehistory, linguistics, biogeography, domestication, and evolution of maize. Elsevier, San Diego, CA, pp 415–428

Childe VG (1929) The most ancient East; the oriental prelude to European prehistory. Knopf, New York

Childe VG (1935) Changing methods and aims in prehistory. Proc Prehistoric Soc 1:1–15

Childe VG (1946) What happened in history, 1st edn. Penguin Books, Baltimore, MD

Childe VG (1950) Prehistoric migrations in Europe. Instituttet for sammenlignende kulturforskning, Serie A: Forelesninger, 20. Harvard University Press, Cambridge

Childe VG (1951a) Social evolution. Meridian Books, Cleveland

Childe VG (1951b) Man makes himself, 2nd edi. The New American Library, New York

Chishom B, Blake M (2006) Diet in prehistoric soconusco. In: Staller JE, Tykot RH, Benz BF (eds) Histories of Maize: Multidisciplinary approaches to the prehistory, linguistics, biogeography, domestication, and evolution of maize. Elsevier, San Diego, CA, pp 161–172

Christenson AJ (2001) Art and society in a highland Maya community – the altarpiece of Santiago Atitlán. University of Texas Press, Austin, TX

Christenson AJ (2003) Popol Vuh: sacred book of the Maya. O Books, London

Christenson AJ (2008) Places of emergence: sacred mountains and cofradía ceremonies. In: Staller JE (ed) Pre-Columbian landscapes of creation and origin. Springer, New York, pp 95–122

Clement RM, Horn SP (2001) Pre-Columbian land-use history in Costa Rica: a 3000-year record of forest clearance, agriculture and fires from Laguna Zoncho. Holocene 11(4):419–426

Cieza de León P (1977) [1551] El Señorio de los Incas, 2nd Editorial Universo, Lima

Cieza de León P (1998[1553]) The discovery and conquest of Peru. Chronicles of the New World encounter. Edited and translated by Alexandra Parma Cook and Noble David Cook. Duke University Press, Durham

Clark JE (1994) The development of early formative rank societies in the soconusco, Chiapas, Mexico. Ph.D. dissertation, University of Michigan, Ann Arbor. Ann Arbor: University Microfilms

Clark RM, Nussbaum Wagler T, Quijada P, Doebley J (2006) A distant upstream enhancer at the maize domestication gene tb1 has pleiotropic effects on plant and inflorescent architecture. Nat Genet 38:594–597

Classen C (1993) Inca cosmology and the human body. University of Utah Press, Salt Lake City, UT

Clavigero FJ (1844) Historia Antigua de México y de su conquista. Ministerio de Educación y Ciencía. México D.F.

Clayton WD, Renvoize SA (1986) Genera graminum: grasses of the World, vol 13, Kew bulletin additional series. HMSO Publications, London

Cobo FrB (1990 [1653]) Inca religion and customs. University of Texas Press, Austin, TX Translated by R. Hamilton (ed). Forward by John H. Rowe

Coe MD (1994a) Mexico, 4th edn. Thames and Hudson, New York

Coe SD (1994b) America's first cuisines. University of Texas Press, Austin, TX

Coe SD, Coe MD (1996) The true history of chocolate. Thames and Hudson Ltd, London

Collier D (1946) Archaeology of Ecuador. In JH Steward (ed) Handbook for South American Indians, vol 2, The Andean Civilizations. Bulletin 143, Bureau of American Ethnology, Smithsonian Institution. Washington DC, pp. 767–784

Collier D, Murra JV (1943) Survey and excavations in Southern Ecuador. Field Museum of Natural History Publication 528, Chicago

Collins GN (1917) Hybrids of *Zea tunicata* and *Zea ramose*. J Agric Res 9:383–396

Collins GN (1921) Teosinte in Mexico. J Hered 12:339–350

Collins GN (1931) The phylogeny of maize. Bull Torrey Bot Club 57:199–210

Colinvaux P (1993) Pleistocene biogeography and diversity in tropical forests of South America. In: Goldblatt P (ed) Biological relationships between Africa and South America. Yale University Press, New Haven, CT, pp 473–499

Colinvaux PA, Bush MB (1991) The rain-forest ecosystem as a resource of hunting and gathering. Am Anthropol 93:153–160

Colinvaux PA, De Oliveira PE, Moreno PE, Miller MC, Bush MB (1996a) A long pollen record from lowland Amazonia: forest and cooling in glacial times. Science 274:85–88

Colinvaux PA, Liu K, De Oliveira P, Bush MB, Miller MC, Steinitz-Kannan M (1996b) Temperature depression in the lowland tropics in glacial times. Clim Change 32:19–33

Colinvaux PA, Bush MB, Steinitz-Kannan M, Miller MC (1997) Glacial and postglacial pollen records from the Ecuadorian Andes and Amazon. Quaternary Res 48:69–78

Columbus C (1930) [1492–1507?] The Voyages of Christopher Columbus being the Journal of his first and third, and the letters concerning his first and last voyages, to which is added the account of his second voyage written by Andres Bernaldez. Edited and translated by Cecil Jane. London: The Agronaut Press. Empire House, 175 Piccadilly

Columbus C (1970) [1492] Capitulaciones del Almirante don Cristóbal Colón y salvoconductos para el descubrimiento del nuevo mundo. 17 April 1492. Ministerio de Educación y Ciencía. Madrid

Columbus C (1990 [1492]) Journal of the first Voyage. Aris & Phillips, Ltd, Warminster, England. Warminster Edited and translated by E. W. Ife

Cook DE, Kovacevich B, Beach T, Bishop RL (2006) Deciphering the inorganic chemical record of ancient human activity using ICP-MS: a reconnaissance study of the late classic soil floors at Cancuén, Guatemala. J Archaeol Sci 33:628–640

Cortés H (1963 [1485-1547?]) Cartas y Documentos. In: Hernández Sánchez-Barba MH (ed) Biblioteca Porrúa, vol 2. Editorial Porrúa, Mexico City

Cortez H (1991 [1519-1526]) Five letters, 1519-1526. Norton, New York Edited by B. Morris

Cummins TBF (2002) Toasts with the Inca: Andean abstraction and colonial images on Quero vessels. University of Michigan Press, Ann Arbor, MI

Cummins TBF (2004) Silver threads and golden needles: the Inca, the Spanish and the sacred World of humanity. In: Phipps E, Hecht J, Martín CE (eds) The colonial Andes: tapestries and silverwork 1530-1830, The Metropolitan Museum of Art, New York. Yale University Press, New Haven, CT, pp 3–15

Cutler HC (1952) A preliminary survey of plant remains of tularosa cave. In: Martin PS, Rinaldo JB, Bluhm E, Cutler HC, Grange R Jr (eds) Mogollon cultural continuity and change. The stratigraphic analysis of tularosa and cordova caves, *Fieldiana*: Anthropology 40. Field Museum of Natural History Museum, Chicago, pp 461–479

Cutler HC (2001) Cultivated plants from Picuris. In: Blake LW, Cutler HC (eds) Plants from the Past. The University of Alabama Press, Tuscaloosa, pp 16–36

Cutler HC, Cardenas M (1947) *Chicha*, a native South American beer. Bot Mus Leafl Harv Univ 12:33–61

Cutler HC, Whitaker TW (1967). Curcurbits from the Tehuacan Caves. In: DS Byers (ed) The prehistory of the Tehuacán valley, vol. 1, Environment and subsistence. University of Texas Press, Austin TX, pp 212–219

Dahlin BH, Ardren T (2002) Modes of exchange and regional patterns: chunchucmil, Yucatan. In: Masson MA, Freidel DA (eds) Ancient Maya political economies. Altimira, New York, pp 249–284

Dahlin BH, Jensen CT, Terry RT, Wright DR, Beach T, Magnoni A (2007) In search of an ancient Maya Market. Lat Am Antiq 18(4):363–384

Dahlin BH, Blair D, Beach T, Moriarty T, Terry R (2009) The dirt on food: ancient feasts and markets among the Lowland Maya. In: Staller JE, Carrasco MD (eds) Pre-Columbian foodways: interdisciplinary approaches to food, culture and markets in Mesoamerica. Springer, New York, pp 191–232

Darwin C (1868) The variation of animals and plants under domestication. John Murray, London

Davidson JR (1981) El Spondylus en la cosmología chimú. Revista del Museo Nacional 45:75–87

Deagan KA (1989) The search for La Navidad Columbus's 1492 settlement. In: Milanich JL, Milbrath S (eds) First encounters: Spanish explorations in the Caribbean and the United States. University of Florida Press/Florida Museum of Natural History, Gainesville, FL, pp 1492–1570

DeBoer WR (2003) Ceramic assemblage variability in the formative of Ecuador and Peru. In: Raymond JS, Burger RL (eds) Archaeology of formative Ecuador. Dumbarton Oaks Research Library and Collections, Washington D.C, pp 289–336

DeNiro MJ (1987) Stable isotopy and archaeology. Am Sci 75:182–191

DeNiro MJ, Epstein S (1981) Influence of diet on the distribution of nitrogen isotopes in animals. Geochim Cosmochim Acta 45:341–351

DeNiro MJ, Schoeninger MJ (1983) Stable carbon and nitrogen isotope ratios of bone collagen: variations within individuals, between sexes, and within populations raised on monotonous diets. J Archaeol Sci 10:199–203

deWet JMJ, Harlan JR (1972) Origin of maize: the tripartite hypothesis. Euphytica 21:271–279

deWet JMJ, Harlan JR (1978) Tripsacum and the origin of maize. In: Walden DW (ed) Maize breeding and genetics. Wiley, New York, pp 129–142

Dennis E, Peacock W (1984) Knob heterochromatin homology in maize and its relatives. J Mol Evol 20:341–350

De Vorsey Jr L (1991) Keys to the encounter: a library of congress resource guide for the age of discovery. Library of Congress, Washington DC

Diamond J (1997) Location, location, location: the first farmers. Science 278:1243–1244

Díaz del Castillo B (1953) [1567-75] The discovery and conquest of Mexico 1517-1521. In: E. Garrett (ed) Abridged by Alfred P. Maudslay. Mexico City: Ediciones Tolteca, Mariano Escobedo 218, México 17, D.F

Dillehay TD, Rossen J, Andres TC, Williams DE (2007) Earliest evidence of peanut, cotton and squash farming found. Science 316:1890–1893

Dixon RMW (1997) The rise and fall of languages. Cambridge University Press, Cambridge, MA

Dobrizhoffer M (1822 [1784]) An account of the Abipones. John Murray, London

Doebley JF (1990a) Molecular data and the evolution of maize. Econ Bot 44(3 Suppl.):6–27

Doebley JF (1990b) Molecular systemics of *Zea*. Maydica 35:143–150

Doebley JF (1994) Morphology, molecules, and maize. In: Johannessen S, Hastorf CA (eds) Corn and culture in the prehistoric New World. Westview Press, Boulder, CO, pp 101–112

Doebley J (2001) George Beadle's other hypothesis: one-gene: one trait. Perspectives. Genet Soc Am 158:487–493

Doebley JF, Iltis HH (1980) Taxonomy of *Zea* (Gramineae): a subgeneric classification with key to taxa. Am J Bot 67(6):982–993

Doebley J, Lukens L (1998) Transcriptional regulators and the evolution of plant form. Plant Cell 10:1075–1082

Doebley JF, Stec A (1991) Genetic analysis of the morphological differences between maize and teosinte. Genetics 129:285–295

Doebley JF, Wang RL (1997) Genetics and the evolution of plant form: an example from maize. Cold Spring Harb Symp Quant Biol 62:361–367

Doebley JF, Goodman MM, Stuber CW (1983) Isozyme variation in maize from the Southwestern United States: taxonomic and anthropological implications. Maydica 28:97–120

Doebley J, Goodman M, Stuber C (1984) Isoenzymatic variation in zea (Gramineae). Syst Bot 93:203–218

Doebley J, Renfroe W, Blanton A (1987) Restriction site variation in the *Zea* chloroplast genome. Genetics 117:139–147

Doebley J, Stec A, Wendel J, Edwards M (1990) Genetic and morphological analysis of a maize teosinte F_2 population: implications for the origin of maize. Proc Natl Acad Sci USA 87:9888–9892

Doebley JF, Stec A, Hubbard L (1997) The evolution of apical dominance in maize. Nature 386:485–488

Doebley JF, Gaut BS, Smith BD (2006) The molecular genetics of crop domestication. Cell 127:1309–1321

Doolittle WE, Frederick CD (1991) Phytoliths as indicators of prehistoric maize (Zea mays subsp. mays, Poaceae) cultivation. Plant Syst Evol 177(3–4):175–184

Dorweiler JE (1996) Genetic and evolutionary analysis of glume development in maize and teosinte. Ph.D. Dissertation, University of Minnesota. University of Michigan Microfilms, Ann Arbor

Dorweiler JE, Doebley J (1997) Developmental analysis of *Teosinte Glume Architecture1*: a key lucus in the evolution of maize (POACEAE). Am J Bot 84(10):1313–1322

Dorweiler J, Stec A, Kermicle J, Doebley J (1993) Teosinte glume architecture 1: a genetic locus controlling a key step in maize evolution. Science 262:233–235

Dull R (2006) The maize revolution: a view from El Salvador. In: Staller JE, Tykot RH, Benz BF (eds) Histories of maize: multidisciplinary approaches to the prehistory, linguistics, biogeography, domestication, and evolution of maize. Elsevier, San Diego, CA, pp 357–365

Dull R (2007) Evidence for forest clearance, agriculture, and human-induced erosion in Precolumbian El Salvador. Ann Assoc Am Geogr 97(1):127–141

Durán PD (1880) [1588?] Historia de las Indias de Nueva-España y islas de Tierra Firme por el padre fray Diego Durán, religioso de la Orden de Predicadores, escritor del siglo XVI; la publica con un atlas de estampas, notas é illustraciones, José F. Ramirez, individuo de varias sociedades literarias nacionales y extranjeras. Imprenta de J.M. Andrade y F. Escalante, México

Durán FD (1964 [1588?]) The Aztecs: history of the Indies of New Spain. Orion Press, New York Translated, with notes by Doris Heyden and Fernando Horcasitas. Introduction by Ignacio Bernal

Durán FD (1971) [d. 1588?] Book of the gods and rites and the ancient calendar. Translated and edited by Fernando Horcasitas and Doris Heyden. Foreword by Miguel León-Portilla. University of Oklahoma Press, Norman

Durán FD (1994 [1588?]) The History of the Indies of New Spain. Translated, annotated, and with an introduction by Doris Heyden, vol 102, The civilization of the American Indian series. University of Oklahoma Press, Norman

Elliot JH (1963) Imperial Spain: 1468–1716. Mentor Books, New York

Estrada VE, Meggers BJ, Evans C Jr (1964) The Jambelí culture of South Coastal Ecuador. In: Proceedings of the U.S. National Museum, vol. 115 (3492). Smithsonian Institution Press, Washington DC, pp. 483–558

Eubanks MW (1995) A cross between two maize relatives: *Tripsacum dactyloides* and *Zea diploperennis* (Poaceae). Econ Bot 49:172–182

Eubanks MW (1997) Molecular analysis of crosses between *Tripsacum dactyloides* and *Zea diploperennis* (Poaceae). Theor Appl Genet 94:707–712

Eubanks MW (1999) Corn in clay: maize paleobotany in pre-Columbian art. University of Florida Press, Gainesville, FL

Eubanks MW (2001a) An interdisciplinary perspective on maize. Lat Am Antiq 12:91–98

Eubanks MW (2001b) The mysterious origins of maize. Econ Bot 55(4):492–514

Eubanks MW (2001c) The origins of maize: evidence for *Tripsacum* ancestry. Plant Breed Rev 20:15–61

Enfield E (1866) Indian corn: its value, culture, and uses. Appleton, New York

Ericson JE (1985) Strontium isotope characterization in the study of prehistoric human ecology. J Hum Evol 14:503–514

Ericson JE (1989) Some problems and potentials of strontium isotope analysis for human and animal ecology. In: Rundel PW, Ehleringer JR, Nagy KA (eds) Stable isotopes in ecological research. New York, Springer, pp 252–259

Eyre-Walker A, Gaut RL, Hilton H, Feldman DL, Gaut BS (1998) Investigation of the bottleneck leading to the domestication of maize. Proc Natl Acad Sci USA 95:4441–4446

Evans RJW (1979) The making of the habsburg monarchy, 1550–1700: an interpretation. Clarendon Press, New York

Evershed RP, Heron C, Charters S, Goad LJ (1992) The survival of food residues: new methods of analysis, interpretation and application. In: Mark Pollard A (ed) New developments in archaeological science. Oxford University Press, New York, pp 187–208

Ferdon EN Jr (1950) The climates of Ecuador. In: Studies in Ecuadorian geography. School of American Research and Museum of New Mexico Bulletin 15, pp. 35–63

Fernández FG, Terry RE, Inomata T, Eberl M (2002) An ethnoarchaeological study of chemical residues in the floors and soil of Q'eqchi' Maya Houses at Las Pozas, Guatemala. Geoarchaeol Int J 17:487–519

Fernández de Oviedo y Valdes G (1950 [1526]) Sumario de la natural historia de las Indias. Fondo de Cultura Económica, México D.F Editor, introduction and notes by José Miranda

Fernández de Oviedo y Valdes G (1959) [1526] Natural History of the West Indies (1526), translated and edited by S.A. Stoudemire, University of North Carolina Studies in the Romance Languages and Literature, No. 32, Chapel Hill: University of North Carolina Press

Fernández de Oviedo y Valdés, G. 1969 [1535]. *De La Natural Hystoria de Las Indias*. A Facsimile Edition issued in Honor of Sterling A. Stoudemire, University of North Carolina Studies in the Romance Languages and Literature, No. 85, Chapel Hill: University of North Carolina Press.

Finan JJ (1950) Maize in the great herbals. Chronica Botanica Company, Waltham, MA With forward by Edgar Anderson

Finucane BC (2009) Maize and sociopolitical complexity in the Ayacucho valley, Peru. Curr Anthropol 50(4):535–545

Flannery KV (1969) Origins and ecological effects of early domestication in Iran and the Near East. In: Ucko PJ, Dimbleby GW (eds) The domestication and exploitation of plants and animals. Duckworth, London, pp 73–100

Flannery KV (1972a) The origins of the village as a settlement type in Mesoamerica and the Near East: a comparative study. In: Ucko PJ, Tringham R, Dimbleby GW (eds) Man, settlement and urbanism. Schenkman Publishing, Cambridge, MA, pp 23–54

Flannery KV (1972b) The cultural evolution of civilizations. Annu Rev Ecol Syst 3:399–426 Palo Alto, California

Flannery KV (1973) The origins of agriculture. Annu Rev Anthropol 2:271–310

Flannery KV (1986a) The research problem. In: Flannery KV (ed) Guilá Naquitz: archaic foraging and early agriculture in Oaxaca. Academic, San Diego, CA, pp 1–18

Flannery KV (1986b) Ecosystem models and information flow. In: Flannery KV (ed) Guilá Naquitz: archaic foraging and early agriculture in Oaxaca. Academic, San Diego, CA, pp 19–28

Flannery KV (1986c) Food procurement and preceramic diet. In: Flannery KV (ed) Guilá Naquitz: archaic foraging and early agriculture in Oaxaca. Academic, San Diego, CA, pp 303–317

Flannery KV (1986d) Adaptation, evolution, and archaeological phases: some implications of Reynolds' simulation. In: Flannery KV (ed) Guilá Naquitz: archaic foraging and early agriculture in Oaxaca. Academic, San Diego, CA, pp 501–507

Flannery KV (2002) The origins of the village revisited: From nuclear to extended households. Am Antiq 67(3):417–433

Ford J (1969) A comparison of formative of cultures in the Americas, vol II, Smithsonian contributions in anthropology. Smithsonian Institution, Washington, DC

Ford RI (1000) Gardening and farming before A.D. 1000: patterns of prehistoric cultivation North of Mexico. J Ethnobiol 1(1):6–27

Ford RI (1985a) Processes of food production in North America. In: Ford RI (ed) Prehistoric food production in North America, Anthropological Papers No. 75. Museum of Anthropology, University of Michigan, Ann Arbor, MI, pp 341–364

Ford RI (1985b) Processes of food production in North America. In: Ford RI (ed) Prehistoric food production in North America, Anthropological Papers No. 75. Museum of Anthropology, University of Michigan, Ann Arbor, MI, pp 1–19

Freidel D, Schele L, Parker J (1993) Maya cosmos: three thousand years on the Shaman's Path. William Morrow and Company, New York

Freidel DA, Reilly FK (2009) The flesh of God: cosmology, food, and the origins of political power in ancient Southeastern Mesoamerica. In: Staller JE, Carrasco MD (eds) Pre-Columbian Foodways: Multidisciplinary approaches to food, culture and markets in Mesoamerica. Springer, New York, pp 635–680

Freitas FO (2001) Estudo genético-evolutivo de amostras modernas e arqueológicas de milho (Zea mays mays, L.) e feijão (Phaseolus vulgaris, L.). Unpublished Ph.D. Dissertation, Escola Superior de Agricultura Luiz de Queiroz. University of São Paulo, Brazil

Freitas FO, Brendel G, Allaby RG, Brown TA (2003) DNA from primitive maize landraces and archaeological remains: implications for the domestication of maize and its expansion into South America. J Archaeol Sci 31:901–908

Fried MH (1967) The Evolution of Political Society: An Essay in Political Anthropology. Random House, New York

Freiwald CR (2009) Dietary diversity in the upper Belize river Valley: a zoo archaeological and isotopic perspective. In: Staller JE, Carrasco MD (eds) Pre-Columbian foodways: interdisciplinary approaches to food, culture and markets in Mesoamerica. Springer, New York, pp 399–420

Fritz GJ (1994) Are the first American farmers getting younger? Curr Anthropol 35:305–309

Fuchs L (1942) De historía stirpium commentarii insignes. Basel, Germany

Fuchs L (1978) [1543] *New Kreüterbůch: in welchem nit allein die gantz Histori, das ist, Namen, Gestalt, Statt und Zeit der Wachsung, Natur, Krafft und Würckung, des meysten Theyls der Kreüter so in teütschen vnnd andern Landen wachsen, mit dem besten Vleiss beschriben, sonder auch aller derselben Wurtzel, Stengel, Bletter, Blůmen, Samen, Frücht, und in summa die gantze Gestalt allso artlich vnd kunstlich abgebildet vnd contrafayt ist, das dessgleichen vormals nie gesehen, noch an tag komen Durch den hochgelerten Leonhart Fuchsen, der Artzney Doctorn vnnd derselbigen zů Tübingen lesern. Mit dreyen nützlichen Registern.* Facsimile reprint of the 1543 German edition Basel ed., printed by Michael Isingrin. Includes indexes. Verlegt von Ernst Battenberg, München

Fussell B (1992) The story of maize. University of New Mexico Press, Albuquerque

Gadacz RR (1982) The language of Ethnohistory. Anthropologica 24(2):147–165

Galinat WC (1954) Corn grass I: corn grass as the possible prototype or a false progenitor of maize. Am Nat 88:101–104

Galinat WC (1964) Tripsacum a possible Amphidiploid of Manisuris and wild maize. Maize Genet Coop News Lett 38:50

Galinat WC (1970) The cupule and its role in the origin and evolution of maize. Massachusetts Agric Exp Stat Bull 585:1–20

Galinat WC (1976) The inheritance of some traits essential to maize and Teosinte. In: Walden DW (ed) Maize breeding and genetics. Wiley, New York, pp 93–111

Galinat WC (1983) The origin of maize as shown by key morphological traits of its Ancestor, Teosinte. Maydica 28:121–138

Galinat WC (1985) Domestication and diffusion of maize. In: Ford RI (ed) Prehistoric food production in North America, Anthropology Papers 75. University of Michigan, Ann Arbor, MI

Galinat WC (1988) The origin of corn. In: Sprague GF, Dudley JW (eds) Corn and corn improvement, 3rd edn. American Society of Agronomy, Madison, WI, pp 1–31

Gallavotti A, Zhao Q, Kyozuka J, Meeley RB, Ritter MK, Doebley JF, Pe ME, Schmidt RJ (2004) The role of barren stalk1 in the architecture of maize. Nature 432(2):630–635

Gebauer AB, Price TD (eds) (1992) Transitions to agriculture in prehistory. Monographs in World Archaeology No. 4. Prehistory Press, Madison, WI

Gerard J (1975) [1633] The Herbal or General History of Plants. The complete 1633 edn. revised by Thomas Johnson. Facsimile reprint of the ed. printed by Islip A. Norton J. Whitakers R London, under title: The Herball, or Generall Historie of Plantes. New York: Dover Publications.

Gerbi A (1985 [1975]) Nature in the New World. University of Pittsburgh Press, Pittsburgh, PA From Christopher Columbus to Gonzalo Fernández de Oviedo. Translated by J. Moyle

Gil AF (2003) Zea mays on the South American periphery: chronology and dietary Importance. Curr Anthropol 44(2):295–300

Gil AF, Tykot RH, Neme G, Shelnut NR (2006) Maize on the frontier; Isotopic and macrobotanical data from central-western Argentina. In: Staller JE, Tykot RH, Benz BF (eds) Histories of Maize: Multidisciplinary approaches to the prehistory, linguistics, biogeography, domestication, and evolution of maize. Elsevier, San Diego, CA, pp 199–214

Godfray HCJ (2007) Linnaeus in the information age. Nature 446(15):259–260

Goldstein DJ, Coleman RC (2004) Schinus molle L. (Anacardiaceae) Chicha production in the Central Andes. Econ Bot 58(4):523–529

Goodman MM (1968) The races of maize II. Use of multivariate analysis of variance to measure morphological similarity. Crop Sci 8:693–698

Goodman MM (1973) Genetic distances: measuring dissimilarity among populations. Yearb Phys Anthropol 17:1–38

Goodman MM (1978) A brief survey of the races of maize and current attempts to infer racial relationships. In: Walden DW (ed) Maize Breeding and Genetics. Wiley, New York, pp 143–158

Goodman MM (1988) The history of evolution of maize. CRC Crit Rev Plant Sci 7(3):197–220

Goodman MM (1994) Racial sampling and identification of maize: quantitative genetic variation versus environmental effects. In: Johannessen S, Hastorf CA (eds) Corn and culture in the prehistoric New World. Westview Press, Boulder, CO, pp 89–100

Goodman MM, McKBird R (1977) The races of maize. IV. Tentative grouping of 219 Latin American races. Econ Bot 31:204–221

Goloubinoff P, Pääbo S, Wilson AC (1993) Evolution of maize inferred from sequence diversity of an Adh2 gene segment from archaeological specimens. Proc Natl Acad Sci USA 90:1997–2001

Goman M, Byrne R (1998) A 5000-year record of agriculture and tropical forest clearance in the Tuxtlas, Veracruz, Mexico. Holocene 8(1):83–89

Gowlett JAJ (1993) Ascent to civilization. The archaeology of early humans. McGraw-Hill, New York

Gould SJ (2002) The structure of evolutionary theory. The Belknap Press of Harvard University Press, Cambridge, MA

Grobman A, Salhuana W, Sevilla R (1961) *Races of maize in Peru, their origins, evolution and classification* [by] and in collaboration with Paul C. Mangelsdorf. National Academy of Sciences-National Research Council, Washington

Guaman Poma de Ayala (Waman Puma) F (1980) [1583-1615]. Nueva Cronica y Buen Gobierno. In: Murra JV, Adorno R, Jorge L. Urioste. 3 vols. Siglo XXI. América Nuestra, Mexico D.F.

Hard RJ, Roney JR (1998) A massive terraced village complex in Chihuahua, Mexico, 3000 Years before present. Science 279:1661–1664

Harlan JR (1975) Crops and man. American Society of Agronomy, Madison, WI

Harlan JR, de Wet JMJ (1973) On the quality of evidence for the origin and dispersal of cultivated plants. Curr Anthropol 14(1–2):51–62

Harlan JR, de Wet JMJ, Price E (1973) Comparative evolution of cereals. Evolution 22:311–325

Harris DR (1972) The origins of agriculture in the tropics. Am Sci 60(2):180–193

Harris DR (1989) An evolutionary continuum of people-plant interaction. In: Harris DR, Hillman G (eds) Foraging and farming: the evolution of plant exploitation. Unwin Hyman, London, pp 11–26

Harshberger JW (1893) Maize: a botanical and economic study. Contributions from the Botanical Laboratory of the University of Pennsylvania. University of Pennsylvania Press, Philadelphia

Harshberger JW (1896a) Fertile crosses of teosinte and maize. Gard Forest 9:522–523

Harshberger JW (1896b) The purpose of Ethnobotany. Am Antiq 17(2):73–81

Hart JP, Thompson RG, Brumbach HJ (2003) Phytolith evidence for early maize (Zea Mays) in the Northern Finger Lakes region of New York. Am Antiq 68(4):619–640

Hart JP, Brumbach HJ, Lusteck R (2007a) Extending the phytolith evidence for early maize (Zea mays ssp. mays) and squash (Cucurbita sp.) in central New York. Am Antiq 72:563–583

Hart JP, Lovis WA, Schulenberg JK, Urquhart GR (2007b) Paloeodietary implications from stable carbon isotope analysis of experimental cooking residues. J Archaeol Sci 35:804–813

Haskin F (1913) The Panama canal. Doubleday, Page and Company, New York

Haslam M (2004) The decomposition of starch grains in soils: implications for archaeological residue analysis. J Archaeol Sci 31:1715–1734

Hastorf CA (1994) The changing approaches to maize research. In: Oyuela-Caycedo A (ed) History of Latin American archaeology. Avebury, Brookfield, CO, pp 139–154

Hastorf CA (1999) Recent research in paleoethnobotany. J Archaeol Res 7(1):55–103

Hastorf CA, DeNiro MJ (1985) New isotopic method used to reconstruct prehistoric plant production and cooking processes. Nature 315:489–491

Hastorf CA, Whitehead WT, Bruno MC, Wright M (2006) The movements of maize into the middle horizon tiwanaku, Bolivia. In: Staller JE, Tykot RH, Benz BF (eds) Histories of Maize: Multidisciplinary approaches to the prehistory, linguistics, biogeography, domestication, and evolution of maize. Elsevier, San Diego, pp 429–448

Hatch MD, Slack CR (1966) Photosynthesis by sugarcane leaves. A new carboxylation reaction and the pathway of sugar formation. Biochem J 101:103–111

Heiser CB Jr (1985) Some botanical considerations of the early domesticated plants north of Mexico. In: Ford RI (ed) Prehistoric Food Production in North America, vol Anthropological Papers No. 75. Museum of Anthropology, University of Michigan, Ann Arbor, pp 57–73

Heiser C (1988) Aspects of unconscious selection and the evolution of domesticated plants. Euphytica 37:77–85

Heron C, Evershed R, Goad L (1991) Effects of migration of soil lipids on organic residues associated with buried potsherds. J Archaeol Sci 18:641–659

Herrera Casasús ML (1989) Presencía y esclavitud del Negro en la Huasteca. Editoríal Purrúa; Universidad Autónoma de Tamaulipas, Instituto de Investigaciones Históricas, Mexico City

Hill JH (2001) Proto-Uto-Aztecan: a community of cultivators in central Mexico? Am Anthropol 1003:913–934

Hill JH (2006) The historical linguistics of maize cultivation in Mesoamerica and North America. In: Staller JE, Tykot RH, Benz BF (eds) Histories of maize: multidisciplinary approaches to the prehistory, linguistics, biogeography, domestication, and evolution of maize. Elsevier, San Diego, CA, pp 631–645

Hillman GC (1996) Late Pleistocene changes in wild plant-foods available to hunters and gatherers in the Fertile Crescent. Possible preludes to cereal cultivation. In: Harris DR (ed) The origins and spread of agriculture and pastoralism in Eurasia. Smithsonian Institution Press, Washington D.C, pp 159–203

Hillman GC, Davies MS (1990) Measured domestication rates in crops of wild type wheats and barley and their archaeological implications. World Prehistory 4:157–222

Hillman GC, Davies MS (1992) Domestication rate in wild wheats and barley under primitive cultivation: Preliminary results and archaeological implications of field measurements of selection coefficient. In: Anderson PC (ed) Prehistoire de l'Agriculture: Nouvelles Approches Expérimentales et Ethnographiques, vol Monographic du CRA No.6. Centre National de la Recherche Scicntifque, Paris, pp 114–158

Hillman GC, Colledge SM, Harris DR (1989) Plant-food economy during the Epipalaeolithic period at Tell Abu Hureyra, Syria: Dietary diversity, seasonality, and modes of exploitation. In: Harris DR, Hillman G (eds) Foraging and farming: the evolution of plant exploitation. Unwin Hyman, London, pp 240–268

Hocquenghem AM (1991) Frontera entre "Areas Culturales" nor y centro Andinas en los valles y la costa del extremeo norte Peruano. Bulletin de l'Institut Français d'Etudes Andines 20(2):309–348

Hocquenghem AM (1993) Rutas de entrada del *mullu* en el extremo norte del Perú. Bulletin del 'Institut Français d'Etudes Andines 22(3):701–719

Hocquenghem AM, Idrovo J, Kaulicke P, Gomis D (1993) Bases del intercambio entre las sociedades norperuanas y surecuadorianas: una zona de transición entre 1500 A.C. and 600 D.C. Bulletin de l'Institut Français d'Etudes Andines 22(2):443–466

Hodell DA, Quinn RL, Brenner M, Kamenov G (2004) Spatial variation of strontium isotopes (87Sr/86Sr) in the Maya region: a tool for tracking ancient human migration. J Archaeol Sci 31:585–601

Holm O (1980) *La Cultura Chorrera Formativo Tardio Apr 1500 – 500 a.C.* Museo Antropológico Banco Central del Eucador, Guayaquil, Ilustre Concejo Municipal de San Lorenzo de Vinces, y colegio Fiscal "10 de Agosto" Guayaquil. Ecuador

Holst I, Moreno JE, Piperno DR (2007) Identification of teosinte, maize, and tripsacum in Mesoamerica by using pollen, starch grains, and phytoliths. Proc Natl Acad Sci USA 104 (45):17608–17631

Homza LA (2006) The Spanish Inquisition, 1478–1614: an anthology of sources. Hackett Publishing Company, Indianapolis

Hoopes JW (1994) Ford revisited: a critical review of the chronology and relationship of the earliest ceramic complexes in the New World, 6000-1500 B.C. J World Prehistory 8(1):1–49

Howie L, White CD, Longstaffe FJ (2009) Potographies and biographies: the role of food in ritual and identity as seen through life histories of selected Maya pots and people. Precolumbian foodways: interdisciplinary approaches to food, culture and markets in Mesoamerica. Springer, New York, pp 369–398

Hopkins NA (2006) The place of maize in Indigenous Mesoamerican Folk Taxonomies. In: Staller JE, Tykot RH, Benz BF (eds) Histories of Maize: Multidisciplinary approaches to the prehistory, linguistics, biogeography, domestication, and evolution of maize. Elsevier, San Diego, CA, pp 611–622

Horn SP (2006) Pre-Columbian maize agriculture in Costa Rica: pollen and other evidence from Lake and Swamp Sediments. In: Staller JE, Tykot RH, Benz BF (eds) Histories of Maize: Multidisciplinary approaches to the prehistory, linguistics, biogeography, domestication, and evolution of maize. Elsevier, San Diego, CA, pp 368–380

Horn S, Kennedy LM (2001) Pollen evidence of maize cultivation 2700 b.p. at La

Selva Biological Station (2001) Costa Rica. Biotropica 33(1):191–196

Huckell BB (1990) Late preceramic farmer-foragers in Southeastern Arizona: A Cultural and Ecological consideration of the Spread of Agriculture into the Arid Southwestern United States. Unpublished Ph.D. Dissertation, Department of Arid Lands Resource Sciences, University of Arizona, Tucson.

Huckell BB, Huckell L, Fish SK (1995) Investigations at Milagro, a Late Preceramic Site in the Eastern Tucson Basin. Technical Report 94-95. Center for Desert Archaeology, Tucson

Huckell L (2006) Ancient maize in the American Southwest: What does it look like and what can it tell us. In: Staller JE, Tykot RH, Benz BF (eds) Histories of Maize: Multidisciplinary approaches to the prehistory, linguistics, biogeography, domestication, and evolution of maize. Elsevier, San Diego, CA, pp 97–107

Hyland S (2003) The Jesuit and the Incas: the extraordinary life of Padre Blas Valera, S.J. The University of Michigan Press, Ann Arbor

Hyslop J (1990) Inka settlement planning. University of Texas Press, Austin, TX

Ife EW (1990) Introduction. In Christopher Columbus: Journal of the first Voyage. Edited and translated by E.W. Ife. Aris & Phillips Ltd, Warminster, England, pp. v–xxvi

Iltis HH (1971) The maize mystique - a reappraisal of the origin of corn. (photo-offset). Botany Department University of Wisconsin, Madison [1985. republished in Contrib. Univ. of Wisconsin-Madison Herbarium 5: 1-4]

Iltis HH (1972) The taxonomy of Zea mays (Gramineae). Phytologia 23:248–249

Iltis HH (1983) From teosinte to maize: the catastrophic sexual transmutation. Science 222:886–893

Iltis HH (2000) Homeotic sexual translocations and the origin of maize (Zea mays, Poaceae): a new look at an old problem. Econ Bot 54(1):7–42

Iltis HH (2004) Domestication of Zea: First for sugar and then for grain? A novel idea with vast implications. Paper presented at the 69th Annual Meeting of the Society for American Archaeology, Montreal

Iltis HH (2006) Polystichy in Maize. In: Staller JE, Tykot RH, Benz BF (eds) Histories of Maize: Multidisciplinary approaches to the prehistory, linguistics, biogeography, domestication, and evolution of maize. Elsevier, San Diego, CA, pp 21–53

Iltis HH, Benz BF (2000) Zea nicaraguensis (Poaceae), a new teosinte from Pacific coastal Nicaragua. Novon 10:382–390

Iltis HH, Doebley JF (1980) Taxonomy of Zea (Gramineae). II. Sub-specific categories in the Zea mays complex and a generic synopsis. Am J Bot 67:994–1004

Iltis HH, Doebley JF, Guzman MR, Pazy B (1979) Zea diploperennis (Gramineae): a new teosinte from Mexico. Science 203:186–188

Innes H (1969) The Conquistadores. Alfred A. Knopf, New York

Islebe GA, Hooghiemstra H, Brenner M, Curtis JH, Hodell DA (1996) A Holocene vegetation history from lowland Guatemala. Holocene 6:265–271

Jaenicke-Després V, Buckler ES, Smith BD, Gilbert TM, Cooper A, Doebley J, Pääbo S (2003) Early allelic selection in maize as revealed by ancient DNA. Science 302:1206–1208

Jaenicke-Després V, Smith BD (2006) Ancient DNA and the integration of archaeological and genetic approaches to the study of maize domestication. In: Staller JE, Tykot RH, Benz BF (eds) Histories of Maize: Multidisciplinary approaches to the prehistory, linguistics, biogeography, domestication, and evolution of maize. Elsevier, San Diego, CA, pp 83–95

Jakes KA (ed) (2002) Archaeological chemistry: materials, methods and meaning. American Chemical Society, Washington, DC

Jane J, Chen J (1992) Effect of amylose molecular size and amylopectin branch chain length on paste properties of starch. Cereal Chem 69:60–65

Jane J, Chen J, Lee LF, McPherson AE, Wong KS, Radosavljevic M, Kasemsuwan T (1999) Effects of amylopectin branch chain length and amylose content on the gelatinization and pasting properties of starch. Cereal Chem 76:629–637

Jimenez Borja A (1953) La comida en el antiquo Perú. Revista de Museo Nacional 22:113–134

Johnson F, MacNeish RS (1972) Chronometric dating. In: Johnson F (ed) The prehistory of the Tehuacán valley, vol 4, Chronology and Irrigation. University of Texas Press, Austin, TX, pp 3–58

Katzenberg MA (2000) Stable isotope analysis: a tool for studying past diet, demography, and life history. In: Katzenberg MA, Saunders SR (eds) Biological anthropology of the human skeleton. Wiley-Liss, New York, pp 305–328

Kaufman T (1994) The native languages of Mesoamerica. In: Moseley C, Asher RE (eds) Atlas of the World's languages. Routledge, London, pp 34–45

Kaplan L (1967) Archaeological Phaseolus from Tehuacan. In: Byers DS (ed) The prehistory of the Tehuacán valley, vol 1, Environment and subsistence. University of Texas Press, Austin, p 201

Kaplan L, Lynch TF (1999) *Phaseolus* (Fabacaea) in archaeology. AMS radiocarbon dates and their significance for Pre-Columbian agriculture. Econ Bot 53:261–272

Kelly ER, Marino AB, DeNiro MJ (1991) Stable isotope ratios of carbon in phytoliths as a quantitative method of monitoring vegetation and climate change. Quaternary Res 35:222–233

Kelly RL (1995) The foraging spectrum: diversity in hunter-gatherer lifeways. Smithsonian Institution Press, Washington, DC

Kellerman WA (1895) Primitive corn. Meehan's Mon 5(44):53

Kellogg E, Birchler J (1993) Linking phylogeny and genetics: *Zea mays* as a tool for phylogenetic studies. Syst Biol 42:415–439

Kimura M (1986) The Neutral theory of molecular evolution. Cambridge University Press, Cambridge

Kirkby MJ, Whyte AV, Flannery KV (1986) The physical environment of the Guilá Naquitz cave group. In: Flannery KV (ed) Guilá Naquitz: archaic foraging and early agriculture in Oaxaca. Academic, San Diego, CA, pp 43–62

Knapp S, Polaszek A, Watson M (2007) Spreading the word. Nature 446(15):261–262

Kolodny K, Luz B, Navon O (1983) Oxygen isotope variations in phosphate of biogenic apatites. Earth Planet Sci Lett 64:398–404

La Barre W (1938) Native American beers. Am Anthropol 40:224–234

Laden G (2006) Toward a biologically based method of phytolith classification. In: Staller JE, Tykot RH, Benz BF (eds) Histories of Maize: Multidisciplinary approaches to the prehistory, linguistics, biogeography, domestication, and evolution of maize. Elsevier, San Diego, CA, pp 123–128

Landa D de (1975) [1566] The Maya: Diego de Landa's Account of the Affairs of Yucatán. Edited and translated by A. R. Pagden. J Philip O'Hara, Chicago

Langham D (1940) The inheritance of intergeneric differences in *Zea Euchlaena* hybrids. Genetics 25:88–107

Las Casas B de (1971) [1527-1565] History of the Indies, Translated and edited by A. Collard. Harper and Row, New York

Lathrap DW (1970) Upper amazon. Thames and Hudson, London

Lathrap DW (1974) The moist tropics, the arid lands, and the appearance of the great art styles of the New World. In: King ME, Traylor I Jr (eds) Art and environment in native America. Texas Tech University Museum, Lubbock, pp 115–158

Lathrap DW, Collier D, Chandra H (1975) Ancient Ecuador culture, clay, and creativity 3000–300 BC. Field Museum of Natural History, Chicago

Latournerie Moreno L, Tuxill J, Yupit Moo E, Arias Reyes L, Cristobal Alejo J, Jarvis DI (2006) Traditional maize storage methods of Mayan farmers in Yucatan, mexico: implications for seed selection and crop diversity. Biodivers Conserv 15(5):1771–1795

Layfield J (1995) [1598] Relación del viaje a Puerto Rico de la expedición de Sir George Clifford, tercer conde de Cumberland, escrita por el Reverendo Doctor John Layfield, capellán de la expedición. (Fragmentos) Año 1598, In: Cronicas de Puerto Rico: desde la Conquista Hasta Nuestros Días (1493-1595), E. Fernandez Mendez (compiled and edited). Editorial Universitaria. Puerto Rico, San Juan

Letts J, Evans J, Fung MG, Hillman C (1994) A chemical method of identifying charred plant remains using infra-red spectroscopy. In: Johannessen S, Hastorf CA (eds) Corn and culture in the prehistoric New World. Westview Press, Boulder, pp 64–89

Lévi-Strauss C (1973) From honey to ashes: introduction to a science of mythology: 2. Translated by J. and D. Weightman, Harper and Row, New York

Lewis HT (1972) The role of fire in the domestication of plants and animals in Southwest Asia. Man 7(2):195–222

Lipp FJ (1991) The mixe of Oaxaca: religion, ritual, and healing. University of Texas Press, Austin, TX

Libby WF (1955) Radiocarbon dating. University of Chicago Press, Chicago

Lindley J (1846) A note upon the wild state of maize or Indian corn. J Hortic Soc Lond 1:114–117

Long A, Fritz GJ (2001) Validity of AMS dates on maize from the Tehuacán Valley: a comment on MacNeish and Eubanks. Lat Am Antiq 12:87–90

Long-Solís J (1986) Capsicum y cultura: La historia del chilli. D. F., Fondo de Cultura Economica, México

Long A, Benz BF, Donahue DJ, Jull AJT, Toolin LJ (1989) First direct AMS dates on early maize from Tehuacán, Mexico. Radiocarbon 31(3):1035–1040

López de Gómara F (1943 [1554]) Historía de la conquista de Mexico, vol 1. Editorial Pedro Robredo, México D.F

Lunniss R (2008) Where the land and ocean meet: the Engoroy phase ceremonial site at Salango, Ecuador. In: Staller JE (ed) Pre-Columbian landscapes of creation and origin. Springer, New York, pp 203–248

Lusteck R (2006) The migrations of maize into the Southeastern United States. In: Staller JE, Tykot RH, Benz BF (eds) Histories of maize: multidisciplinary approaches to the prehistory, linguistics, biogeography, domestication, and evolution of maize. Elsevier, San Diego, CA

Luz B, Kolodny Y, Horowitz M (1984) Fractionation of oxygen isotopes between mammalian bone-phosphate and environmental drinking water. Geochim Cosmochim Acta 48:1689–1693

MacNeish RS (1947) Preliminary report on coastal Tamaulipas, Mexico. Am Antiq 13:1–15

MacNeish RS (1958) Preliminary Archaeological Investigations in the Sierra de Tamaulipas, Mexico, vol 48. American Philosophical Society, Transactions, Philadelphia, PA Part 6

MacNeish RS (1961) First Annual Report of the tehuacan archaeological-botanical project. Publication of the Robert S, Peabody Foundation for Archaeology Phillips Academy, Andover, MA

MacNeish RS (1962) Second Annual Report of the tehuacan archaeological-botanical project. Publication of the Robert S, Peabody Foundation for Archaeology Phillips Academy, Andover, MA

MacNeish RS (1967a) Introduction. In: Byers DS (ed) The prehistory of the Tehuacán valley, vol 1, Environment and subsistence. University of Texas Press, Austin, pp 1–13

MacNeish RS (1967b) An interdisciplinary approach to an archaeological problem. In: Byers DS (ed) The Prehistory of the Tehuacán Valley, vol 1, Environment and subsistence. University of Texas Press, Austin, pp 14–24

MacNeish RS (1967c) A summary of the subsistence. In: Byers DS (ed) The Prehistory of the Tehuacán Valley, vol 1, Environment and subsistence. University of Texas Press, Austin, pp 290–309

MacNeish RS (1978) The Science of Archaeology? Duxbury Press, North Scituate, MA

MacNeish RS (1985) The archaeological record on the problem of the domestication of corn. Maydica 30:171–178

MacNeish RS (1992) The origins of agriculture and settled life. University of Oklahoma Press, Norman

MacNeish RS (2001a) Mesoamerican Chronology. In Oxford Encyclopedia of Mesoamerican Cultures. Oxford University Press, New York

MacNeish RS (2001b) A response to Long's radiocarbon determinations that attempt to put acceptable chronology on the fritz. Lat Am Antiq 12(1):99–104

MacNeish RS, Eubanks MW (2000) Comparative analysis of the Río Balsas and Tehuacán models for the origin of maize. Lat Am Antiq 11(1):3–20

MacNeish RS, Cook A, Lumbreras L, Vierra R, Nelken-Terner A (1981) Prehistory of the Ayacucho Basin, Peru, vol 2. University of Michigan Press, Ann Arbor

MacCormack S (2004) Religion and society in Inca and Spanish Peru. In: Phipps E, Hecht J, Martín CE (eds) The colonial andes: tapestries and silverwork 1530–1830, The Metropolitan Museum of Art, New York. New Haven, Yale University Press, pp 101–112

Madariaga S de (1947) The rise of the Spanish American empire. The Free Press, New York

Maltby W (2002) The reign of Charles V. Palgrave Publishers, New York

Mangelsdorf PC (1948) The role of pod corn in the origin and evolution of maize. Ann Missouri Bot Gard 35:377–406

Mangelsdorf PC (1958) Ancestor of corn. A genetic reconsideration yields clues to the nature of an extinct wild ancestor. Science 128:1313–1320

Mangelsdorf PC (1974) Corn: its origin evolution and improvement. The Belknap Press of Harvard University, Cambridge

Mangelsdorf PC (1983) The mystery of corn: new perspectives. Proc Am Philos Soc 127:215–247

Mangelsdorf PC (1986) The origin of corn. Sci Am 255:80–86

Mangelsdorf PC, Reeves RG (1931) Hybridization of maize, tripsacum, and euchlaena. J Hered 22:329–343

Mangelsdorf PC, Reeves RG (1938) The origin of maize. Proc Natl Acad Sci USA 24:303–312

Mangelsdorf PC, Reeves RG (1939) The origin of Indian corn and its relatives. Texas Agricultural Experiment Station. Bulletin No. 547 College Station, Texas.

Mangelsdorf PC, Reeves RG (1959a) The origin of corn. I. Pod corn, the ancestral form. Bot Mus Lealf Harv Univ 18:329–356

Mangelsdorf PC, Reeves RG (1959b) The origin of corn. IV. Place and time of origin. Bot Mus Lealf Harv Univ 18:413–427

Mangelsdorf PC, Galinat WC (1964) The tunicate locus in maize dissected and reconstituted. Proc Natl Acad Sci USA 51:147–150

Mangelsdorf PC, MacNeish RS, Galinat WC (1967) Prehistoric wild and cultivated maize. In: Byers DS (ed) The prehistory of the Tehuacán valley, vol 1, Environment and Subsistence. University of Texas Press, Austin, pp 178–200

Mangelsdorf PC, Barghoorn ES, Banerjee UC (1978) Fossil pollen and the origin of corn. Bot Mus Lealf Harv Univ 26:238–255

Manzanilla L (1987) Cobá. Análisis De Dos Unidades Habitacionales Mayas. Universidad Nacional Autónoma de México, Mexico, Quintana Roo

Manzanilla L (1996) Corporate groups and domestic activities at Teotihuacan. Lat Am Antiq 7:228–246

Manzanilla L, Barba L (1990) The study of activities in classic households: two case studies from Cobá and Teotihuacán. Anc Mesoamerica 1:41–49

Martire P. d' Anghiera (1907 [1516]) De orbe novo de Pierre Martyr Anghiera. Les huit décades tr. du latin, avec notes et commentaires par Paul Gaffarel. E. Leroux, Paris

Matson RG (2003) The Spread of Maize Agriculture into the U.S. Southwest. In: Bellwood P, Renfrew C (eds) Examining the Farming/Language Dispersal Hypothesis, McDonald Institute for Archaeological Research. University of Cambridge, Cambridge, pp 341–356

Matsuoka Y, Vigouroux Y, Goodman MM, Sanchez J, Buckler E, Doebley J (2002) A single domestication for maize shown by multilocus microsatellite genotyping. Proc Natl Acad Sci USA 99:6080–6084

Mayr E (1942) Systematics and the Origin of Species. Columbia University Press, New York

McClintock B (1960) Chromosome constitutions of Mexican and Guatemalan races of maize. Carnegie Inst Wash Year B 59:461–472

McClintock B (1976) Significance of chromosome constitutions in tracing the origin and migration of races of maize in the Americas. In: Walden DW (ed) Maize Breeding and Genetics. Wiley, New York, pp 159–189

McEwan C, van der Guchte M (1992) Ancestral time and sacred space in inca state ritual. In: Townsend RE (ed) The Ancient Americas Art from Sacred Landscapes. Art Institute of Chicago, Chicago, pp 359–371

McNeil CL (2006) Traditional Cacao Use in Modern Mesoamerica. In: McNeil CL (ed) Chocolate in Mesoamerica: a cultural history of Cacao. University Press of Florida, Gainesville, FL, pp 341–366

McNeil CL (2009) Death and chocolate: the significance of cacao offerings in ancient maya tombs and caches at Copan, Honduras. In: Staller JE, Carrasco MD (eds) Pre-Columbian Foodways: interdisciplinary approaches to food, culture and markets in Mesoamerica. Springer, New York, pp 293–314

McNeil CL, Hurst WJ, Sharer RJ (2006) The use and representation of cacao during the classic period at Copan, Honduras. In: McNeil CL (ed) Chocolate in Mesoamerica: a cultural history of Cacao. University Press of Florida, Gainesville, FL, pp 224–252

Meggers BJ (1966) Ecuador. Ancient peoples and places series 49, edited by Glynn Daniel. Praeger Publication, NewYork

Meggers BJ, Evans C Jr, Estrada VE (1965) Early formative of Coastal Ecuador: the valdivia and machalilla phases. Smithsonian Contributions in Anthropology, vol. 1. Washington DC

Memorias para la historia de Sinaloa. Carta Anua 1593. MS 227. University of California, Berkeley Bancroft Library. Berkeley

Milanich JL, Milbrath S (eds) (1989) First encounters: Spanish explorations in the Caribbean and the United States, 1492–1570. University of Florida Press. Florida Museum of Natural History, Gainesville, FL

Miller M, Taube K (1993) The Gods and symbols of ancient Mexico and the Maya. Thames and Hudson, London

Mitchem JM (1989) Artifacts of exploration: archaeological evidence from Florida. In: JL. Milanich, S. Milbrath (eds) First Encounters: Spanish explorations in the Caribbean and the United States, 1492-1570.University of Florida Press. Florida Museum of Natural History, Gainesville, FL, pp 99–109

Molina C de (El Cuzqueño) (1989) [ca. 1575]. Relación de los ritos y fábulas de los Ingas. In: Urbano H, Duvoils P (eds) Fábulas y mitos de los Incas. Crónicas de América series. Historia 16, Madrid, pp. 47–134

Monsalve J (1985) A pollen core from the Hacienda Lusitania. In: Bray W (ed) Pro Calima: Archälogische-ethnologisches Projekt im Westlichen Columbien/Sudamerika No. 5. Vereingung Pro Calima, Bern, pp 40–44

Moore JD (1989) Pre-Hispanic beer in coastal Peru: technology and social context of prehistoric production. Am Anthropol 91:682–695

Morris C (1979) Maize beer in the economics, politics, and religion of the Inca state. In: Gastineau C, Darby W, Turner T (eds) The role of fermented beverages in nutrition. Academic, New York, pp 21–34

Morris C (1993) The wealth of a Native American state. Value, investment, and mobilization in Inka economy. In: Henderson J, Netherly PJ (eds) Configurations of Power. Cornell University Press, Ithaca, pp 36–50

Morton JD, Schwarcz HP (2004) Palaeodietary implications from stable isotopic analysis of residues on prehistoric Ontario ceramics. J Archaeol Sci 31(5):503–517

Motolinía F, de Benavente T (1979[1528]) Historia de los Indios de la Nueva España. Editorial Porrúa, México

Morris C, Thompson DE (1985) Huánuco pampa: an inca city and its hinterlands. Thames and Hudson, New York

Mt Pleasant J (2006) The science behind the three sisters mound system: an agronomic assessment of an indigenous agricultural system in the Northeast. In: Staller JE, Tykot RH, Benz BF (eds)

Histories of Maize: Multidisciplinary approaches to the prehistory, linguistics, biogeography, domestication, and evolution of maize. Elsevier, San Diego, pp 529–537

Mulholland S, Rapp G (1992) A morphological classification of grass silica bodies. In: Rapp G Jr, Mulholland SC (eds) Phytolith Systematics: Emerging Issues, vol 1, Advances in archaeological and museum science. Plenum Press, New York

Murra JV (1975 [1972]) El control vertical de un máximo de pisos ecológicos en los sociedades andinas. In: Murra JV (ed) Formaciones Económicas y Politicas del Mundo Andino. Instituto de Estudios Peruanos, Lima, pp 59–115

Murra JV (1973) Rite and crop in the Inca State. In: DR Gross (ed) Peoples and cultures of South America. Natural History Press, Doubleday, Garden City NJ, pp 377–389

Murra JV (1980) Economic Organization of the Inca State. JAI Press, Greenwich, Connecticut

Murra JV (1982) The mit'a obligation of ethnic groups to the Inca State. In: Collier G, Rosaldo R, Wirth J (eds) The Inca and Aztec States, 1400–1800. Academic, New York, pp 237–264

Murua M de (1962 [1590]) Historia general del Perú, origen y descendencia de los Incas. Biblioteca Americana Vetus, Madrid

Myrick H (1904) The book of corn; a complete treatise upon the culture, marketing and uses of maize in America and elsewhere, for farmers, dealers, manufacturers and others–a comprehensive manual upon the production, sale, use and commerce of the world's greatest crop. O. Judd Company, New York

Neff H, Arroyo B, Jones JG, Pearsall DM, Freidel, D. E. 2002. Nueva evidencia pertinente a la ocupación temprana del sur de Mesoamérica. Paper Presented at the XII Encuentro Internacional: Los Investigadores de la Cultura Maya, Campeche, November 10-14, 2002. University of Campeche, Mexico

Niederberger C (1979) Early sedentary economy in the basin of Mexico. Science 203:131–142

Newsom LA, Deagan KA (1994) Zea mays in the West Indies: the archaeological and early historic record. In: S. Johannessen, C.A. Hastorf (eds) Corn and culture in the prehistoric New World. Westview Press, Boulder, CO, pp. 203–217

Newsom LA (1996) The population of the Amazon basin in 1492: a view from the Ecuadorian headwaters. Trans Inst Br Geogr 21(1):5–26

Newsom LA (2006) Caribbean maize: first farmers to Columbus. In: Staller JE, Tykot RH, Benz BF (eds) Histories of Maize: Multidisciplinary approaches to the prehistory, linguistics, biogeography, domestication, and evolution of maize. Elsevier, San Diego, CA, pp 325–335

Nicholson GE (1960) Chicha maize types and chicha manufacturing in Peru. Econ Bot 14:290–299

Ochoa Salas L, Jaime Riverón O (2005) The cultural mosaic of the Gulf Coast during the Pre-Hispanic Period. In: Sandstrom AR, García Valencía EH (eds) Native peoples of the Gulf Coast of Mexico. University of Arizona Press, Tucson, pp 22–44

Olmsted I (1993) Wetlands of Mexico. In: Whigham DF, Dykyjová D, Slavomil H (eds) Wetlands of the World: inventory, ecology, and management, Handbook of Vegetation Science, pt. 15/2. Kluwer, Dordrecht, pp 637–677

Ortiz A, Barba L (1993) La Química En Los Estudios De Áreas De Actividad. In: Manzanilla L (ed) Anatomia De Un Conjunto Residencial Teotihuacano En Oztoyahualco, vol 2. Universidad Nacional Autónoma de México, Mexico D.F, pp 617–660

Otero GA (1951) La piedra mágica. Instituto Indiginista Americano, Mexico D.F

Parker PL (1964) The biochemistry of the stable isotope fractionation during photosynthesis. Geochim Cosmochim Acta 28:1155–1164

Parker III TA, Carr JL (eds) (1992) Status of Forest Remnants in the Cordillera de la Costa and Adjacent Areas of Southwestern Ecuador Rapid Assessment Program (RAP Working Papers 2) Conservation International October 1992

Parnell JJ, Terry RE, Golden G (2001) The use of in-field phosphate testing for the rapid identification of Middens at Piedras Negras, Guatemala. Geoarchaeol Int J 16:855–873

Parnell JJ, Terry RE, Nelson Z (2002a) Soil chemical analysis applied as an interpretive tool for ancient human activities at Piedras Negras, Guatemala. J Archaeol Sci 29:379–404

Parnell JJ, Terry RE, Sheets PD (2002b) Soil chemical analysis of ancient activities in Cerén, El Salvador: A case study of a rapidly abandoned Site. Lat Am Antiq 13:331–342

Parsons JR (2006) The last pescadores of chimalhuacán, mexico: an archaeological ethnography. Anthropological Papers Museum of Anthropology, University of Michigan, Ann Arbor, MI, Number 96

Parsons JR (2009) The Pastoral Niche in Pre-Hispanic Mesoamerica. In: Staller JE, Carrasco MD (eds) Pre-Columbian Foodways: Interdisciplinary approaches to food, culture and markets in Mesoamerica. Springer, New York, pp 108–136

Parsons JR, Parsons M (1990) Maguey utilization in highland central Mexico: an archaeological ethnography, appendix by Sandra L. Dunavan. Anthropological Papers, Museum of Anthropology, University of Michigan, Ann Arbor, MI, Number 96

Paulsen AC (1974) The Thorny Oyster and the voice of God: *Spondylus* and *Strombus* in Andean Prehistory. Am Antiq 39(4):597–607

Pearsall DM (1978) Phytolith analyses of archaeological soils: Evidence of maize cultivation in Formative Ecuador. Science 199:177–178

Pearsall DM (1979) The application of ethnobotanical techniques to the problem of subsistence in the Ecuadorian formative. Ph.D. dissertation, Department of Anthropology, University of Illinois, Urbana

Pearsall DM (1989) Paleoethnobotany: a handbook of procedures. Academic, San Diego, CA

Pearsall DM (1992) The origins of plant cultivation in South America. In: C. Wesley Cowan, Patty Jo Watson, with the assistance of N. L. Benco (eds) The origins of agriculture: an International perspective. Smithsonian Institution Press, Washington DC, pp. 173–205

Pearsall DM (1994) Issues in the analysis and interpretation of archaeological maize in South America. In: Johannesson S, Hastorf CA (eds) Corn and Culture in the Prehistoric New World. Westview Press, Boulder, pp 245–272

Pearsall DM (1999) The impact of maize on subsistence systems in South America: an example from the Jama River Valley, Coastal Ecuador. In: Gosden C, Hather J (eds) The prehistory of food: appetites for change. Routledge, London, pp 419–437

Pearsall DM (2000) Paleoethnobotany: a handbook of procedures, 2nd edn. Academic, San Diego, CA

Pearsall DM (2002) Maize is *still* ancient in Prehistoric Ecuador: a view from Real Alto, with comments on Staller and Thompson. J Archaeol Sci 29(1):51–55

Pearsall DM (2003) Plant food resources of the ecuadorian formative: an overview and comparison to the central andes. In: Raymond JS, Burger RL (eds) Archaeology of formative ecuador: a symposium at Dumbarton Oaks, 7 and 8 October 1995. Dumbarton Oaks Research Library and Collection, Washington D.C, pp 213–257

Pearsall DM, Piperno DR (1990) Antiquity of maize cultivation in Ecuador: summary and reevaluation of the evidence. Am Antiq 55:324–337

Pearsall DM, Piperno DR (1993) The nature and status of phytolith analysis. In: Pearsall DM, Piperno DR (eds) Current research in phytolith analysis: applications in archaeology and paleoecology. MASCA, University of Pennsylvania Museum, Philadelphia, pp 9–18

Pearsall DM, Chandler-Ezell K, Chandler-Ezell A (2003) Identifying maize in Neotropical sediments and soils using cob phytoliths. J Archaeol Sci 30:611–627

Perales H, Brush SB, Qualset C (2003a) Maize landraces of central Mexico: an altitudinal transect. Econ Bot 57:7–20

Perales H, Brush SB, Qualset C (2003b) Dynamic management of maize landraces in central Mexico. Econ Bot 57:21–34

Perales H, Benz B, Brush S (2005) Maize diversity and ethnolinguistic diversity in Chiapas, Mexico. Proc Natl Acad Sci USA 102:949–952

Pérez J (2005) The Spanish inquisition: a history. Yale University Press, Translated by J. Lloyd. New Haven

Perry L, Sandweiss DH, Piperno DR, Rademaker K, Malpass MA, Umire A, Vera P, Umire A (2006) Early maize agriculture and interzonal interaction in southern Peru. Nature 440(2):76–79

Perry L, Dickau R, Zarrillo S, Holst I, Pearsall DM, Piperno DR, Berman MJ, Cooke RG, Rademaker K, Ranere AJ, Raymond JS, Sandweiss DH, Scaramelli F, Tarble K, Zeidler JA (2007) Starch Fossils and the Domestication and Dispersal of Chili Peppers (Capsicum spp. L.) in the Americas. Science 315:986–988

Pillsbury J (1996) The Thorny Oyster and the origins of Empire: implications of recently uncovered *Spondylus* imagery from Chan Chan, Peru. Lat Am Antiq 7(4):313–340

Piperno DR (1984) A comparison and differentiation of phytoliths from maize (Zea mays L.) and wild grasses: use of morphological criteria. Am Antiq 49:361–383

Piperno DR (1985) Phytolithic analysis of geological sediments from Panama. Antiquity 59:13–19

Piperno DR (1988) Phytolith analysis: an archaeological and geological perspective. Academic, San Diego, CA

Piperno DR (1991) The status of phytolith analysis in the American tropics. J World Prehistory 5(2):155–191

Piperno DR (1994) On the emergence of agriculture in the New World. Curr Anthropol 35 (5):637–643

Piperno DR (1999) The origins and development of food production in Pacific Panama. In: Blake M (ed) Pacific Latin America in prehistory: The evolution of Archaic and Formative cultures. Washington State University Press, Pullman, pp 123–134

Piperno DR, Holst I, Ranere AJ, Hansell P, Stothert K (2001) The occurrence of genetically controlled phytoliths from maize cobs and starch grains from maize kernels on archaeological stone tools and human teeth, and in archaeological sediments from southern Central America and Northern South America. Phytolith Bull Soc Phytolith Res 13(2&3):1–7

Piperno DR (2006) Phytoliths: a comprehensive guide for archaeologists and paleoecologists. AltaMira Press, Lanham MD

Piperno DR, Flannery KV (2001) The earliest archaeological maize (*Zea mays* L.) from highland Mexico: new accelerator mass spectrometry dates and their implications. Proc Natl Acad Sci USA 98:2101–2103

Piperno DR, Jones JG (2003) Paleoecological and archaeological implications of a Late Pleistocene/Early Holocene Record of vegetation and climate from the Pacific Coastal plain of Panama. Quaternary Res 59:79–87

Piperno DR, Pearsall DM (1993) Phytoliths in the reproductive structures of maize and teosinte: Implications for the study of maize evolution. J Archaeol Sci 17:665–677

Piperno DR, Pearsall D (1998) The origins of agriculture in the lowland neotropics. Academic, San Diego, CA

Piperno DR, Bush MB, Colinvaux PA (1990) Paleoenvironments and human occupation in late-glacial Panama. Quaternary Res 33:108–116

Piperno DR, Clarey KH, Cooke RG, Ranere AJ, Weiland D (1985) Preceramic maize in central Panama: Phytolith, Pollen evidence. Am Anthropol 87:871–878

Piperno DR, Ranere AJ, Moreno JE, Iriarte J, Holst I, Lachniet M (2004) Preliminary results of investigations into maize history in the central Balsas watershed. Paper presented at an invited symposium, "The Stories of Maize I-IV." Organized by John P. Hart, Michael Blake, John E. Staller, and Robert G. Thompson at the 69th Annual Meeting of the Society for American Archaeology, Montreal, Canada

Piperno DR, Ranere AJ, Moreno JE, Iriarte J, Holst I, Lachniet M, Jones JG, Ranere AJ, Castanzo R (2007) Late Pleistocene and Holocene environmental history of the Iguala Valley, Central Balsas Watershed of Mexico. Proc Natl Acad Sci USA 104:11874–11881

Piperno DR, Ranere AJ, Holst I, Iriarte J, Dickau R (2009) Starch grain and phytolith evidence for early ninth millennium B.P. maize from the Central Balsas River Valley, Mexico. Proc Natl Acad Sci USA 106:5019–5024

Pizarro P (1978[1571]) Relación del Discubrimiento y Conquista del Perú. Pontífica Universidad Católica del Perú, Lima

Pizarro P (1921[1571]) Relation of the discovery and conquest of the kingdoms of Peru. Translated & edited by P. Ainworth Means. The Cortés Society, New York

Pohl MED, Pope KO, Jones JG, Jacob J, Piperno D, de France S, Lentz DL, Gifford J, Danforth M, Josserand JK (1996) Early agriculture in the Maya Lowlands. Lat Am Antiq 7(4):355–372

Pohl MED, Piperno DR, Pope KO, Jones JG (2007) Microfossil evidence for pre-Columbian maize dispersals in the neotropics from San Andrés, Tabasco, Mexico. Proc Natl Acad Sci USA 104:6870–6875

Pope KO, Dahlin BH (1989) Ancient Maya wetland agriculture: new insights from ecological and remote sensing research. J Field Archaeol 16:87–106

Pope KO, Pohl M, Jones JG, Lentz DL, von Nagy C, Varga FJ, Quitmyer IR (2001) Origin and environmental settings of ancient agriculture in the Lowlands of Mesoamerica. Science 292:1370–1373

Price TD (ed) (2000) Europe's First Farmers. Cambridge University Press, Cambridge

Price TD, Manzanilla L, Middleton WD (2000) Immigration and the ancient city of Teotihuacan in Mexico: a study using strontium isotope ratios in human bone and teeth. J Archaeol Sci 27:903–913

Price TD, Burton JH, Bentley RA (2002) The characterization of biologically available strontium isotope ratios for the study of prehistoric migration. Archaeometry 44:117–135

Provine WB (2004) Ernst Mayr: genetics and speciation. Genetics 167:1041–1046

Rahman A, Wong K, Jane J, Myers AM, James MG (1998) Characterization of SU1 isoamylase, a determinant of storage starch structure in maize. Plant Physiol 117:425–435

Rainey KD, Spielmann KA (2006) Protohistoric and contact period salinas pueblo maize: trend or departure. In: Staller JE, Tykot RH, Benz BF (eds) Histories of Maize: Multidisciplinary approaches to the prehistory, linguistics, biogeography, domestication, and evolution of maize. Elsevier, San Diego, CA, pp 487–496

Ramirez ER, Nicholson Calle GE, Anderson E, Brown WL (1960) Races of maize in Bolivia. National Academy of Sciences, National Research Council, Washington

Randolph L (1975) Contributions of the wild relatives of maize to evolutionary history to domesticated maize: a synthesis of divergent hypotheses I. Econ Bot 30:321–345

Ranere AJ, Piperno DR, Holst I, Dickau R, Iriarte J (2009) The cultural and chronological context of early Holocene maize and squash domestication in the Central Balsas River Valley, Mexico. Proc Natl Acad Sci USA 106:5014–5018

Rankin RL (2006) Siouan tribal contacts and dispersions evidenced in the terminology for *maize* and other cultigens. In: Staller JE, Tykot RH, Benz BF (eds) Histories of Maize: Multidisciplinary approaches to the prehistory, linguistics, biogeography, domestication, and evolution of maize. Elsevier, San Diego, CA, pp 563–575

Ransom SL, Thomas M (1960) Crassulacean acid metabolism. Annu Rev Plant Physiol 11:81–110

Raven PH (2005) Transgenes in Mexican maize: desireability or inevitability. Proc Natl Acad Sci USA 102:13003–13004

Raymond JS, DeBoer WR (2006) Maize on the Move. In: Staller JE, Tykot RH, Benz BF (eds) Histories of Maize: Multidisciplinary approaches to the prehistory, linguistics, biogeography, domestication, and evolution of maize. Elsevier, San Diego, CA, pp 337–342

Raynor GS, Ogden EC, Hayes KV (1972) Dispersion and deposition of corn pollen from experimental sources. Agron J 64:420–427

Reber EA (2006) Hard row to hoe: changing maize use in the American bottom and surrounding areas. In: Staller JE, Tykot RH, Benz BF (eds) Histories of Maize: Multidisciplinary approaches to the prehistory, linguistics, biogeography, domestication, and evolution of maize. Elsevier, San Diego, pp 236–248

Reber EA, Evershed RP (2004) Identification of maize in absorbed organic residues: a cautionary tale. J Archaeol Sci 31:399–410

Reber EA, Dudd SN, van der Merwe NJ, Evershed RP (2004) Direct detection of maize processing in archaeological pottery through compound-specific stable isotope analysis of *n*-dotriacontanol in absorbed organic residues. Antiquity 78(301):682–691

Redman CL (1999) Human impacts on ancient environments. University of Arizona Press, Tucson

Reinhard J (1985) Chavín and Tiahuanaco: a new look at two Andean Ceremonial Centers. Natl Geogr Res 1:395–422

Renfrew C (2000) Towards a population prehistory of Europe. In: Renfrew C, Boyle K (eds) Archaeogenetics: DNA and the population history of Europe. McDonald Institute for Archaeological Research, Cambridge, pp 3–11

Richards M (2003) The Neolithic invasion of Europe. Annu Rev Anthropol 32:135–162

Rindos D (1984) The origins of agriculture; an evolutionary perspective. With forward by R.C. Dunnell. Academic, Orlando

Rivera MA (2006) Prehistoric maize from Northern Chile. In: Staller JE, Tykot RH, Benz BF (eds) Histories of Maize: Multidisciplinary approaches to the prehistory, linguistics, biogeography, domestication, and evolution of maize. Elsevier, San Diego, CA, pp 403–413

Rodriquez MF, Aschero CA (2000) Archaeological evidence of Zea mays L. (POACEAE) in the Southern Argentinean Puna (Antofagasto de la Sierra, Catamarca). J Ethnobiol 27:256–271

Rose F (2008) Intra-community variation in diet during the adoption of a new staple crop in the Eastern Woodlands. Am Antiq 73(3):413–439

Rostworowski de Diez Canseco M (1986) Estructuras andinas del poder. Ideología religiosa y política, 2nd edn. Instituto de Estudios Peruanos, Lima

Rostworowski de Diez Canseco M (1999) History of the Inca Realm. Cambridge University Press, Translated by Harry B. Iceland. Cambridge

Rossen J, Dillehay TD, Ugent D (1996) Ancient cultigens or modern intrusions? Evaluating plant remains in an Andean case study. J Archaeol Sci 23:391–407

Rouse I, Cruxent JM (1963) Venzuelan Archaeology. Yale University Press, New Haven

Roush W (1997) Squash seeds yield new view of early American farming. Science 276 (5314):894–895

Roth BJ (1989) Late Archaic settlement and subsistence in the Tucson Basin. Ph.D. Dissertation, Department of Anthropology, University of Arizona, Tucson

Rovner I (1971) Potential of opal Phytoliths for use in Paleoecological reconstruction. Quaternary Res 1:343–359

Rovner I (1983) Plant Opal Phytolith Analysis: Major Advances in Archaeobotanical Research. In: Schiffer M (ed) Advances in archaeological method and theory, vol 6. Academic, New York, pp 225–266

Rovner I (1995) Mien, mean, and meaning. The limits of typology in phytolith analysis. Paper presented at the 60th Annual Meetings of the Society for American Archaeology, Minneapolis, Minn.

Rovner I (1999) Phytolith analysis. Science 283:488–489

Rovner I (2004) On transparent blindfolds: comments on identifying maize in Neotropical sediments and soils using cob phytoliths. J Archaeol Sci 31:815–819

Rovner I, Gyuli F (2007) Computer-assisted morphometry: a new method for assessing and distinguishing morphological variation in wild and domestic seed populations. Econ Bot 61 (2):154–172

Rovner I, Russ J (1992) Darwin and design in phytolith systematics: morphometric methods for mitigating redundancy. In: Rapp GR Jr, Mulholland SC (eds) Phytolith systematics: emerging issues, vol 1, Advances in Archaeological and Museum Science. Plenum Press, New York, pp 253–276

Russ JC, Rovner I (1989) Stereological identification of opal phytolith populations from wild and cultivated zea. Am Antiq 54:784–792

Rowe JH (1944) An introduction to the archaeology of Cuzco. Pap Peabody Mus Am Archaeol Ethnol 27(2):3–69

Rowe JH (1961) Stratigraphy and seriation. Am Antiq 26(3):324–330

Rue DJ (1989) Archaic Middle American agriculture and settlement: recent pollen data from Honduras. J Field Archaeol 16(2):177–184

Rust WF, Leyden BW (1994) Evidence of maize use at Early and Middle Preclassic La Venta Olmec sites. In: Johannessen S, Hastorf CA (eds) Corn and Culture in the Prehistoric New World. Westview Press, Boulder, CO, pp 181–202

Sachse F (2008) Over distant waters: origin places and creation in colonial k'iche'an sources. In: Staller JE (ed) Pre-Columbian landscapes of creation and origin. Springer, New York, pp 123–160

Sage RF, Wedin DA, Meirong L (1999) The Biogeography of C_4 Photosynthesis: Patterns and Controlling Factors. In: Sage RF, Monson RK (eds) C_4 Plant Biology. Academic, San Diego, CA, pp 313–371

Sahagún Fr B de (1963) [1590] Florentine Codex: General History of the Things of New Spain. Book 11: Earthly Things, translated and edited by C.E. Dibble and A.J. Anderson. Monographs of the School of American Research and the Museum of New Mexico, Santa Fe. Salt Lake City: University of Utah Press

Salisbury JH (1848) Maize or Indian corn. Trans N Y Agric Soc 8:678–873

Sandstrom AR (1991) Corn is our Blood: Cultural and Ethnic Identity in a contemporary Aztec Village. University of Oklahoma Press, Norman

Sarmiento de Gamboa P (1942) 1572. Historia de los Incas. Buenos Aires, Emece Editores

Sauer CO (1950) Cultivated plants of South and Central America. In: Steward JH (ed.) Handbook of South American Indians. Physical Anthropology, Linguistics and Cultural Geography of South American Indians, vol. 6. Smithsonian Institution Bureau of American Ethnology Bulletin 143. U.S. Government Printing Office, Washington DC, pp 487–543

Sauer CO (1952) Agricultural origins and dispersals. American Geographic Society, New York

Sauer CO (1969) The Early Spanish Main. The University of California Press, Berkeley

Scheffler TE (2002) El Gigante Rock Shelter: archaic Mesoamerica and Transitions to Settled Life. Reports Submitted to FAMSI. www.famsi.org/reports/00071/index.html

Schiebinger L (2004) Plants and empire: colonial bioprospecting in the atlantic World. Harvard University Press, Cambridge

Schiebinger L, Swan C (2005) Introduction. In: Schiebinger L, Swan C (eds) Colonial botany: science, commerce, and politics in the early modern World. University of Pennsylvania Press, Philadelphia, pp 1–16

Schoeninger MJ, Kohn MJ, Valley JW (2000) Tooth oxygen isotope ratios as paleoclimate monitors in arid ecosystems. In: Ambrose SH, Katzenberg MA (eds) Biogeochemical approaches to paleodietary analysis. Plenum, New York, pp 117–140

Schoenwetter J, Smith LD (1986) Pollen analysis of the oaxaca archaic. In: Flannery KV (ed) Guilá Naquitz, Archaic Foraging and early agriculture in Oaxaca, Mexico. Academic, Orlando, pp 179–218

Scholes FV, Roys RL (1968) The Maya chontal Indians of Acalan-Tixchel. a contribution to the history and ethnography of the yucatan peninsula. University of Oklahoma Press, Norman with the assistance of Eleanor B. Adams and Robert S. Chamberlain

Schwarcz HP (2006) Stable carbon isotope analysis and human diet: a synthesis. In: Staller JE, Tykot RH, Benz BF (eds) Histories of Maize: Multidisciplinary approaches to the prehistory, linguistics, biogeography, domestication, and evolution of maize. Elsevier, San Diego, CA, pp 315–321

Schwartz SB (2000) Victors and vanquished: Spanish and Nahua Views of the Conquest of Mexico. Bedford/St. Martin's, Boston, MA

Gines de Sepulveda J, de Las Casas B (1975 [c. 1540]) Apología. In: Gines de Sepulveda J, de las Casas Bartolome F (eds) Traducción castellana de los textos originales latinos, introduccion, notas e indices por Angel Losada. Editorial Nacional, Madrid

Sevilla R (1994) Variation in modern Andean maize and its implications for prehistoric patterns. In: Johannessen S, Hastorf CA (eds) Corn & culture in the prehistoric new World. Westview Press, Boulde, pp 219–244

Service ER (1975) Origins of the state and civilization: the process of cultural evolution. W. W. Norton, New York

Serra M, Lazcano JC (2009) The drink mescal its origins and ritual uses. In: Staller JE, Carrasco MD (eds) Pre-Columbian Foodways: Interdisciplinary approaches to food, culture and markets in Mesoamerica. Springer, New York, pp 137–156

Siemens AH, Puleston DE (1972) Ridged Fields and associated features in Southern Campeche: new perspectives on the Lowland Maya. Am Antiq 37(7):228–239

Shady R (2006) Caral-Supe and the North-Central area of Peru: the history of maize in the land where civilization came into being. In: Staller JE, Tykot RH, Benz BF (eds) Histories of Maize: Multidisciplinary approaches to the prehistory, linguistics, biogeography, domestication, and evolution of maize. Elsevier, San Diego, CA, pp 381–402

Simmons AH (1986) New evidence for the use of early cultigens in the American Southwest. Am Antiq 51(1):73–89

Sluyter A (1997) Analysis of maize (*Zea mays* subsp. *mays*) pollen: normalizing the effects of microscopic – slide mounting media on diameter determinations. Palynology 21:35–39

Sluyter A, Dominguez G (2006) Early maize (*Zea mays* L.) cultivation in Mexico: dating sedimentary pollen records and its implications. Proc Natl Acad Sci USA 103:1147–1151

Smalley J, Blake TM (2003) Sweet beginnings: stalk sugar and the domestication of maize. Curr Anthropol 44(5):675–703

Smith BD (1977) Archaeological inference and inductive confirmation. Am Anthropol 79(3):598–617

Smith BD (1997a) The initial domestication of Curcurbita pepo in the Americas 10, 000 years ago. Science 276:932–934

Smith BD (1997b) Reconsidering the Ocampo caves and the era of incipient cultivation in Mesoamerica. Lat Am Antiq 8:342–383

Smith BD (1998) The emergence of agriculture. Scientific American Library Publication. W. H. Freedman and Company, New York

Smith BD (2000) Guilá Naquitz revisited: agricultural origins in Oaxaca, Mexico. In: Fienman G, Manzanilla L (eds) Cultural evolution, contemporary viewpoints. Plenum Press, New York, pp 15–59

Smith BD (2001) Documenting plant domestication: the consilience of biological and archaeological approaches. Proc Natl Acad Sci USA 98:1324–1326

Smith BD (2005a) Reassessing Coxcatlan cave and the early history of domesticated plants in Mesoamerica. Proc Natl Acad Sci USA 102:9438–9944

Smith BD (2005b) Low level food production and the Northwest Coast. In: Duer D, Turner N (eds) Traditions of plant use and cultivation in the Northwest Coast. University of Washington Press, Seattle, pp 37–66

Smith BD (2006) Documenting plants in the archaeological record. In: Zeder MA, Bradley DG, Emshwiller E, Smith BD (eds) Documenting domestication: new genetic and archaeological paradigms. University of California Press, Berkeley, pp 15–24

Smith BD, Yarnell RA (2009) Initial formation of an indigenous crop complex in eastern North America at 3800 B.P. Proc Natl Acad Sci USA 106:6561–6566

Smith CE Jr (1967) Plant remains. In: Byers DS (ed) The prehistory of the Tehuacán valley, vol 1, Environment and subsistence. University of Texas Press, Austin, TX, pp 220–255

Smith CE Jr (1986) Preceramic plant remains from Guilá Naquitz. In: Flannery KV (ed) Guilá Naquitz: archaic foraging and early agriculture in Oaxaca. Academic, San Diego, CA, pp 265–274

Spencer HJ (1864) The classification of the sciences: which are added reasons for dissenting from the philosophy of M. Comte. D. Appleton and Company, New York

Spencer HJ (1897) The principles of sociology. Appleton, New York

Spinden HJ (1917) The origin and distribution of agriculture in America. In: Proceedings of the19[th] International Congress of Americanists (1915), Washington DC, pp 269–276

Spores R (1980) New World Ethnohistory and archaeology (1970–1980). Annu Rev Anthropol 9:575–603

Staller JE (1994) Late Valdivia Occupation in El Oro Province Ecuador: Excavations at the Early Formative Period (3500-1500 B.C.) site of La Emerenciana, Ph.D. dissertation. Department of Anthropology, Southern Methodist University, Dallas, Texas. University Microfilms, Ann Arbor

Staller JE (1996) El sitio Valdivia tardio de La Emerenciana en la costa sur del Ecuador y su significación del desarollo de complejidad en la costa de sudamerica. Cuadernos de Historia y

Arqueología 48-50: 65-118. Edición en homenaje a Olaf Holm. Publicación de la Casa la Cultura Ecuatoriana. Benjamin Carrion, Nucleo del Guayas, Guayaquil

Staller JE (2000) Political and prehistoric frontiers: How history influences our understanding of the past. In: Boyd M, Erwin JC, Hendrickson M (eds) The entangled past: integrating history and archaeology, pp. 242–258. Proceedings of the 30th Annual Chacmool Conference Calgary, Alberta. The Archaeological Association of The University of Calgary. Alberta, Canada

Staller JE (2001a) Reassessing the chronological and developmental relationships of the Formative of coastal Ecuador. J World Prehistory 15(2):193–255

Staller JE (2001b) The Jelí Phase Complex at La Emerenciana, a late Valdivia site in southern El Oro Province, Ecuador. Andean Past 6:117–174. Cornell University, Latin American Studies Program

Staller JE (2003) An examination of the paleobotanical and chronological evidence for an early introduction of maize (Zea mays L.) into South America: a response to Pearsall. J Archaeol Sci 30(3):373–380

Staller JE (2004) Domesticación vs. Domesticando: Revaluando los Componentes Primarios del "Formativo" en el Centro & NO Sudamérica. Revista Arqueologia del Area Intermedia 6:51–82. Instituto Colombiano de Antropologia e Historia, & Sociedad Colombiana de Arqueología. Bogotá, Colombia

Staller JE (2005) La agricultura de la irrigación en el Período Formativo del desierto de Atacama. In: Rivera MA (ed.) Arqueología del Desierto de Atacama, Chile: La Etapa Formativa en el Area de Ramaditas/Guatecondo, Santiago: Ediciones Universidad Bolivarana. Chile, pp 89–102

Staller JE (2006a) An introduction to the histories of maize. In: Staller JE, Tykot RH, Benz BF (eds) Histories of Maize: Multidisciplinary approaches to the prehistory, linguistics, biogeography, domestication, and evolution of maize. Elsevier, San Diego, CA, pp xxi–xxv

Staller JE (2006b) The social, symbolic and economic significance of zea mays l. in the late horizon period. In: Staller JE, Tykot RH, Benz BF (eds) Histories of Maize: Multidisciplinary approaches to the prehistory, linguistics, biogeography, domestication, and evolution of maize. Elsevier, San Diego, CA, pp 449–467

Staller JE (2006c) Domesticación de Paisajes: ¿Cuáles son los componentes primarios del Formativo? Revista Estudios Atacameños 32: 43–58. Universidad Católica del Norte, Chile

Staller JE (2007a) Una aproximación interdisciplinaria para nuestra comprensión de la introducción y el rol temprano del maíz (Zea mays L.) en los Andes Occidentales. In: Oliva F, Grandis de N, Rodríguez J (eds) Arqueología Argentina en los Inicios de un Nuevo Siglo, pp. 23-38. Tomo I. Publicación del XIV. Rosario: Congreso Nacional de Arqueología, Sept. 17–21, 2001. (Keynote Address)

Staller JE (2007b) Un reevaluación del papel de la ideología en el intercambio de larga distancia temprano y a los orígenes de la civilización andina. In F. García (ed.) II Congreso Ecuatoriano de Antropologia y Arqueologia: Balance de la última década: aportes, retos y nuevos temas, pp. 511-548, Tomo 1. Abya Yala, Quito

Staller JE (2008a) Introduction to Pre-Columbian landscapes of creation and origin. In: Staller JE (ed) Pre-Columbian landscapes of creation and origin. Springer, New York, pp 1–9

Staller JE (2008b) Dimensions of place: the significance of centers to the development of andean civilization: an examination of the ushnu concept. In: Staller JE (ed) Pre-Columbian landscapes of creation and origin. Springer, New York, pp 269–313

Staller JE (2009) Ethnohistoric sources on foodways, feasts and festivals in Mesoamerica. In: Staller JE, Carrasco MD (eds) Pre-Columbian Foodways: interdisciplinary approaches to food, culture and markets in Mesoamerica. Springer, New York, pp 23–69

Staller JE, Carrasco MD (2009) Pre-Columbian foodways in Mesoamerica. In: Staller JE, Carrasco MD (eds) Pre-Columbian foodways: interdisciplinary approaches to food, culture and markets in Mesoamerica. Springer, New York, pp 1–20

Staller JE, Thompson RG (2000) Reconsiderando la Introdución del Maíz en el Occidente de America del Sur. Bulletin de l'Institut Français d'Etudes Andines 30(1):123–156. Lima

Staller JE, Thompson RG (2002) A multidisciplinary approach to understanding the initial introduction of maize into coastal Ecuador. J Archaeol Sci 29(1):33–50

Staller JE, Tykot RH, Benz BF (eds) (2006) Histories of Maize: Multidisciplinary approaches to the prehistory, linguistics, biogeography, domestication, and evolution of maize. Elsevier, San Diego, CA

Staller JE, Tykot RH, Benz BF (eds) (2009) Histories of Maize in Mesoamerica: Multidisciplinary Approaches. Left Coast Press, Walnut Creek, CA

Stiner MC (2001) Thirty years on the "broad spectrum revolution" and Paleolithic demography. Proc Natl Acad Sci USA 98:6993–6996

Steward JH (1955) The theory of culture change: the methodology of multilinear evolution. University of Illinois Press, Urbana

Stross B (2006) Maize in word and image in Southeastern Mesoamerica. In: Staller JE, Tykot RH, Benz BF (eds) Histories of Maize: Multidisciplinary approaches to the prehistory, linguistics, biogeography, domestication, and evolution of maize. Elsevier, San Diego, CA, pp 577–598

Stross B (2009) This world and beyond: food practices and the social order in mayan religion. In: Staller JE, Carrasco MD (eds) Pre-Columbian Foodways: Interdisciplinary approaches to food, culture and markets in Mesoamerica. Springer, New York, pp 553–576

Stothert KE (1985) The preceramic Las Vegas culture of Coastal Ecuador. Am Antiq 50(3):613–637

Strurtevani EL (1899) Varieties of corn. U.S.D.A. Offprint of the Expeditionary State Bulletin. 57

Stuessey TF (1990) Plant Taxonomy. Columbia University Press, New York

Swadesh M (1951) Diffusional cumulation and archaic residue as historical explanations. Southwest J Anthropol 7:1–21

Swadesh M (1959a) Linguistics as an instrument of prehistory. Southwest J Anthropol 15:20-35

Swadesh M (1959b) Mapas de Clasificación Lingüística de México y las Américas. Universidad Nacional Autonoma de Mexico, Mexico D.F

Tagg MD (1996) Early cultigens from fresnal shelter, Southeastern New Mexico. Am Antiq 61 (2):311–324

Talbert L, Doebley J, Larson S, Chandler V (1990) The ancestry of *Tripsacum andersonii* is a natural hybrid involving *Zea* and *Tripsacum*: molecular evidence. Am J Bot 77:722–726

Tanksley SD, McCouch SR (1997) Seed banks and molecular maps: unlocking genetic potential from the wild. Science 277:1063–1066

Taube KA (1996a) The olmec maize God: the face of corn in formative mesoamerica. *Res* – Anthropol Aesthetics 29–30:39–81

Taube K (1996b) Lightning celts and corn fetishes: the Formative Olmec and the development of maize symbolism in Mesoamerica and the American Southwest. In: Clark JE, Pye ME (eds) Olmec art and archaeology in Mesoamerica. National Gallery of Art and Yale University Press, Washington, D.C, pp 297–337

Tedlock D (1985) Popol Vuh. Simon and Schuster, New York

Terrell JE, Hart J, Barut S, Cellinese N, Curet A, Denham T, Kusimba C, Latinis K, Oka R, Palka J, Pohl M, Pope K, Williams P, Haines H, Staller JE (2003) Domesticated landscapes: the subsistence ecology of plant and animal domestication. J Archaeol Method Theor 10(4):323–367

Terry RE, Fernández FG, Parnell JJ, Inomata T (2004) The Story in the floors: chemical signatures of ancient and modern Maya activities at Aguateca, Guatemala. J Archaeol Sci 31:1237–1250

Thomas J (1993) Discourse, totalisation and the 'Neolithic'. In: C.H. Tilley (ed.) Interpretative archaeology. Berg, London, pp. 357–394

Thomson JA (2007) Seeds for the future: the impact of genetically modified crops on the environment. Comstock Publication, Ithaca, NY

Thompson JES (1970) Maya history and religion. University of Oklahoma Press, Norman

Thompson RG (2005) Phytolith analysis of food residues from coprolites and a pottery sherd recovered at Ramaditas, Chile. In: Rivera MA (ed) Arqueologia del Desierto de Atacoma: La Etopa Formativa en el Area de Ramaditas/Guatacondo. Ediciones Universidad Bolivarana, Santiago, pp 211–230

Thompson RG (2006) Documenting the presence of maize in central and south america through phytolith analysis of food residues. In: Zeder MA, Bradley DG, Emshwiller E, Smith BD (eds) Documenting domestication: new genetic and archaeological paradigms. University of California Press, Berkeley, pp 82–95

Thompson RG (2007) Tracing the movement of maize through the analysis of phytoliths recovered from food residues in prehistoric pottery. Ph.D. dissertation, Department of Anthropology, University of Minnesota, Minneapolis

Thompson RG, Mulholland S (1994) The identification of corn in food residues on utilized ceramics at the Shea Site (32CS101). Phytolith News Lett 8(2):7–11

Thompson RG, Staller JE (2001) An analysis of opal phytoliths from food residues of selected sherds and dental calculus from excavations at the site of La Emerenciana, El Oro Province, Ecuador. Phytolith Bull Soc Phytolith Res 13(2&3):8–16

Thompson RG, Kluth RA, Kluth DW (1995) Brainerd ware pottery function explored through opal phytolith analysis of food residues. Minn Archaeol 53:86–95

Timothy DH, Harry D (1963) Races of maize in Ecuador. National Academy of Sciences-National Research Council, Washington, DC

Timothy D, Bertulfo Pena VH, Ramirez ER (1961) Races of maize in Chile. In: collaboration with William L. Brown and Edgar Anderson. National Academy of Sciences-National Research Council, Washington

Yupanqui TC, de Castro DD (1985[1570]) Ynstrucción del Ynga Don Diego de Castro Titu Cusi Yupanqui. Edición facsímil de Luis Millones. Ediciones El Virrey, Lima

Toledo de F (1940) [1571] Información hecha en el Cuzco por orden del Virrey Toledo, Cuzco, 13–18 marzo 1571. In: R. Levillier (ed.) Don Francisco de Toledo, supremo organizador del Perú: Su vida, su obra (1515–1582), vol. 2, Sus informaciones sobre los incas (1570–1572). Espasa-Calpe, Buenos Aires, pp. 65–98

Tozzer AM (1907) A comparative study of the Mayas and the Lacandones. Archaeo- logical Institute of America, New York

Traboulay DM (1994) Columbus and Las Casas. The conquest and christianization of America, 1492-1566. Lanham, New York

Tudela J (1977) [1541] Relación de Michoacán. Relación de las ceremonias y ritos y población y gobierno de los indios de la Provincia de Michoacán: reproducción facsímil del Ms. ç.IV.5 de El Escorial; estudio preliminar por José Corona Núñez. México: Balsal Editores

Tuxill J, Reyes LA, Latournerie L, Cob V, Jarvis DI (2009) All maize is not equal: maize variety choices and mayan foodways in Rural Yucatan, Mexico. In: Staller JE, Carrasco MD (eds) Pre-Columbian foodways: interdisciplinary approaches to food, culture and markets in Mesoamerica. Springer, New York, pp 466–486

Tykot RH (2002) Contributions of stable isotope analysis to our understanding dietary variation among the Maya. In: Jakes KA (ed) Archaeological chemistry: materials, methods and meaning. Washington D.C, American Chemical Society, pp 214–230

Tykot RH (2006) Stable isotope analysis and human diet. In: Staller JE, Tykot RH, Benz BF (eds) Histories of Maize: Multidisciplinary approaches to the prehistory, linguistics, biogeography, domestication, and evolution of maize. Elsevier, San Diego, CA, pp 131–142

Tykot RH, Staller JE (2002) On the importance of early maize agriculture in coastal Ecuador: new data from the Late Valdivia Phase site of La Emerenciana. Curr Anthropol 43(4):666–677

Tykot RH, van der Merwe NJ, Burger RL (2006) The importance of maize and marine foods to Initial Period/Early Horizon subsistence in highland and coastal Peru. In: Staller JE, Tykot RH, Benz BF (eds) Histories of Maize: Multidisciplinary approaches to the prehistory, linguistics, biogeography, domestication, and evolution of maize. Elsevier, San Diego, CA, pp 187–197

Tylor EB (1889) Anthropology: an introduction to the study of man and civilization. Appleton and Company, New York

Ubelaker DH, Katzenberg MA, Doyon L (1995) Status and diet in precontact highland Ecuador. Am J Phys Anthropol 97:403–411

Ubelaker DH, Bubniak Jones E (2002) Formative period human remains from coastal ecuador: La Emerenciana Site (OOSrSr-42). J of the Washington Academy of Sciences 88(2):59–72

Ugent D (1994) Chemosystematics in archaeology: A preliminary study of the use of chromatography and spectrophotometry in the identification of four prehistoric root crop species from the desert coast of Peru. In: Hather JG (ed) Tropical archaeobotany: applications and new developments. Series on World Archaeology, New York, pp 215–226

Ugent D, Pozorski S, Pozorski T (1982) Archaeological potato tuber remains from the Casma Valley of Peru. Econ Bot 36:182–192

Ugent D, Pozorski S, Pozorski T (1984) New evidence for ancient cultivation of *Canna edulis* in Peru. Econ Bot 38:417–432

Ugent D, Pozorski S, Pozorski T (1986) Archaeological manioc (*Manihot*) from Coastal Peru. Econ Bot 40:78–102

Urton G (1985) Animal metaphors and the life cycle in an Andean community. In: Gary Urton (ed) Animal Myths and Metaphors in South America. University of Utah Press, Salt Lake City, pp 251–284

Urton G (1990) The history of a myth: Pacariqtambo and the origin of the Inka. University of Texas Press, Austin, TX

Urton G (1999) Inca Myths. University of Texas Press, Austin, TX

Valera B (1968) 1594. Relación de las costumbres antiguas del Peru. With an introduction by F. Loayza, Lima

Valcárcel LE (1946) Indian markets and fairs in Peru. In: Steward JH (ed) Handbook of South American Indians. The Andean civilizations, vol 2, Smithsonian Institution Bureau of American Ethnology Bulletin 143. U.S. Government Printing Office, Washington D.C, pp 477–482

van der Merwe NJ, Vogel JC (1978) ^{13}C content of human collagen as a measure of prehistoric diet in woodland North America. Nature 276:815–816

van der Merwe NJ (1982) Carbon isotopes, photosynthesis, and archaeology. Am Sci 70 (1982):596–606

van der Merwe NJ, Lee-Thorp JA, Raymond JS (1993) In: Lambert JB, Grup G (eds) Prehistoric human bone archaeology at the molecular level. Springer, New York, pp 63–97 Light, stable isotopes and the subsistence base of formative cultures at Valdivia, Ecuador

Vavilov NI (ed) (1926) Centers of origin of cultivated plants. In: Origin and geography of cultivated plants, translated by Doris Löve. Cambridge University Press, New York

Vega, "El Inca" Garcilaso de la (1966) [1609]. Royal Commentaries of the Incas and General History of Peru. Parts I-II. Translated by H.V. Livermore, forward by A.J. Toynbee. University of Texas Press, Austin, TX

Vierra BJ, Ford RI (2006) Early maize agriculture in the Río Grande Valley, New Mexico. In: Staller JE, Tykot RH, Benz BF (eds) Histories of Maize: Multidisciplinary approaches to the prehistory, linguistics, biogeography, domestication, and evolution of maize. Elsevier, San Diego, CA, pp 497–510

Vigouroux Y, McMullen M, Hittinger CT, Houchins K, Schulz L, Kresovich S, Matsouka Y, Doebley J (2002) Identifying genes of agronomic importance in maize by screening microsatellites for evidence of selection during domestication. Proc Natl Acad Sci USA 99 (15):9650–9655

Vigouroux Y, Matsuoka Y, Doebley J (2003) Directional evolution for microsatellite size in maize. Mol Biol Evol 20(9):1480–1483

Villalba M (1988) *Cotocollao.* Miscelanea Antropógica Ecuatoriana 2. Quito, Ecuador

Vogel JC, van der Merwe NJ (1977) Isotopic evidence for early maize cultivation in New York State. Am Antiq 42:238–242

Wallace HA, Brown WL (1956) Corn and its early fathers. The Michigan State University Press, Kalamazoo

Wang RI, Stec J, Hey J, Lukens L, Doebley J (1999) The limits of selection during maize domestication. Nature 398:236–239

Ward SAW, Prothero GW, Mordaunt S, Benians EA (1912) The Cambridge modern history atlas. London, Cambridge University Press

Watson RA, Watson PJ (1969) In: Ucko PJ, Dimbleby GW (eds) The domestication and exploitation of plants and animals. Duckworth, London, pp 397–405 Early cereal cultivation in China

Watson S (1891) Contributions to American Botany, 3 Upon a wild species of Zea from Mexico. Proc Am Acad Arts Sci 26:108–161

Weatherwax P (1954) Indian corn in Old America. MacMillan, New York

Wellhausen EJ, Roberts LM, Hernandez X E (1952) Races of maize in Mexico; their origin, characteristics and distribution. In: collaboration with Paul C. Mangelsdorf. Bussey Institution of Harvard University, Cambridge

Wellhausen EJ, Fuentes OA, Hernández Corzo A (1957) Races of maize in Central America. In: collaboration with Paul C. Mangelsdorf. National Academy of Sciences, National Research Council, Washington

Wells CE, Terry RE, Parnell JJ, Hardin PJ, Jackson MW, Houston SD (2000) Chemical analyses of ancient anthrosols in residential areas at Piedras Negras, Guatemala. J Archaeol Sci 27:449–462

Whitaker TW, Cutler HC, MacNeish RS (1957) Curcurbit materials from the three caves near Ocampo, Tamaulipas. Am Antiq 22(4):352–358

White C, Schwarcz HP (1989) Ancient Maya diet as inferred from isotopic and chemical analyses of human bone. J Archaeol Sci 16:451–474

White CD, Spence MW, Longstaffe FJ, Law KR (2000) Testing the nature of Teotihuacán imperialism at Kaminaljuyú using phosphate oxygen-isotope ratios. J Anthropol Res 56:535–558

White CD, Longstaffe FJ, Schwarcz HP (2006) Past and future directions for isotopic anthropology in mesoamerican maize research. In: Staller JE, Tykot RH, Benz BF (eds) Histories of Maize: Multidisciplinary approaches to the prehistory, linguistics, biogeography, domestication, and evolution of Maize. Elsevier, San Diego, CA, pp 143–159

Whitehead DR, Langham EJ (1965) Measurement as a means of identifying fossil maize pollen. Bull Torrey Bot Club 92:7–20

Whitehead DR, Sheehan MC (1971) Measurement as a means of identifying fossil maize pollen, II: the effect of slide thickness. Bull Torrey Bot Club 98:268–271

Whitt SR, Wilson LM, Tenaillon MI, Gaut BS, Buckler ES IV (2002) Genetic diversity and selection in the maize starch pathway. Proc Natl Acad Sci USA 99:12595–12562

Whorf BL (1954) Language, meaning, and maturity. Harper and Bros, New York

Wilkes HG (1977) Hybridization of Teosinte in Mexico and Guatemala and its improvement of maize. Econ Bot 31:254–293

Wilkes HG (1967) Teosinte: the closest relative of maize. Harvard University, Cambridge

Wilkes HG (1989) Maize: domestication, racial evolution, and spread. In: Harris DR, Hillman GC (eds) Foraging and farming: the evolution of plant exploitation. Unwin Hyman, London, pp 440–455

Willey GR (1964) An introduction to North American ARchaeology, vol 1, North America. Prentice-Hall, New York

Willey GR (1971) An introduction to South American archaeology, vol 2, South America. Prentice-Hall, New York

Willey GR, Sabloff JA (1980) A history of American archaeology. W. H. Freeman and Company, San Francisco

Wills WH (1988) Early prehistoric agriculture in the American Southwest. School of American Research, Santa Fe

Wilson DJ (1981) Of maize and men; a critique of the maritime hypothesis of state origins on the Coast of Peru. Am Anthropol 83:93–120

Wright HE Jr, Mann DH, Glasner PH (1984) Piston corers for peat and lake sediments. Ecology 65:657–659

Wright SI, Vroh Bi I, Schroeder SI, Yamasaki M, Doebley JF, Mullen MD, Gaut BS (2005) The effects of artificial selection on the maize genome. Science 308:1310–1314

Xerex F de (1985) [1534] Verdaderas relación de la conquista del Perú, edited by C. Bravo. Historia 16. Madrid

Zarrillo S, Pearsall DM, Raymond JS, Tisdale MA, Quon DJ (2008) Directly dated starch residues document early formative maize (Zea mays L.) in tropical Ecuador. Proc Natl Acad Sci USA 105(13):5006–5011

Zeder MA, Emshwiller E, Smith BD, Bradley DG (2006) Documenting domestication: the intersection of genetics and archaeology. Trends Genet 22:139–155

Zevallos JMP (2005) The ethnohistory of Southern Veracruz. In: Sandstrom AR, García Valencía EH (eds) Native peoples of the Gulf Coast of Mexico. University of Arizona Press, Tucson, pp 66–99

Zevallos-Menéndez C, Galinat WC, Lathrap DW, Leng ER, Marcos J, Klumpp K (1977) The San Pablo corn kernel and its friends. Science 196:385–389

Zohary D (1996) The mode of domestication of the founder crops of Southwest Asian agriculture. In: Harris DR (ed) The origins and spread of agriculture and pastoralism in Eurasia. UCL Press, London, pp 142–158

Zohary D (2004) Unconscious selection and the evolution of domesticated plants. Econ Bot 58:5–10

Zohary D, Hopf M (1993) Domestication of Plants in the Old World, 2nd edn. Oxford, Oxford University Press

Zuidema RT (1973) Kinship and ancestor cult in three Peruvian communities: Hernández Príncipe's account in 1622. Bulletin, Institut Français d'Études Andines 2:16-23. Lima.

Zuidema RT (2002) Inca religion: its foundations in the central Andean context. In: Sullivan LE (ed) Native Religions and Cultures of Central and South America. Continuum, London, pp 236–253

Zuidema RT (2008) The astronomical significance of ritual movements in the Calendar of Cuzco. In: Staller JE (ed) Pre-Columbian landscapes of creation and origin. Springer, New York, pp 249–268

Zvelebil M (2000) The social context of the agricultural transition in Europe. In: Renfrew C, Boyle K (eds) Archaeogenetics: DNA and the Population History of Europe. McDonald Institute for Archaeological Research, Cambridge, pp 57–79

Index